ADVANCES IN CHEMICAL PHYSICS

VOLUME XLIII

EDITORIAL BOARD

Advances in CHEMICAL PHYSICS

EDITED BY

I. PRIGOGINE

University of Brussels
Brussels, Belgium
and
University of Texas
Austin, Texas

AND

STUART A. RICE

Department of Chemistry
and
The James Franck Institute
The University of Chicago
Chicago, Illinois

VOLUME XLIII

AN INTERSCIENCE® PUBLICATION

JOHN WILEY & SONS

NEW YORK • CHICHESTER • BRISBANE • TORONTO

An Interscience® Publication

Library of Congress Catalog Number: 58-9935

ISBN 0-471-05741-X

Printed in the United States of America

10 9 8 7 6 5 4 3 2 1

CONTRIBUTORS TO VOLUME XLIII

BRUCE L. CLARKE, Department of Chemistry, University of Alberta, Edmonton, Alberta, Canada

PETER H. RICHTER, Department of Chemistry, Massachusetts Institute of Technology, Cambridge, Massachusetts

ITAMAR PROCACCIA, Department of Chemical Physics, Weizmann Institute of Science, Rehovot, Israel

JOHN ROSS, Department of Chemistry, Massachusetts Institute of Technology, Cambridge, Massachusetts

INTRODUCTION

Few of us can any longer keep up with the flood of scientific literature, even in specialized subfields. Any attempt to do more, and be broadly educated with respect to a large domain of science, has the appearance of tilting at windmills. Yet the synthesis of ideas drawn from different subjects into new, powerful, general concepts is as valuable as ever, and the desire to remain educated persists in all scientists. This series, *Advances in Chemical Physics*, is devoted to helping the reader obtain general information about a wide variety of topics in chemical physics, which field we interpret very broadly. Our intent is to have experts present comprehensive analyses of subjects of interest and to encourage the expression of individual points of view. We hope that this approach to the presentation of an overview of a subject will both stimulate new research and serve as a personalized learning text for beginners in a field.

ILYA PRIGOGINE

STUART A. RICE

CONTENTS

STABILITY OF COMPLEX REACTION NETWORKS

BRUCE L. CLARKE

Department of Chemistry
University of Alberta
Edmonton, Alberta, Canada

CONTENTS

INDEX OF EXAMPLES

INDEX OF SYMBOLS

The symbols that are used most frequently throughout the chapter are listed below. More specialized uses may have subscripts or superscripts. A "vector" means a matrix with one column. Inner products are written as matrix multiplication between the transpose of a column matrix and another column matrix. For example, $\mathbf{x}'\mathbf{y}$.

$\mathbf{a(j)}$	another steady-state concentration for given $\mathbf{j}$
c	coefficient in a polynomial
d	rank of $\underline{\boldsymbol{\nu}}$
$\mathbf{e}$	$(1, 1, \ldots, 1)'$
f	number of extreme currents
$f(\mathbf{X})$	a function
$g(\pi)$	tree function for a graph
$g(\mathbf{X})$	exponomial of $f(\mathbf{X})$
$\mathbf{h}$	reciprocal steady-state concentration vector
$\mathbf{j}$	coordinates of a current relative to the frame of $\mathcal{C}_v$
$\mathbf{k}$	rate constant vector
l	number of simplices in a decomposition of Π_v or $\mathcal{C}_v$
n	number of species
$\mathbf{p}$	parameter vector
$\mathbf{q}$	secondary parameter vector
r	number of reactions
s_i	ith principal minor of a matrix
t	time; term in a polynomial
$\mathbf{v}$	reaction velocity vector; $\mathbf{v}^0$ steady state
$\mathbf{x}$	dimensionless concentration vector; variable in an exponomial
$\mathbf{y}$	a complex; an exponent vector of a term in a polynomial
A	Routh-Hurwitz array
A_j	arrow on $\mathfrak{D}_C$
$B(\mathrm{X}_i, A_j)$	number of barbs on arrow A_j of $\mathfrak{D}_C$ at X_i
B	bifurcation set for steady states
$\mathbf{C}$	concentration vector of chemical components
C	number of thermodynamic components
$C(\mathbf{a})$	cycle on $\mathfrak{D}_{CM}$ involving species $\mathrm{X}_{a(1)}, \ldots, \mathrm{X}_{a(k)} \cdots$
D	a parameter domain
E	extreme current matrix
$\mathbf{E}_i$	extreme current vector of the ith extreme current

E	extreme subnetwork; an edge of a polytope
F	number of thermodynamic degrees of freedom; a face of a polytope
F_i	set of functions for the generalized Routh-Hurwitz–Liénard-Chipart conditions
G	a graph; Gibbs free energy
I	identity matrix
$\mathbf{J}$	flux vector
$\mathbf{J}_E$	equilibrium flux vector
$\mathbf{J}_N$	nonequilibrium flux vector
K	a knot
L	a Lyapunov function
$L(\mathrm{X}_i, A_j)$	number of left feathers on the arrow A_j of $\mathfrak{D}_C$ at X_i
M	matrix of dynamics linearized about steady state
N	a network
P	an exponent polytope; a projection operator
Q	matrix of a quadratic Lyapunov function
R	real numbers; R_+ positive real numbers; $\overline{R}_+$ nonnegative real numbers
R	matrix in Lyapunov's matrix equation
R_i	ith pseudoreaction
S	a subspace; see Table I
$\mathsf{S}^{(i)}$	interaction matrix for the extreme current $\mathbf{E}_i$
T	set of terms in a polynomial
$T(\mathrm{X}_i, A_j)$	number of tails on arrow A_j of $\mathfrak{D}_C$ at X_i
U	any matrix (sign stability)
V	a vertex of a polytope
$\mathsf{V}(\mathbf{j})$	defined in (III.10)
$\mathbf{X}$	concentration vector; variables in a polynomial; $\mathbf{X}^0$ steady-state concentration vector
X_i	chemical species
Y	set of complexes; set of exponent vectors of a polynomial
Y	matrix of complex vectors
$\alpha_i(\mathbf{p})$	coefficient of the characteristic equation
$\beta_i(\gamma, \mathbf{j})$	coefficient of $\mathbf{h}$ in the expansion of $\alpha_i(\mathbf{h}, \mathbf{j})$
$\underline{\gamma}$	conservation matrix

γ	vector of integers representing subscripts of a set of species
δ	deficiency of a network
ζ	deviation of concentrations from steady state
η	deviation of reaction extents from steady state
$\underline{\kappa}$	kinetic order matrix
λ	eigenvalue of M; distance along a ray
μ_i	chemical potential of X_i
$\mu(\lambda, \mathbf{p})$	minimal polynomial of M
$\underline{\nu}$	stochiometric matrix
ξ	extent of reaction vector
π	a point on a graph
$\rho(\mathbf{k}, \mathbf{C})$	number of steady states for given $\mathbf{k}$ and $\mathbf{C}$
σ	entropy production density
ϕ	mapping between parameter domains
$\chi(\lambda, \mathbf{p})$	relevant factor of the characteristic polynomial
$\boldsymbol{\chi}$	thermodynamic force vector
ψ	mapping from S_ξ to S_X
Γ_k	defined by (VI.33)
Δ_i	Hurwitz determinant
Λ	set of relevant eigenvalues of M; a ray
$\Xi(N, \mathbf{E}_i)$	mapping from a network N to its subnetwork corresponding to $\mathbf{E}_i$
Π_v	current polytope
$\Pi_X(\mathbf{C})$	concentration polyhedron
$\Pi_\xi(\mathbf{C})$	extent of reaction polyhedron
Π_E	equilibrium simplex
Σ	unit sphere; a simplex in a decomposition of Π_v; a sign matrix set; a set of species on a current cycle
Σ^0	outer sphere of vertices of the exponent polytope of Δ_i
$\mathcal{A}$	any affine space
$\mathcal{C}_E$	equilibrium (current) cone
$\mathcal{C}_v$	current cone
$\mathcal{C}_0(V)$	cone of dominance of the vertex V of an exponent polytope
$\mathcal{D}_{ASP}$	diagram of all stabilizers and polygons
$\mathcal{D}_C$	current diagram
$\mathcal{D}_{CM}$	current matrix diagram

$\mathcal{D}_{EC}$	exteme current diagram
$\mathcal{D}_N$	network diagram
$\mathcal{D}_{NOSP}$	diagram of nonoverlapping stabilizers and polygons
$\mathcal{E}$	set of all extreme networks
$\mathcal{K}$	set of knots of a network
$\mathcal{N}$	set of all networks; see Section III.B
$\mathcal{S}$	set of all real symmetric positive definite matrices
$\mathcal{S}_0$	set of all real symmetric positive semidefinite matrices
$\mathcal{U}$	set of all sign matrix sets
$\mathcal{X}$	set of species of a network
$\square$	the empty complex (blank side of a chemical reaction)
$\emptyset$	the null (empty) set

I. INTRODUCTION

A. The Major Problems

Instability of the steady states of chemical reaction systems always leads to exotic dynamical phenomena, such as switching between multiple steady states, explosions, sustained oscillations, or even more complicated phenomena, which are known by mathematicians as motion on an attracting k-torus and strange attractors (chaos). The oscillatory phenomena in chemistry[1,2] that are currently of intense interest are not the damped oscillations that sometimes occur as a system approaches steady state, but sustained oscillations that appear spontaneously in systems that seem to be at steady state initially. These oscillations are often called "self-oscillations" (in chemistry) and "limit cycles" (in mathematics).

It is important for the chemist to be able to identify chemical mechanisms that can exhibit exotic dynamics. Stability analysis is by far the most effective technique because it uses linearized equations of motion; the direct identification of limit cycles requires a tedious analysis of the complete nonlinear equations of motion, which is possible only for simple systems. Using stability analysis to detect oscillatory chemical mechanisms has an inherent danger. In principle, exotic dynamics may occur in systems of differential equations even when no steady states are unstable. Stability analysis is thus not a direct test for the existence of exotic dynamics; however the differential equations for chemical systems appear to be special. So far, no remotely plausible model has been constructed of a chemical network with oscillations and only stable steady states. Hence linear stability analysis may be used with some confidence to determine which chemical models are capable of exotic dynamics in cases that are far too complex for the direct identification of limit cycles.

We shall make an important distinction between a chemical system and a chemical network. A *chemical system* is a set of reaction *stoichiometries*, which specify the numbers of molecules produced and consumed in each reaction, and a function $v_i(\mathbf{X})$ for each reaction, known as the *rate law*, giving the reaction rate in terms of the chemical concentration vector $\mathbf{X}$. The stoichiometries and rate laws completely determine a system of differential equations that may be studied numerically. Chemical rate laws are rational polynomials and hence are determined by two types of constants. First, the form of the polynomial is determined by constants called *exponents* (in mathematics) and *orders of kinetics* (in chemistry); these are frequently small, exact integers, arising from the bimolecular nature of molecular collisions. The remaining constants are the coefficients of the terms and are called *rate constants*. These constants are often poorly known experimentally and can range over very many orders of magnitude. They also cannot change sign, so that the sign of a term may be regarded as part of the form of the polynomial and the rate constants may always be considered positive. Thus they may be represented as an r-tuple $\mathbf{k} \in R^r_+$, the r-dimensional positive orthant of the real numbers. The rate law $v_i(\mathbf{X})$ is thus a function of $\mathbf{k}$ and some exponents $\underline{\boldsymbol{\kappa}}$; hence we sometimes write $v_i(\mathbf{X}, \mathbf{k})$ or $v_i(\mathbf{X}, \mathbf{k}, \underline{\boldsymbol{\kappa}})$.

When a chemist specifies a *mechanism*, he gives a set of reaction stoichiometries; the form of the rate laws $v_i(\mathbf{X})$ is determined by the stoichiometries if the mechanism is *elementary*; otherwise, the form is given explicitly. However the mechanism does not specify the rate constants $\mathbf{k}$, which are left as experimental parameters. Thus a mechanism is not a chemical system, but a set of chemical systems, one system for each choice of $\mathbf{k} \in R^r_+$.

We shall use the term "network" in the same sense as "mechanism." A *chemical network* is a set of chemical systems, one system for each value of a parameter vector $\mathbf{p} \in R^q$ (analogous to $\mathbf{k}$), which lies in a specified domain D of parameter space.

A chemical network is *stable* if every steady state of every corresponding chemical system is stable. Otherwise the network is *unstable*. The term "qualitative stability" is also used to refer to stability for all values of the parameters. Every unstable network must exhibit exotic dynamics for some parameter values. Every network so far constructed that exhibits exotic dynamics is unstable. The first problem of interest is

The Network Stability Problem. **Find necessary and sufficient conditions for a chemical network to be stable that are practical to apply to complex**

reaction networks. A solution to this problem would help the chemist choose plausible mechanisms for oscillating chemical systems.

If a network is unstable, the physically relevant domain D of parameter space may be divided into two mutually exclusive subsets D_u and D_s, where the corresponding chemical systems have an unstable steady state or only stable steady states, respectively. This partitioning of parameter space is a *stability diagram*, which is similar to the chemist's phase diagram. The parameter $\mathbf{p}$ is analogous to temperature or pressure; the regions D_u and D_s are analogous to one-phase regions such as solid or gas; and the boundary separating D_u and D_s is called the *bifurcation set for steady states* and is analogous to a phase coexistence curve (e.g., a vapor pressure curve). The bifurcation set for steady states is also known as the *catastrophe set* or the *boundary of stability*. The second major problem is

The Stability Diagram Problem. **Devise an efficient method of calculating the bifurcation set for steady states of an unstable chemical reaction network.** A solution would enable a chemist who knows the rate constants to predict when a mechanism would be on the verge of oscillation or other exotic dynamics. Conversely, if the chemist does not know the rate constants, the experimentally determined bifurcation set could be used to obtain some of them.

This chapter carries the theoretical analysis of these two problems far beyond the level represented by the current literature, which is so sparse that it is briefly summarized in Section I.B. Section I.C outlines how this approach (called stoichiometric network analysis) has been able to circumvent the problems that others have encountered. These two sections are written for theoretical chemists; hence readers who are unfamiliar with the literature will probably find them impossible to understand. They need not despair, because a purely deductive approach to the subject begins from first principles in Section II. Most of the results discussed in Section I.B. eventually will be derived from our approach.

The subject matter is a highly specialized branch of mathematics that has practical applications in chemistry. Since this is the first time that this subject has been developed in a coherent fashion in print, I feel compelled to give complete proofs of every step in the main line of development. Consequently the material reads like mathematics. I have given definitions and examples of chemical terminology so that it can be read by theoreticians with little knowledge of chemistry. There is much in this chapter that would be of practical value to experimental chemists if they could only understand it. Since few chemists will have the mathematical background to work through this chapter in a reasonable length of time, I suggest that chemists attempt to read Sections II.A, II.E, II.F, II.H, and IV.G, then try to work through Section V. Those interested primarily in how instability occurs will find that they can understand the main results of Section V after having studied Section II and the first three sections of Section III.

B. Summary of Previous Work on Stability Theorems for Chemical Networks

Although most chemical networks are stable, the literature of attempts to prove stability is not large, probably because stability was generally taken for granted until recently. This section outlines two lines of development that prove stability by finding a function, called a *Lyapunov function*, with certain properties such that it can only exist when the steady state is asymptotically stable. These approaches both assume that the rate law has the form known as *mass action kinetics*. A third approach, which is based on the Hurwitz determinants, is discussed later.

One line begins with Hearon,[3,4] who showed that systems of unimolecular reactions (linear rate laws) have a linear equation of motion; the matrix of this equation has real eigenvalues, and hence the possibility of damped oscillations is excluded. Hearon's proof rests on the symmetrizability of the matrix. Using a method of diagramming chemical networks devised by Delattre,[5] Hyver[6] showed that the same symmetrization technique could be used on the linearized dynamics of more general networks whose diagrams corresponded to *tree graphs*. Tree graph networks therefore have real eigenvalues. This work has been extended beyond the tree graph structure by Solimano et al.[7,8] to include some networks with cycles coming from enzymes. Recently it has been proved that a kind of symmetrizability, called *D-symmetrizability*, is a sufficient condition for the existence of a Lyapunov function for these systems.[9] Hence tree networks are stable.

The second line of development obtains Lyapunov functions for the full nonlinear equations, not just for their linearization. The Lyapunov function is closely related to the thermodynamic free energy of a mixture. The theorems proved identify a class of systems that (1) have only one stoichiometrically accessible steady state for each set of physically possible initial conditions, and (2) always approach this steady state. Such systems are *quasithermodynamic*. The first proof in this approach is due to Shear[10] (an incorrect proof was given by Shear earlier[11,12]) for systems whose rate constants **k** made the steady states satisfy the detailed balance condition; that is, forward and reverse reaction rates are equal at steady state. He proved that detailed balanced networks are quasithermodynamic.

This line of development continued in a series of very thorough and elegant papers by Horn, Feinberg, and Jackson (HFJ)[14–19] which show that the same function used by Shear is a Lyapunov function for a much broader class of systems that satisfy a condition called *complex balancing*. This condition is easily understood. The expressions, such as 2X or X + Y, which appear on the sides of a chemical reaction, such as $2X \rightarrow X + Y$, are called *complexes*. Complex balancing means that all the complexes are at steady state; that is, the sum of rates of reactions producing the complex must equal the sum of the rates of reactions consuming the same complex. Since for unimolecular reaction systems, the complexes are the chemical species, the species are at steady state if and only if the complexes are at steady state. All unimolecular systems are therefore complex balanced, and it follows from the theorems of HJF that unimolecular networks are quasithermodynamic. If a system is detailed balanced, forward and reverse reactions consume and produce the same pair of complexes at the same rate; thus the system is also complex balanced and hence quasithermodynamic. Thus we see how Shear's result is a special case. Complex balancing holds for all steady states of many networks that are not detailed balanced. More generally, complex balancing holds when the rate constants satisfy certain equations. HJF give a simple method of calculating the number of independent equations that must be satisfied by the rate constants for complex balancing to hold. This number is called a network's *deficiency* δ. All steady states of networks with zero deficiency are complex balanced. Hence zero-deficiency networks are quasithermodynamic and consequently stable.

Many of the tree networks have zero deficiency. It has been shown by Clarke[20] that tree networks have $\delta > 0$ only when a sufficient number of the species undergo a pseudo-first-order exchange with species that do not appear in the dynamics, such as diffusive exchange with a reservoir. The same paper thoroughly discusses the relationship between these two lines of development.

Section IV treats a new Lyapunov function that enables one to prove stability for a much broader class of networks than is covered by the HJF theorems. In fact, complex balanced networks all have this Lyapunov function and hence are a special case. We also develop an approach based on Hurwitz determinants, which appears to resolve the network stability problem in every case. This approach grew out of my work and is discussed fully later.

Thermodynamic formulations of the stability problem do not consider the detailed network structure, with the exception of Oster, Perelson, and Katchalsky's *network thermodynamics*.[21,22] This theory regards the concentrations of the species to be analogous to electrical charges stored in various capacitors, one for each species. The reaction $X + Y \rightarrow Z$ is analogous to an electric current that discharges "capacitors" corresponding to X and Y and charges a capacitor corresponding to Z. The chemical potential is analogous to the electrical potential. The analogy with electricity is not perfect because the potential of a capacitor is linear in the charge, whereas the chemical potential of a species is not linear in its concentration. Network thermodynamics linearizes where necessary to make nonequilibrium thermodynamics formally identical with electrical network analysis so that the latter's powerful formalism can be used. The Glansdorff-Prigogine stability criterion can be obtained. Section II demonstrates that thermodynamic formulations use a system of parameters that is not well suited to stability analysis. The parameters of nonequilibrium thermodynamics are chemical potentials and thermodynamic fluxes; the matrix of the dynamics linearized about steady state is not a sufficiently simple function of these parameters. Neither the Glansdorff-Prigogine stability criterion nor network thermodynamics has yet led to a proof that networks with certain general structural features are stable or unstable.

We now turn from the problem of proving stability to the problem of proving instability. Unstable chemical networks are relatively uncommon, and most theoretical work has been based on studies of a few famous models, most of which are noted more for their simplicity than their applicability to chemistry. Some of the most important models are the Lotka-Volterra model,[23,24] the Turing model,[25] the Brusselator (also known as the trimolecular model or Prigogine-Lefever model),[3] the Selkov model of glycolysis,[26] the Schlögl model,[27,28] the Oregonator (Field-Noyes model),[29,30] the Edelstein model,[31,32] and the Goodwin model.[33,34] The stability diagram has been calculated for many of these networks.

Since oscillatory reaction systems are usually much more complicated than these models, research on instabilities in complex reaction systems is urgently required. I first approached this question with numerical calculations on a complicated model.[35] Following the observation that the main features of the results could easily be calculated by making approximations based on a scheme involving a convex polytope, a "graph theoretic" approach was formulated.[36–39] Further experience in treating realistic complex reaction systems was gained during an analysis of the Field-Noyes-Körös mechanism for the Belousov-Zhabotinski (BZ) reaction system.[40,41] The concepts used in the analysis were general, and the analysis illustrates an approach that can be applied to any complex reaction system. This chapter presents this approach after much further theoretical refinement.

C. Stoichiometric Network Analysis

In presenting a comprehensive theory of stability and instability in chemical networks, we develop a formalism, called *stoichiometric network analysis*, which makes it possible to treat any network, no matter how complex, using one approach. The approach is called "stoichiometric network analysis" because it applies to all physical systems where the dynamical variables satisfy a system of linear constraints called stoichiometry. Such constraints occur when the dynamical variables are chemical concentrations, populations of species in ecology, or commodities in economics. The formalism elucidates a mathematical structure that is primarily a consequence of stoichiometry.

The rate laws for the "reactions" $v_j(\mathbf{X}, \mathbf{k}, \underline{\boldsymbol{\kappa}})$ can have any form. A common form in chemistry is a *power law*

$$v_j(\mathbf{X}, \mathbf{k}, \underline{\boldsymbol{\kappa}}) = k_j X_1^{\kappa_{1j}} \cdots X_n^{\kappa_{nj}} \tag{I.1}$$

Differentiating at steady state $\mathbf{X} = \mathbf{X}^0$ yields

$$\kappa_{ij} = \frac{\partial \log v_j(\mathbf{X}^0, \mathbf{k}, \underline{\boldsymbol{\kappa}})}{\partial \log X_i^0} \tag{I.2}$$

The function $\mathbf{v}(\mathbf{X}, \mathbf{k}, \underline{\boldsymbol{\kappa}})$ affects the stability only through this derivative; hence when $\mathbf{v}$ is a power law, the constancy of this derivative simplifies the problem. For more general rate laws, we take (I.2) to be the definition of an *effective power function*. Since the concentration $\mathbf{X}^0$ may be considered to be a function of $\mathbf{k}$, the effective power function is written $\kappa_{ij}(\mathbf{k})$, where all required parameters are considered components of $\mathbf{k}$. The solution for constant κ_{ij} can then be extended over the range of the effective power function.

Like other mathematical formalisms in physics (thermodynamics, statistical mechanics, quantum mechanics), stoichiometric network analysis has an elaborate mathematical structure that has emerged from solving the mathematical problems associated with the basic stoichiometric constraint. Some of the key problems affecting the formalism are as follows.

1. The first step in the usual approach to linear stability analysis of a model is to calculate the steady-state concentrations $\mathbf{X}^0$ as a function of the rate constants $\mathbf{k}$ to obtain the function $\mathbf{X}^0(\mathbf{k})$. If this can be done, the matrix of the dynamics linearized about steady state is then a function of $\mathbf{k}$, that is, $\mathsf{M}^\dagger(\mathbf{k})$. This method fails for complex systems because $\mathbf{X}^0(\mathbf{k})$ is the solution to a system of nonlinear equations that is rarely soluble in closed form. The formalism avoids this obstacle entirely by using a different set of parameters $\mathbf{p}$ that are equivalent to $\mathbf{k}$. Hence we obtain the matrix function $\mathsf{M}(\mathbf{p})$ and a mapping $\mathbf{p} = \phi_\kappa(\mathbf{k})$, such that $\mathsf{M}(\phi_\kappa(\mathbf{k})) = \mathsf{M}^\dagger(\mathbf{k})$. $\mathsf{M}(\mathbf{p})$ is stable for all physically meaningful $\mathbf{p}$ if and only if $\mathsf{M}^\dagger(\mathbf{k})$ is stable for all $\mathbf{k} \in R_+^r$. The mapping ϕ_κ can be chosen so that $\mathsf{M}(\mathbf{p})$ has a simple form in general.

2. The parameter vector $\mathbf{p} = (\mathbf{h}, \mathbf{j})$, where $\mathbf{h}$ is a vector whose components are the reciprocals of the steady-state concentrations, and $\mathbf{j}$ is a vector with the following property. A set of reaction velocities $\mathbf{v}$ ($v_i > 0$) satisfies the steady-state condition if and only if one may write

$$\mathbf{v} = \mathsf{E}\mathbf{j} \tag{I.3}$$

for a fixed matrix E determined by stoichiometry, and $j_i > 0$. The theory behind this equation was developed in connection with flows in economic systems and transportation networks. From this theory comes the idea of an *extreme network* that is fundamental to our entire theory.

3. The necessary and sufficient conditions that the matrix $\mathsf{M}(\mathbf{p})$ be stable for all $\mathbf{p}$ lead to polynomials in h_i and j_i that have very many terms and would normally be considered intractable. A new method[42] for determining where such polynomials change sign is indispensable to this formalism. The method is an n-dimensional generalization of an idea called *Newton's polygon*, which has had a long and important history in its two-dimensional version.

4. Instability can be proved by showing that the coefficient of a single term in a polynomial is negative. The coefficients of terms are complicated determinants; however a method of diagramming the network enables one to find negative signs by looking for certain simple but very rare diagrammatic features.

5. Networks can be proved to be stable by verifying that all the coefficients of every polynomial are positive; usually one constructs the polynomial on a computer. This method is not practical for large networks because the polynomials have too many terms. To circumvent this obstacle, a Lyapunov function has been found for a broad class of extreme networks. This class of *mixing stable extreme networks* includes all extreme subnetworks of complex balanced networks, most extreme subnetworks of tree networks, and very many other extreme networks. *The same function is also a Lyapunov function for all mixtures of these networks.* Hence the stability of networks is proved for all networks that can be decomposed into mixing stable extreme networks. This result, along with the idea of making a simplicial decomposition of the cone of steady states, produces a great simplification of the network stability and stability diagram problems for networks in cases that must be treated by computer.

6. The computations that are sometimes necessary require efficient algorithms for finding the vertices, edges, and higher-dimensional faces of a convex polyhedron expressed in the internal representation. Such algorithms have been recently found.[43]

The formalism of stoichiometric network analysis is very closely related to a problem in mathematics known as the *sign stability* or *qualitative stability* problem for matrices. In this problem a matrix function $\mathbf{M}(\mathbf{p})$ is given and one must determine necessary and sufficient conditions for $\mathbf{M}(\mathbf{p})$ to be the matrix of an unstable linear dynamical system for some $\mathbf{p} \in R^r_+$. The only difference between this problem and the network stability problem is that now $\mathbf{M}(\mathbf{p})$ has a considerably simpler structure. The sign stability problem has been solved recently by Jefferies, Klee, and Van den Driessche,[44] after an error was found by Jefferies[45] in the original solution by Quirk and Ruppert.[46] The method of solution is strikingly similar to the partial solution of the network stability problem given here. Instability was proved[46] by showing that a negative term of a certain type exists in a polynomial whenever a graph (diagram) had cycles of certain types. Stability was proved by constructing a quadratic Lyapunov function in the remaining cases, where another type of diagram had a tree structure. Despite its great simplicity, the sign stability problem required some complicated additional criteria to decide between stability and instability in borderline cases.

The chemical network stability problem still has a sizable gap between those networks that can be proved unstable and those that can be proved stable. Within this gap, the stability always can be decided by constructing polynomials on the computer, provided the network is not too large. At present, the limit is about 9 parameters; however this number could be increased to possibly 13 using a reasonable conjecture.

This formalism has been used to examine the exact solutions to a number of network stability problems that had been "solved" previously by making approximate models. Making such models is far more dangerous than generally realized. I have seen models whose original network was unstable, whose first approximation was stable, and whose second approximation contained an instability entirely unrelated to the original source of instability. From this theory of instability will emerge some insight into how to safely approximate complex networks.

Stoichiometric network analysis may also be used in ecology or economics, where populations of plants, animals, or commodities are analogous to chemical concentrations. Stoichiometry enters when plant or animal reproduction or a technological process uses resources in approximately fixed ratios to produce a new individual or commodity. The ratios do not have to be precisely fixed; rather, the possible steady states must be constrained to be a convex set. Variability in the ratios can be treated within this formalism; the role of stoichiometry is only to limit the range of the variability. The range should be narrow because species and industries that survive in a competitive system cannot be more efficient at reproduction than is possible for the given level of evolution or technology, nor can they lag too far behind the level of efficiency of their competitors. We have discussed how the rate laws $v_j(\mathbf{X}, \mathbf{k}, \underline{\kappa})$ can have any form and that the stability problem depends only on the effective power function $\kappa_{ij}(\mathbf{k})$. In ecology, economics, and chemistry the range of this function should be narrow, for it is unreasonable to suppose that a power law approximation to the actual rate law would have an exponent that differed by many orders of magnitude from unity. If the range of κ_{ij} were to become this large, the theory would degenerate into the sign stability problem for matrices, which was originally proposed in an economic context by Quirk and Ruppert.[46] The sign stability problem has been discussed in ecology by May,[47] and in chemistry by Clarke[38] and Tyson.[48]

The sign stability problem is not really useful in any of these contexts because the system appears much more unstable than it really is, owing to the assumption that $\kappa_{ij}(\mathbf{k})$ can vary over many orders of magnitude. When the theorems on sign stability state that a system can be unstable, the instability often occurs only for parameter values that are equivalent to $\kappa_{ij}(\mathbf{k})$ being many orders of magnitude different from unity. Hence the present theory should be used in ecology and economics in place of the theorems on sign stability. Stoichiometric network analysis is also more appropriate because it is based on a decomposition of all possible flows in the network into extreme flows, as is done in economics. Sign stability omits this complication.

Stoichiometric network analysis is an outgrowth of the "graph theoretic approach" I developed earlier.[36,37] The earlier approach did not use extreme networks, although they were used in the analysis of the Belousov-Zhabotinski reaction system.[40,41] This chapter reconstructs the earlier theory using extreme networks, generalizes it to include non-power-law kinetics, and extends it considerably with new theorems for proving both stability and instability.

II. THE NETWORK AND ITS ACCESSIBLE STATES

A. The Network

The nonitalic symbol X is used in chemistry for the species whose concentration is represented by the italic symbol X. Chemical species will be divided into two types. The concentrations of *internal species* have a significant time variation on the time scale of interest and therefore are dynamical variables. The concentrations of *external species* remain sufficiently close to a constant to be omitted as dynamical variables. For example, omitting the external species (conventionally represented by letters at the beginning of the alphabet) in the reaction

$$A + X \rightarrow B + Y$$

with rate law $v = k'AX$, gives the *pseudoreaction*

$$X \rightarrow Y \tag{II.1}$$

with the rate law $v = kX$, where $k = k'A$. The external species A provides a means whereby the rate constant k of the pseudoreaction may be adjusted throughout the range $0 < k < \infty$ for any fixed $k' > 0$. Physical limits on the concentration of A (e.g., solubility) plus the detailed dependence of the rate laws on the external species determine the *physically accessible domain* of **k**-space. To avoid considering a complicated domain in **k**-space, we consider all of **k**-space to be of interest; this generalization of the problem allows us to omit external species and consider only pseudoreactions between the internal species. We do not include $k_i = 0$ in the accessible domain because when a reaction vanishes, the system is a network with fewer reactions.

The possibility of external species justifies the mathematical treatment of many mechanisms that are unsound chemically and justifies regarding all the rate constants as being adjustable parameters. One may also use interactions with radiation in a similar way and then, for a fixed radiation field, photons of different energies should be regarded as different external species.

When the omission of external species leaves one side of a pseudoreaction blank, the symbol □ appears. This symbol is analogous to the number 0, the vector **0**, and the empty set Ø; however a new symbol is required because the expressions that chemists write on each side of a reaction arrow are not numbers, vectors, or sets. Some authors use Ø, however this choice can lead to confusion with the null set Ø. A commonly occurring pseudoreaction is

$$X \rightleftharpoons \square$$

which could represent diffusive exchange with an external reservoir, a phase equilibrium with an unreactive phase, or production of X from a species with a high (therefore almost constant) concentration.

The set of n-tuples **x** of real numbers will be called R^n. The *positive orthant* of R^n is $\{\mathbf{x} \in R^n | x_i > 0, \text{ i} = 1, \ldots, n\}$ and is denoted R^n_+. Its closure is $\bar{R}^n_+$.

A *chemical network* is a set of n internal species $X_1, \ldots, X_n$, a set of r pseudoreactions $R_1, \ldots, R_r$, a matrix of net reaction stoichiometries $\underline{\boldsymbol{\nu}}$, and a vector function $\mathbf{v}(\mathbf{X}, \mathbf{k})$ whose jth component gives the velocity of the jth reaction when the concentration vector is $\mathbf{X} \in R^n_+$ and the rate constants are $\mathbf{k} \in R^r_+$. Each element ν_{ij} of the net stoichiometric matrix $\underline{\boldsymbol{\nu}}$ is the difference between the number of units of X_i on the right and on the left sides of R_j. The network's *kinetic equation*

$$\frac{d\mathbf{X}}{dt} = \underline{\boldsymbol{\nu}}\mathbf{v}(\mathbf{X}, \mathbf{k}) \tag{II.2}$$

determines the dynamics of a set of chemical systems, one system for each $\mathbf{k} \in R^r_+$.

Example II.1. The Oregonator pseudoreactions and rate laws are as follows:

R_1	$Y \to X$	$v_1 = k_1 Y$
R_2	$X + Y \to \square$	$v_2 = k_2 XY$
R_3	$X \to 2X + 2Z$	$v_3 = k_3 X$
R_4	$2X \to \square$	$v_4 = k_4 X^2$
R_5	$2Z \to fY$	$v_5 = k_5 Z$

This reaction scheme is set up in the formalism by defining

$$\mathbf{v}(\mathbf{k}, \mathbf{X}) \equiv (v_1, \ldots, v_5)^t$$

$$\mathbf{X} \equiv (X_1, X_2, X_3)^t \equiv (X, Y, Z)^t$$

$$\underline{\boldsymbol{\nu}} \equiv \begin{pmatrix} 1 & -1 & 1 & -2 & 0 \\ -1 & -1 & 0 & 0 & f \\ 0 & 0 & 2 & 0 & -2 \end{pmatrix}$$

B. The Concentration Polyhedron

If an atom or other subunit appears only in the internal species, the reaction stoichiometries must conserve the subunit. Let there be γ_{ki} subunits of type k in species X_i; then the total concentration of subunit k is

$$\sum \gamma_{ki} X_i = C_k \tag{II.3}$$

which is a conserved quantity during any motion. The net production of subunit k due to the production of X_i in R_j is $\gamma_{ki}\nu_{ij}$. Since a reaction cannot destroy a conserved subunit, summing over species yields

$$\sum_i \gamma_{ki} \nu_{ij} = 0 \tag{II.4}$$

Since this is true for all j, we may replace ν_{ij} in this equation with $\boldsymbol{\nu}_j$, the jth column of $\underline{\boldsymbol{\nu}}$; hence the columns of $\underline{\boldsymbol{\nu}}$ are linearly dependent. If d is the rank of the $n \times r$ matrix $\underline{\boldsymbol{\nu}}$, there are d linearly independent columns, and each of the remaining $n - d$ linearly dependent columns determines one independent relationship among the columns of $\underline{\boldsymbol{\nu}}$. Corresponding to each such independent relationship there is an independent conserved subunit.

Thus there are $n - d$ conserved subunits, and the *conservation matrix* $\underline{\gamma}$ has $n - d$ linearly independent rows and n columns. The *conservation constraint* vector is $\mathbf{C} \in \bar{R}_{+}^{n-d}$. In matrix notation (II.3) and (II.4) become

$$\underline{\gamma}\mathbf{X} = \mathbf{C} \tag{II.5}$$

$$\underline{\gamma}\,\underline{\nu} = 0 \tag{II.6}$$

For each $\mathbf{C} \in \bar{R}_{+}^{n-d}$ (we call $\bar{R}_{+}^{n-d}$ the *conservation constraint cone*), the set of $\mathbf{X}$ that are stoichiometrically compatible with $\mathbf{C}$ and physically meaningful (nonnegative) is

$$\Pi_X(\mathbf{C}) \equiv \left\{\mathbf{X} \in \bar{R}_{+}^{n} \mid \underline{\gamma}\mathbf{X} = \mathbf{C}\right\} \tag{II.7}$$

This is the set of feasible solutions to a linear programming problem and is in general a (possibly unbounded) *convex polyhedron*.

Few physical scientists are familiar with the n-dimensional geometry of convex sets, which are used in numerous applications throughout this chapter. An introductory text such as those of Hadley[49, 50] should be consulted. A good reference book is by Rockafellar.[51]

Since $\Pi_X(\mathbf{C})$ lies in the d-dimensional subspace S_X, there are d independent dynamical variables and the remaining $n - d$ species are determined by (II.5). From a thermodynamic viewpoint, the d variables are considered internal degrees of freedom and the remaining $n - d$ variables are determined by the thermodynamic state. Hence the number of components in the sense of the Gibbs phase rule is $C = n - d$. We will call $\Pi_X(\mathbf{C})$ the *concentration polyhedron*. The term *reaction simplex* is widely used, but it is a misnomer because $\Pi_X(\mathbf{C})$ is rarely a simplex. The term is also too vague because several other polyhedra can be associated with a reaction system.

The representation (II.7) of $\Pi_X(\mathbf{C})$ is called an *external representation*. It gives a test for determining whether any arbitrarily chosen $\mathbf{X}^*$ is in the set. We now pass over to an *internal representation*, which by definition is an algorithm for constructing points in the set directly.

If the concentration is initially $\mathbf{X}$, a new stoichiometrically accessible concentration $\mathbf{X}^*$ may be obtained by shifting each reaction R_j by an amount ξ_j, called the *extent of reaction*. Then

$$\mathbf{X}^* = \mathbf{X} + \underline{\nu}\xi \tag{II.8}$$

Substitution in (II.5) shows that every $\underline{\xi} \in R^r$ determines $\mathbf{X}^*$ satisfying $\underline{\gamma}\mathbf{X}^* = \underline{\gamma}\mathbf{X}$ and this is a point of $\Pi_X(\underline{\gamma}\mathbf{X})$ provided $\mathrm{X}_i^* \geqslant 0$. Hence

$$\Pi_X(\underline{\gamma}\mathbf{X}) = \{\mathbf{X} + \underline{\nu}\xi \mid \xi \in R^r, \mathbf{X} + \underline{\nu}\xi \geqslant 0\} \tag{II.9}$$

This is only partly internal because the representation still contains a test (inequality). For some $k > 0$, an $n \times k$ matrix H exists such that

$$\Pi_X(\underline{\gamma}\mathbf{X}) = \left\{\mathsf{H}\,\underline{\boldsymbol{\eta}} \mid \underline{\boldsymbol{\eta}} \in \bar{R}^k_+,\ \sum \eta_i = 1\right\}$$

and the representation is internal.

Example II.2. Suppose that a network containing reactions among the species CH_4, C_2H_6, C_2H_4, and C_2H_2 conserves hydrogen and carbon. In this example $n = 4$, $C = n - d = 2$, so $d = 2$ and $\Pi_X(\mathbf{C})$ is two-dimensional. Indexing species in the order mentioned gives

$$\underline{\gamma} = \begin{pmatrix} 1 & 2 & 2 & 2 \\ 4 & 6 & 4 & 2 \end{pmatrix}$$

One may show that $\Pi_X(\mathbf{C})$ is a triangle (simplex) if $(2, -1)\mathbf{C} \geqslant 0$ or if $(3, -1)\mathbf{C} \geqslant 0$, and is otherwise a quadrilateral. Also, when $(5, -2)\mathbf{C} = 0$, the quadrilateral $\Pi_X(\mathbf{C})$ can be represented in the completely internal form $\{\mathsf{H}\underline{\boldsymbol{\eta}} \mid \underline{\boldsymbol{\eta}} \in \bar{\mathrm{R}}^4_+,\ \sum_i \eta_i = 1\}$, where

$$\mathsf{H} = \begin{bmatrix} 0 & 0 & 4 & 2 \\ 3 & 2 & 0 & 0 \\ 0 & 2 & 0 & 0 \\ 1 & 0 & 2 & 0 \end{bmatrix}$$

A method of proof appears in the next section.

C. The Current Cone and Polytope

We now turn our attention from n-dimensional concentration space to r-dimensional reaction velocity space. The forward and reverse velocities of reversible reactions are given by two nonnegative numbers, and the corresponding columns of $\underline{\nu}$ sum to zero. Hence physical velocities satisfy the constraint

$$v_i(\mathbf{X}, \mathbf{k}) \geqslant 0 \tag{II.10}$$

From (II.2) and the definition of the steady state, $d\mathbf{X}/dt = 0$, we obtain the steady-state condition for reaction velocities

$$\underline{\nu}\mathbf{v}^0(\mathbf{X}, \mathbf{k}) = 0 \tag{II.11}$$

The steady-state condition is similar to Kirchhoff's current law for electrical networks; hence solutions $\mathbf{v}^0$ to (II.10) and (II.11) are called *currents*. The set of all currents is a set called the *current cone*

$$\mathcal{C}_v \equiv \left\{\mathbf{v}^0 \in \bar{R}^r_+ \mid \underline{\nu}\mathbf{v}^0 = 0\right\} \tag{II.12}$$

This is an external representation of $\mathcal{C}_v$ and can be converted to an internal representation whose form is

$$\mathcal{C}_v = \left\{\mathsf{E}\mathbf{j} \mid \mathbf{j} \in \bar{R}^f_+\right\} \tag{II.13}$$

where E is an $r \times f$ matrix. Let us briefly outline the mathematical justification for this step. The details may be found in Hadley,[50] although they are not organized as we need them. One may rearrange (II.12) to obtain

$$\mathcal{C}_v = \{\lambda \mathbf{v}^0 \mid \mathbf{v}^0 \in \Pi_v, \lambda \geq 0\} \tag{II.14}$$

$$\Pi_v \equiv \{\mathbf{v}^0 \in \bar{R}^r_+ \mid \underline{\boldsymbol{\nu}} \mathbf{v}^0 = 0, \mathbf{e}_r^t \mathbf{v}^0 = 1\} \tag{II.15}$$

where $\mathbf{e}_r = (1, \ldots, 1)^t \in R^r$. The representation conforms to the general definition of a convex polyhedral cone given in Section 2-22 of Ref. 50. The two restrictions on $\mathbf{v}^0$ in (II.15) may be combined into the equation

$$\mathsf{B}\mathbf{v}^0 = \mathbf{b} \tag{II.16}$$

where B is obtained by adding a bottom row of 1's to d linearly independent rows chosen from $\underline{\boldsymbol{\nu}}$, and $\mathbf{b} = (0, 0, 0, \ldots, 0, 1)^t \in R^{d+1}$.

Then (II.15) is identical in form to (II.14) and conforms to Hadley's definition of the set of feasible solutions of a linear programming problem. Hadley's[50] Section 2-20 proves that these solutions form a closed bounded convex set. Hence we call Π_v the *current polytope*. Then Theorem IV in Section 2-21 may be used to express Π_v as a convex combination of its extreme points, which we write in matrix notation as

$$\Pi_v = \{\mathsf{E}\mathbf{j} \mid \mathbf{j} \in \bar{R}^f_+, \mathbf{e}_f^t \mathbf{j} = 1\} \tag{II.17}$$

where $\mathbf{e}_f = (1, 1, \ldots, 1)^t \in R^f$. Dropping the second restriction on $\mathbf{j}$ and combining with (II.14) gives the internal representation (II.13) for $\mathcal{C}_v$.

In linear programming the feasible solutions form a convex polytope whose extreme points are called *basic feasible solutions*. The columns of the matrix E consist of all basic feasible solutions. Here the columns will be called *extreme currents*. The normalization conditions on $\mathbf{v}^0$ and $\mathbf{j}$ in (II.15) and (II.17) cause each column of E to sum to 1. The property is not needed because (II.13) is still valid when each column of E is multiplied by an arbitrary positive number. Hence any positive multiple of a basic feasible solution will also be called an extreme current. The extreme currents are vectors in $\bar{R}^r_+$ that point along the edges of the current cone and form the cone's *frame*.

Performing stability analysis on chemical networks frequently calls for the calculation of E. Little linear programming literature exists on this topic (but see Hadley,[50] Section 2-16) because this calculation is not required in that subject. Hence there follows a detailed discussion of how this is done.

Every $\mathbf{v}^0 \in \mathcal{C}_v$ may be represented in the form

$$\mathbf{v}^0 = \mathsf{E}\mathbf{j} \tag{II.18}$$

for some $\mathbf{j} \in \bar{R}^f_+$, and every so representable point is also in $\mathcal{C}_v$. Substituting into (II.11) gives $\underline{\boldsymbol{\nu}}\mathsf{E}\mathbf{j} = 0$ for all $\mathbf{j} \in \bar{R}^f_+$, hence

$$\underline{\boldsymbol{\nu}}\mathsf{E} = 0 \tag{II.19}$$

If $\boldsymbol{\nu}_i^\dagger$ is the ith row of $\underline{\boldsymbol{\nu}}$ and $\mathbf{E}_j$ is the jth column of E, (II.19) says that the inner product $\boldsymbol{\nu}_i^{\dagger\prime}\mathbf{E}_j = 0$ for all i and j. Therefore $\mathbf{E}_j$ is orthogonal to every row of $\underline{\boldsymbol{\nu}}$. Since $\underline{\boldsymbol{\nu}}$ has d linearly independent rows, $\mathbf{E}_j$ must lie in the $(r-d)$-dimensional linear subspace S_v orthogonal to the rows. This is called the *null space* of $\underline{\boldsymbol{\nu}}$. The restriction $\mathbf{v}^0 \in \bar{R}^r_+$ in (II.12) says that $\mathbf{E}_j$ must also lie in the closed positive orthant. Thus $\mathbf{E}_j \in S_v \cap \bar{R}^r_+$. If $S_v \cap R^r_+ = \emptyset$, reactions are present that have zero velocity for all steady states. After eliminating these superfluous reactions $S_v \cap R^r_+ \neq \emptyset$. Then, since R^r_+ is open, $\mathcal{C}_v$ must have the same dimension as S_v; hence $\dim \mathcal{C}_v = r - d$. A cone of this dimension cannot be spanned by fewer than $r - d$ frame vectors, so

$$f \geqslant r - d \tag{II.20}$$

Viewed from within S_v, $\mathcal{C}_v$ has no boundary other than the boundary of $\bar{R}^r_+$. This boundary consists entirely of the coordinate hyperplanes $v_i^0 = 0$, $i = 1, \ldots, r$. If $\mathcal{C}_v$ intersects the ith hyperplane other than at the origin, we can investigate this intersection by setting $v_i^0 = 0$ in (II.11). The resulting equations plus (II.10) are satisfied on cone $\mathcal{C}_{v2}$, which is a *facet* of $\mathcal{C}_v$. This cone lies within $S_{v2} \equiv S_v \cap \{\mathbf{v}^0 \in \bar{R}^r_+ \mid v_i = 0\}$ and is $(r-d-1)$-dimensional. If $\mathcal{C}_{v2}$ lies entirely within the coordinate hyperplanes of some $v_j, j \neq i$, we can drop the corresponding variables at this stage to obtain an r_2-dimensional space containing S_{v2} and $\mathcal{C}_{v2}$ [which are still $(r-d-1)$-dimensional]. Viewed from within S_{v2}, $\mathcal{C}_{v2}$ has no boundary other than the boundary of $\bar{R}^{r_2}_+$, which consists of the remaining coordinate hyperplanes; that is, we exclude $v_i^0 = 0$ and any that were dropped because they contain $\mathcal{C}_{v2}$. If $\mathcal{C}_{v2}$ intersects the jth hyperplane other than at the origin, we can investigate this intersection as we did with $\mathcal{C}_v$ by setting $v_j^0 = 0$ to obtain S_{v3} and $\mathcal{C}_{v3}$. This process may be continued until we obtain a cone that does not intersect any coordinate hyperplane except at the origin. At the kth stage the cone $\mathcal{C}_{vk}$ has dimension $r - d - k + 1$. If the dimension of $\mathcal{C}_{vk}$ is greater than one, S_{vk} must intersect some coordinate axes and the process can always be continued. However when $k = r - d$, $\mathcal{C}_{vk}$ is a one-dimensional cone or *half-line*, and after dropping all components v_l^0

containing the half-line, S_{r-d} will not intersect the coordinate hyperplanes of any remaining component v_i^0 except at the origin. This half-line is a frame vector of $\mathcal{C}_v$, and any point $\mathbf{v}^0 \neq 0$ on it is an extreme current.

It should be clear from the preceding discussion that an extreme current can be found by setting a certain number of components of $\mathbf{v}^0$ equal to zero and then solving (II.16). One must set to zero enough components to give (II.16) a unique solution for the remaining components. If the final row of B were a linear combination of the remaining rows, every solution of (II.16) would satisfy $\mathbf{e}_r^t\mathbf{v}^0 = 0$ and (II.16) would be inconsistent. Since $\mathbf{e}_r^t\mathbf{v}^0 = 1$, B has rank $d+1$. To find an extreme current, we select $d+1$ columns of B and construct a square matrix that usually may be inverted to obtain $\mathbf{v}^0$. All possible extreme currents are found by selecting the $d+1$ columns in all possible ways (see Ref. 50, Section 2-16) and then keeping only solutions satisfying (II.10).

It often happens that for certain choices of $d+1$ columns, the resulting matrix has rank less than $d+1$. Then (II.16) cannot be solved because the chosen columns of B form a singular matrix. For these choices of columns the calculation may be abandoned because every extreme solution of (II.10) and (II.11) can be obtained from (II.16) with a linearly independent set of columns.

Example II.3. Describe $\mathcal{C}_v$ and Π_v when $r = 3$ and $d = 1$; S_v is a two-dimensional plane through the origin. If there are no reactions that vanish for all steady states, S_v intersects R_+^3 and then $\mathcal{C}_v$ is a wedge; Π_v must be a line segment.

Example II.4. Find E when $\underline{\nu} = (1, 1, -1, -1)$; B is a 2×4 matrix and two columns may be chosen in six ways. Of these, the two positive and two negative ones give unphysical solutions. The remaining four choices give

$$\mathsf{E} = \begin{pmatrix} 1 & 1 & 0 & 0 \\ 0 & 0 & 1 & 1 \\ 0 & 1 & 1 & 0 \\ 1 & 0 & 0 & 1 \end{pmatrix}$$

$\mathcal{C}_v$ is a three-dimensional pyramid-shaped cone with a square cross-section.

The dimension of Π_v is $r - d - 1$. If $\dim \Pi_v = 0$, Π_v is a point, and there is only one extreme current. If $\dim \Pi_v = 1$, Π_v is a line segment and there are two extreme currents. If $\dim \Pi_v \geqslant 3$ the number of vertices of Π_v may be more than $r - d$ and the inequality sign in (II.20) may apply.

Example II.5. Find a set of extreme currents for the Oregonator, (Example II.1) with $f = 1$. The rows of $\underline{\nu}$ are independent, so $d = \operatorname{rank} \underline{\nu} = 3$, $r = 5$, $n = 3$, and $\dim \Pi_v = r - d - 1 = 1$. Hence there are two extreme currents. The matrix B in (II.16) is

$$\begin{pmatrix} 1 & -1 & 1 & -2 & 0 \\ -1 & -1 & 0 & 0 & 1 \\ 0 & 0 & 2 & 0 & -2 \\ 1 & 1 & 1 & 1 & 1 \end{pmatrix}.$$

We can choose four columns in five ways, but only two solutions to (II.16) will be nonnegative. Inspection of the third row helps find the proper choice of columns. Normalizing the extreme currents to have integer components gives

$$\mathsf{E} = \begin{pmatrix} 0 & 1 \\ 1 & 0 \\ 1 & 1 \\ 0 & 1 \\ 1 & 1 \end{pmatrix}$$

The extreme currents can easily be found by a computer algorithm that tests all possible choices of columns. The algorithm tests fewer columns if reverse reactions are deleted from $\underline{\boldsymbol{\nu}}$. To accommodate reverse reactions, solutions with $v_i^0 < 0$ are considered to be physical and are reinterpreted as positive velocities for the reverse reactions. When the matrix chosen from B is singular, as discussed previously, one may abandon the calculation. Extreme currents with fewer than $d+1$ nonzero components can be generated in several ways; hence the algorithm must also check for redundancy. To the extreme currents generated by this algorithm, one must add one extreme current for each reversible reaction. Such currents consist of the forward reaction and its reverse.

A polytope with the minimum number of vertices possible for its dimension is called a *simplex*. If a particular set of $n+1$ points can be represented by $n+1$ linearly independent vectors, the points are the vertices of an n-dimensional simplex. Hence an n-dimensional convex polytope is a simplex if and only if it has $n+1$ vertices. The first few simplices for $n = 0, 1, 2, 3$ are the point, the line segment, the triangle, and the tetrahedron. A corollary of Caratheodory's theorem[51] states that every n-dimensional convex polytope can be expressed as the union of a finite number of n-dimensional simplices, whose intersection has dimension less than n and whose vertices are vertices of the polytope. For example, one constructs the *simplicial decomposition* of a square ($n = 2$) by drawing a diagonal to form two closed triangles whose union is the square and whose intersection is their common edge (one-dimensional). Two planes can divide an octahedron into four (irregular) tetrahedra.

We are now coming to the crux of this discussion. We want to find a set of parameters that can represent an element of the set of all currents as simply as possible. These parameters should have the simplest possible domain and be as few as possible. Since $\mathcal{C}_v$ is $(r-d)$-dimensional, we might use $r-d$ reaction velocities, the remaining d being dependent. Better yet, we could set up an orthogonal basis for S_v. Since the basis would have $r-d$ vectors, $r-d$ parameters would be necessary. Unfortunately, inequality (II.10) would make the domain of these parameters very complicated. It will be important later to have all parameters nonnegative. To make them nonnegative we could use a skewed basis for S_v so that $\mathcal{C}_v$

lay entirely in the domain of nonnegative parameters, but then equality (II.10) could still be an undesirably complicated constraint. Alternatively, we could use $\mathbf{j}$ as a parameter vector. Recall that $\mathbf{v}^0 \in \mathcal{C}_v$ if and only if $\mathbf{v}^0 = \mathsf{E}\mathbf{j}$ and $\mathbf{j} \in \bar{R}^f_+$. The domain of $\mathbf{j}$ is simple, but there are f parameters instead of $r-d$. If Π_v is a simplex, $f = r-d$, and this is the best parametrization because the number of parameters is a minimum and their domain is the simplest possible. Otherwise, $f > r-d$, and there are more parameters than necessary. We therefore make a simplicial decomposition of Π_v into l simplices. To each simplex Σ_i corresponds a cone $\mathcal{C}_{vi} \equiv \{\lambda\mathbf{v}^0 \mid \lambda \geqslant 0, \mathbf{v}^0 \in \Sigma_i\}$; hence $\mathcal{C}_v$ has been decomposed into a set of l *simplicial cones*. Each point in $\mathcal{C}_v$ can then be specified by a vector $\mathbf{j}^* \in \bar{R}^{r-d}_+$ and an integer i in the range $1 \leqslant i \leqslant l$ that specifies the simplex Σ_i. Then

$$\mathbf{v}^0 = \mathsf{E}^*_i\mathbf{j}^* \tag{II.21}$$

where E^*_i is the $r \times (r-d)$ matrix made from the $r-d$ columns of E that have been scaled such that $\mathbf{e}^t\mathsf{E}^*_i = \mathbf{e}^t$ and chosen to be vertices of Σ_i.

If $\mathbf{v}^0$ is in the relative interior of $\mathcal{C}_{vi}$, i and $\mathbf{j}^*$ are uniquely determined by (II.21) and the requirement $\mathbf{j} \geqslant 0$. However if $\mathbf{v}^0$ is on the boundary of $\mathcal{C}_{vi}$ it may also lie in other cones, and i is not uniquely determined. To obtain a one-to-one mapping between $\mathbf{v}^0$ and $(\mathbf{j}^*, i)$, the cones $\mathcal{C}_{vi}$ are chosen closed along arbitrary parts of their boundaries and open along the remainder of their boundaries such that

$$\mathcal{C}_v = \bigcup_{i=1}^{l} \mathcal{C}_{vi} \tag{II.22}$$

$$\mathcal{C}_{vi} \cap \mathcal{C}_{vj} = \varnothing \qquad \text{for all } i, j \in [1, l], i \neq j \tag{II.23}$$

Then each $\mathbf{v}^0 \in \mathcal{C}_v$ lies in a unique $\mathcal{C}_{vi}$ and has a unique $\mathbf{j}^*$ in the domain

$$\tilde{R}_i^{+(r-d)} \equiv \{\mathbf{j}^* \in \bar{R}^{r-d}_+ \mid \mathsf{E}^*_i\mathbf{j}^* \in \mathcal{C}_{vi}\} \tag{II.24}$$

If $\mathcal{C}_{vi}$ is closed, $\tilde{R}_i^{+(r-d)} = \bar{R}^{r-d}_+$, and if $\mathcal{C}_{vi}$ is open, $\tilde{R}_i^{+(r-d)} = R^{r-d}_+$. Otherwise $\tilde{R}_i^{+(r-d)}$ is a partly open, partly closed orthant, and in general satisfies $R^{r-d}_+ \subset \tilde{R}_i^{+(r-d)} \subset \bar{R}^{r-d}_+$. Then (II.21) is the one-to-one mapping

$$\phi : \{(\mathbf{j}^*, i) \mid i \in [1, l], \mathbf{j}^* \in \tilde{R}_i^{+(r-d)}\} \to \mathcal{C}_v \tag{II.25}$$

The existence of this mapping is important; however, we will never have to construct it in detail. Practical problems require only the extreme currents.

D. The Extent of Reaction Polyhedron

R^n is the sum of two subspaces S_C and S_X, the former having dimension $n-d$ and being spanned by the $n-d$ linearly independent rows of $\underline{\gamma}$, the latter having dimension d and being spanned by the d linearly independent columns of $\underline{\nu}$. The orthogonality of these subspaces follows from (II.6). Note that for any $\mathbf{Y} \in R^{n-d}$, $\underline{\gamma}'\mathbf{Y}$ is a linear combination of the rows of $\underline{\gamma}$, and hence lies in S_C; similarly, for any $\xi \in R^r$, $\underline{\nu}\xi$ is a linear combination of the columns of $\underline{\nu}$, hence lies in S_X. Thus every $\mathbf{X} \in R^n$ may be decomposed into components in these subspaces by writing

$$\mathbf{X} = \underline{\gamma}^t\mathbf{Y} + \underline{\nu}\xi \tag{II.26}$$

Multiplying on the left by $\underline{\gamma}$ and using (II.5) and (II.6) gives

$$\mathbf{C} = \underline{\gamma}\,\underline{\gamma}^t\mathbf{Y} \tag{II.27}$$

The linear independence of the rows of $\underline{\gamma}$ makes $\underline{\gamma}\underline{\gamma}'$ nonsingular, so that (II.27) may be solved for $\mathbf{Y}$, and (II.26) becomes

$$\mathbf{X} = \underline{\gamma}^t\left(\underline{\gamma}\,\underline{\gamma}^t\right)^{-1}\mathbf{C} + \underline{\nu}\xi \tag{II.28}$$

Differentiating and comparing with (II.2) gives

$$\underline{\nu}\,\frac{d\xi}{dt} = \underline{\nu}\mathbf{v}$$

which has many solutions, one of which is

$$\xi(t) = \int_0^t \mathbf{v}(t)\,dt \tag{II.29}$$

so ξ is called the *extent of reaction vector*.

If ξ is changed by a vector $\Delta\xi \in S_v$, the null space of $\underline{\nu}$, the concentration vector $\mathbf{X}$ is not affected because $\underline{\nu}\Delta\xi = 0$ and $\Delta\xi$ contributes nothing to (II.28). Hence without loss of generality, we may assume that ξ is orthogonal to S_v, that is, $\xi \in S_\xi$, the d-dimensional subspace of R^r orthogonal to S_v. Equation (II.29) does not satisfy this convention, but $\mathsf{P}_\xi\xi(t)$ does, where P_ξ is the matrix that projects R^r onto S_ξ.

Equation (II.28) is a mapping ψ from S_ξ to $\gamma^t(\gamma\gamma^t)^{-1}\mathbf{C} + S_X$, the d-dimensional affine subspace containing $\Pi_X(\mathbf{C})$. Consequently, there is a d-dimensional polyhedron, called the *extent of reaction polyhedron*,

$$\Pi_\xi(\mathbf{C}) \equiv \{\xi \in S_\xi \,|\, \psi(\xi) \geqslant 0\}$$

in S_ξ, which must have the same facet structure as $\Pi_X(\mathbf{C})$ because ψ is a nonsingular linear mapping between two d-dimensional subspaces and the conditions $\mathbf{X} \geqslant 0$ and $\psi(\xi) \geqslant 0$ are equivalent. Hence there is a one-to-one correspondence between the accessible concentration states $\mathbf{X} \in \Pi_X(\mathbf{C})$ and the accessible extents of reaction $\xi \in \Pi_\xi(\mathbf{C})$. The accessible steady-state velocities are points $\mathbf{v}^0 \in \mathcal{C}_v$, but the orthogonality of S_v and S_ξ implies that

$$\boldsymbol{\xi}'\mathbf{v}^0 = 0 \tag{II.30}$$

This orthogonality relation between extents of reaction and currents is related to the orthogonality relation between voltage and currents in electrical networks (Tellegen's theorem). Since Π_v and $\Pi_\xi(\mathbf{C})$ lie in orthogonal subspaces and have unrelated dimensionality, we have proof that there cannot be any general relationship between the facet structures of these polyhedra. The properties of the subspaces and polyhedra discussed so far are summarized in Table I. Note that $\boldsymbol{\nu}_i^\dagger$ and $\gamma_i^\dagger$ are the rows of $\underline{\boldsymbol{\nu}}$ and $\underline{\gamma}$; $\boldsymbol{\nu}_i$ and $\mathbf{E}_i$ are the columns of $\underline{\boldsymbol{\nu}}$ and E; and $\mathbf{v}^0$ is a steady-state reaction velocity.

TABLE I
Properties of Subspaces and Polyhedra

Spaces	Species concentration $\mathbf{X} \in R^n$		Reaction velocity $\mathbf{v} \in R^r$	
Subspaces	$\Delta\mathbf{X} \in S_X$	$\mathbf{C} \in S_C$	$\boldsymbol{\xi} \in S_\xi$	$\mathbf{v}^0 \in S_v$
Dimensions	d	$n-d$	d	$r-d$
Spanned by	$\boldsymbol{\nu}_i$	$\gamma_i^\dagger$	$\boldsymbol{\nu}_i^\dagger$	$\mathbf{E}_i$
Cones		$\bar{R}_+^{r-d}$		$\mathcal{C}_v$
Polyhedron	$\Pi_X(\mathbf{C})$		$\Pi_\xi(\mathbf{C})$	Π_v
Names	Concentration polyhedron		Extent of reaction polyhedron	Current polytope

E. Convex Parameters for the Steady States

The stability of all steady states of the network can best be studied when an elegant set of steady-state parameters is available. In this section we examine three parametrizations: the usual parameters called *kinetic parameters*, the parameters of stoichiometric network analysis called *convex parameters*, and finally the *thermodynamic parameters* of nonequilibrium thermodynamics. The parameters' physical domains are, respectively, the sets D_K, D_C, and D_T. Therefore a given steady state can be represented in any of the three parametrizations as points $\mathbf{p}_K \in D_K$, $\mathbf{p}_C \in D_C$, or $\mathbf{p}_T \in D_T$. To convert between parametrizations we will later define two mappings $\phi_{KC}: D_K \to D_C$ and $\phi_{CT}: D_C \to D_T$. These mappings will be one-to-one and will enable us to convert between any pair of parametrizations. The discussion is slightly simpler if we assume *power law kinetics*; that is, that there exists a matrix $\underline{\boldsymbol{\kappa}}$ such that

$$v_j(\mathbf{X}, \mathbf{k}) = k_j \prod_{i=1}^{n} X_i^{\kappa_{ij}} \tag{II.31}$$

In the kinetic parameters, the steady state is considered a function of the rate constants $k_i \geqslant 0$. If k_i vanishes, the ith reaction can be deleted from the network and a new network results. Hence without loss of generality, the domain of $\mathbf{k}$ for the original network can be the open orthant R_+^r. Similarly, physical concentrations satisfy $X_i \geqslant 0$. If $X_i = 0$, reactions with power law kinetics that consume X_i have zero velocity and the network may be considered to be a different network with fewer species and reactions. Without loss of generality we may therefore assume that $\mathbf{X}$ is in the open orthant R_+^n, and consequently no component of $\mathbf{C}$ can vanish; thus $\mathbf{C} \in R_+^{n-d}$.

In chemical evolution the first appearance of molecule X_i produces a transition from a network where X_i is missing to a network where X_i is present and $X_i > 0$. It is possible for the original network to be stable and the new network to be unstable for small X_i. Hence if we were to take $\mathbf{X}$ in the closed orthant, the network's stability properties could change discontinuously at the orthant boundary, thus leading to mathematical difficulties. This is a possible interpretation of the difficulties that occur if a component of $\mathbf{k}$, $\mathbf{X}$, or $\mathbf{C}$ vanishes. There is no loss of generality in assuming that all components are positive because the remaining cases can be treated as different networks.

The accessible states of a network must now be redefined as open sets. If T is any closed set, the boundary of T ($\operatorname{bdy} T$) is the set of points of T whose every neighborhood contains points outside of T in the affine subspace of minimal dimension containing T. The relative interior of T is $\operatorname{ri} T \equiv T \backslash \operatorname{bdy} T$, where \ means set theoretic subtraction or difference; that is, $A \backslash B$ means the points of A that are not in B. Hence for $\mathbf{C} \in R_+^{r-d}$, the accessible concentration states of the network are in $\operatorname{ri} \Pi_X(\mathbf{C})$, the accessible extents of reactions are in $\operatorname{ri} \Pi_\xi(\mathbf{C})$, and the accessible currents are in $\operatorname{ri} \mathcal{C}_v$, whose cross-section is the open current polytope $\operatorname{ri} \Pi_v$. A simplicial decomposition of $\operatorname{ri} \mathcal{C}_v$ is $\{\hat{\mathcal{C}}_{vi} \mid i = 1, \ldots, l\}$, and we assume that these cones satisfy (II.22) and (II.23). In analogy with (II.24), the domain of $\mathbf{j}^*$ such that $\mathbf{E}_i^* \mathbf{j}^* \in \hat{\mathcal{C}}_{vi}$ will be called $\hat{R}_i^{+(r-d)}$ and as before $R_+^{r-d} \subset \hat{R}_i^{+(r-d)} \subset \overline{R}_+^{r-d}$. Then (II.25) is modified to the one-to-one mapping

$$\hat{\phi} : \left\{ (\mathbf{j}^*, i) \mid i \in [1, l], \mathbf{j}^* \in \hat{R}_i^{+(r-d)} \right\} \rightarrow \operatorname{ri} \mathcal{C}_v$$

For each $\mathbf{k} \in R_+^r$ and $\mathbf{C} \in R_+^{n-d}$, assume that there is a countable number $\rho(\mathbf{k}, \mathbf{C})$ of steady-state solutions $\mathbf{X} \in R_+^n$ for the system of polynomial equations

$$\underline{\nu} v(\mathbf{X}, \mathbf{k}) = 0 \tag{II.32}$$

for which $\mathbf{X}$ is consistent with $\underline{\nu}$ and $\mathbf{C}$ in the sense of (II.5) and (II.6). Section V.I discusses the case of an uncountable number of solutions. Of course a suitable matrix $\underline{\gamma}$ must have been previously chosen to give $\mathbf{C}$ meaning. For $\mathbf{k}$ and $\mathbf{C}$ such that $\rho(\mathbf{k}, \mathbf{C}) = 0$, there is no steady state to parametrize. Otherwise, for each $\mathbf{k}$ and $\mathbf{C}$, label the steady state with an index i, $1 \leqslant i \leqslant \rho(\mathbf{k}, \mathbf{C})$. A steady state can now be specified by $r + n - d + 1$ numbers $\mathbf{p}_K \equiv (\mathbf{k}, \mathbf{C}, i)$. Hence there is a one-to-one correspondence between physical steady states and the elements of

$$D_K \equiv \{(\mathbf{k}, \mathbf{C}, i) | \mathbf{k} \in R_+^r, \mathbf{C} \in R_+^{n-d}, 1 \leqslant i \leqslant \rho(\mathbf{k}, \mathbf{C})\}. \tag{II.33}$$

For each $\mathbf{p}_K \in D_K$, the network has a matrix $\mathsf{M}_K(\mathbf{p}_K)$ that is obtained by linearizing the dynamics about the steady state (see Section III). The *kinetic parametrization* is very poor for stability analysis because $\mathsf{M}_K(\mathbf{p}_K)$ is an extremely complicated function of $\mathbf{p}_K$.

The *convex parameters* are elements of the set

$$D_C \equiv \{(\mathbf{h}, \mathbf{j}^*, i) | \mathbf{h} \in R_+^n, \mathbf{j}^* \in \tilde{R}_i^{+(r-d)}, \qquad 1 \leqslant i \leqslant l\} \tag{II.34}$$

where i is an integer, and l is the number of simplicial cones $\mathcal{C}_{vi}$ in the decomposition of the current cone. The physical steady state corresponding to an element $\mathbf{p}_C \in D_C$ will be defined by specifying the mapping $\phi_{KC}: D_K \to D_C$ and proving this mapping to be one to one. For each $\mathbf{p}_K \in D_K$ there is a positive vector $\mathbf{X}^0$ representing the steady-state concentrations. The first n components of $\mathbf{p}_C \equiv \phi_{KC}(\mathbf{p}_K) \in D_C$ are defined by

$$h_i = 1/X_i^0 \tag{II.35}$$

This equation defines a unique $\mathbf{h}$ for every $\mathbf{p}_K \in D_K$ because D_K does not include steady states where any components of $\mathbf{X}$ vanish. For each $\mathbf{p}_K \in D_K$ there is also a unique positive vector $\mathbf{v}^0$ representing the steady-state reaction rates, hence $\mathbf{v}^0 \in \mathcal{C}_v$. Each such $\mathbf{v}^0$ is an element of one simplicial cone $\mathcal{C}_{vi}$. A unique vector $\mathbf{j}^* \in \tilde{R}_i^{+(r-d)}$ exists so that $\mathbf{v}^0$ can be expressed as in (II.21). Hence $\mathbf{p}_K$ determines a unique $\mathbf{j}^*$ and i that are the remaining $r - d + 1$ components of $\mathbf{p}_C$. Recall that if $\mathbf{v}^0$ is on the boundary of more than one simplicial cone, the cone to be used is specified by convention; consequently the domain $\tilde{R}_i^{+(r-d)}$ frequently is partly open and partly closed. The definition of $\phi_{KC}(\mathbf{p}_K)$ is now complete, and it is clear that $\phi_{KC}(D_K) \subset D_C$.

To prove that ϕ_{KC} is an isomorphism, we show that each $\mathbf{p}_C \in D_C$ is the image of a unique $\mathbf{p}_K \in D_K$. The first n components of each $\mathbf{p}_C \in D_C$ form the vector **h**, which determines a set of n nonzero steady-state concentrations via (II.35). The final $r - d + 1$ components of $\mathbf{p}_C$ determine $\mathbf{j}^*$ and i, and thus a unique $\mathbf{v}^0$ via (II.21). A unique physical steady state is specified by $\mathbf{X}^0$ and $\mathbf{v}^0$. To determine its kinetic parameters, we solve (II.31) for k_i to obtain

$$k_j = v_j^0 \prod_{i=1}^{n} X_i^{0^{-\kappa_{ij}}} \tag{II.36}$$

and use (II.5) to obtain **C**. The vectors **k** and **C** have $\rho(\mathbf{k}, \mathbf{C})$ steady states and the one with $\mathbf{X}^0$ and $\mathbf{v}^0$ has, by convention, a particular value of i. Hence every $\mathbf{p}_C \in D_C$ is the image of a unique $\mathbf{p}_K \in D_K$, so ϕ_{KC} is an isomorphism.

As an aid to understanding the relationship between stoichiometric network analysis and thermodynamic approaches, we now introduce the thermodynamic parameters. First let us compare the number of parameters in the general open system, and in a similar closed system in thermodynamic equilibrium. For the equilibrium system, the Gibbs phase rule gives the number of degrees of freedom

$$F = C - P + 2 = n - d + 1 \tag{II.37}$$

where one phase is assumed ($P = 1$), and the number of components is $C = n - d$. As parameters for this system, choose the reciprocal temperature $\beta = 1/RT$, and the dimensionless product $\beta\mu_j$, $j \in [1, n - d]$, where μ_j is the *chemical potential* of species j. These parameters are the fundamental variables of the *entropy representation* of thermodynamics (Callen) and are the most profound parameters because the partition functions of the statistical mechanical ensembles all take the same general form only when expressed in these parameters. We now have a set of F thermodynamic parameters; however we must add one more parameter i to count the number of steady states, even though we know $i = 1$. In the general open system there are $n + r - d + 1$ parameters, which is $r - 1$ more parameters than in the equilibrium case. To explain why $r - 1$ additional parameters are needed, note that there are r independent rate constants in the general case (using D_K), whereas in the equilibrium case the rate constants can be varied along a curve by varying T. Hence the accessible region of D_K has $r - 1$ fewer dimensions in the equilibrium case. Nonequilibrium thermodynamic parameters could be constructed by keeping $\beta\mu_j$, $j \in [1, n - d]$ and replacing β with r new parameters. Instead we will use $\beta\mu_j$, $j \in [1, n]$ and then only $r - d$ new parameters are needed.

In nonequilibrium thermodynamics the net reaction rates (fluxes) are parameters. To define these, assume that all reverse reactions are present, that the first $r/2$ reactions are *forward*, and that the last $r/2$ reactions are *reverse*. Then $\boldsymbol{\nu}_i + \boldsymbol{\nu}_{i+r/2} = 0$. Define the *flux vector* $\mathbf{J}$ by

$$\mathbf{J} = \mathsf{F}\mathbf{v}, \qquad \mathsf{F} \equiv \begin{pmatrix} \mathsf{I} & \mathsf{I} \\ \mathsf{I} & -\mathsf{I} \end{pmatrix} \tag{II.38}$$

where I is the identity matrix of order $r/2$. Let $\mathbf{J} = (\mathbf{J}_E, \mathbf{J}_N)$, where $\mathbf{J}_E \in R^{r/2}$, $\mathbf{J}_N \in R^{r/2}$. The net reaction rates are contained in $\mathbf{J}_N$, which is called the *nonequilibrium flux vector*; $\mathbf{J}_E$ is called the *equilibrium flux vector* for reasons that will be clear in the next section. Note that $\mathsf{F}^{-1} = \frac{1}{2}\mathsf{F}$; hence

$$\mathbf{v} = \tfrac{1}{2}\mathsf{F}\mathbf{J} = \tfrac{1}{2}(\mathsf{F}_E\mathbf{J}_E + \mathsf{F}_N\mathbf{J}_N) \tag{II.39}$$

where F_E and F_N are the first and last $r/2$ columns of F. Since $\dim \mathcal{C}_v = r - d$, and the mapping (II.38) is nonsingular, only $r - d$ components of $\mathbf{J}$ are independent at steady state. These will be the $r - d$ new independent parameters.

To find the domain of $\mathbf{J}$ corresponding to $\mathcal{C}_v$, multiply (II.39) on the left by $\underline{\boldsymbol{\nu}}$. Since $\underline{\boldsymbol{\nu}}\mathsf{F}_E = 0$, we obtain $\underline{\boldsymbol{\nu}}\mathsf{F}_N\mathbf{J}_N = 0$. The steady-state condition is satisfied for all $\mathbf{J}_E \in R^{r/2}$, however, $\mathbf{v}^0 > 0$ if and only if $\mathbf{J}_E \in R_+^{r/2}$. The vector $\mathbf{J}_N$ is restricted to a subspace of dimension $r/2 - d$, and the condition $\mathbf{v}^0 > 0$ determines a complicated domain $D^{r/2-d}(\mathbf{J}_E)$, which depends on $\mathbf{J}_E$. The domain of the thermodynamic parameters will be

$$D_T \equiv \left\{ (\beta\boldsymbol{\mu}, \mathbf{J}) \mid \beta\boldsymbol{\mu} \in R^n, \mathbf{J} = (\mathbf{J}_E, \mathbf{J}_N), \mathbf{J}_E \in R_+^{r/2}, \mathbf{J}_N \in D^{r/2-d}(\mathbf{J}_E) \right\} \tag{II.40}$$

The set D_T has the proper dimension; however there are too many parameters, so the components of $\mathbf{J}_N$ are interdependent. These parameters are poor when $\mathbf{J}_N \neq 0$.

The mapping $\phi_{CT}: D_C \to D_T$ is easy to construct for an ideal mixture where at steady state

$$\beta\boldsymbol{\mu} = \beta\boldsymbol{\mu}^0 + \ln \mathbf{X}^0 = \beta\boldsymbol{\mu}^0 - \ln \mathbf{h} \tag{II.41}$$

Here μ_i^0 is the chemical potential when $X_i^0 = 1$, and

$$\ln \mathbf{X}^0 \equiv (\ln X_1^0, \ldots, \ln X_n^0).$$

Then ϕ_{CT} is defined by

$$\phi_{CT}(\mathbf{h}, \mathbf{j}^*, i) = (\beta\boldsymbol{\mu}^0 - \ln \mathbf{h}, \mathsf{F}\mathsf{E}_i^*\mathbf{j}^*)$$

Both components of ϕ_{CT} are one to one, so ϕ_{CT} establishes a one-to-one correspondence between D_T and D_C, which has already been proved to correspond to the physical steady states.

A set of parameters that is almost as useful as the convex parameters consists of *redundant convex parameters* whose domain is

$$D_R \equiv \{(\mathbf{h},\mathbf{j}) | \mathbf{h} \in R^n_+, \mathbf{j} \in R^f_+\} \tag{II.42}$$

Each $(\mathbf{h},\mathbf{j}) \in D_R$ determines $\mathbf{v}^0 \in \mathrm{ri}\,\mathcal{C}_v$ by (II.18), which determines i and $\mathbf{j}^*$ by (II.21) and hence a point $(\mathbf{h},\mathbf{j}^*,i) \in D_C$. Conversely every $(\mathbf{h},\mathbf{j}^*,i) \in D_C$ determines $\mathbf{v}^0 \in \mathrm{ri}\,\mathcal{C}_v$ and a set of vectors $\mathbf{j} \in R^f_+$ satisfying (II.18) and hence a set of points of D_R. The mapping between the steady states and D_R is one-to-many (except when $\mathrm{ri}\,\mathcal{C}_v$ is simplicial). There are also more parameters than necessary. These disadvantages are offset because we do not have to construct a simplicial decomposition of $\mathrm{ri}\,\mathcal{C}_v$ and use many matrices E^*_i.

Example II.6. We now give the mapping ϕ_{KR} for the general unimolecular network. Since every reaction has the form $X_i \to X_j$, every column of $\underline{\boldsymbol{\nu}}$ is a permutation of $(1,-1,0,0,\ldots,0)^t$. Take mass action kinetics. Without loss of generality, the network may be assumed to be connected, and then $\underline{\gamma} = (1,1,\ldots,1)$. Kirchhoff has developed a diagrammatic solution[52] to the master equation, which was applied to unimolecular reaction networks by King and Altman.[53] Hill rediscovered and refined the approach; he has recently made extensive use of it for the analysis of complex enzyme networks. The form of ϕ_{KR} is given using Hill's terminology; consult Hill[54] for clarification of the results we now state. The steady-state concentrations are given by the "diagrammatic" formula

$$h_i^{-1} = \frac{\text{the sum of all directional diagrams of species } i}{\Sigma/C} \tag{II.43}$$

where $C = \underline{\gamma}\mathbf{X}$ and Σ is the sum of all directional diagrams. The rate of reaction R_k minus the rate of its reverse R_{-k} (i.e., the "net" rate) at steady state is

$$v_k^{\text{net}} = \frac{\text{sum of all}\begin{pmatrix} \text{cycle diagrams containing } R_k \\ -\text{ cycle diagrams containing } R_{-k}\end{pmatrix}}{\Sigma/C} \tag{II.44}$$

We will now show that this formula is consistent with the following elegant choice for the redundant current parameters:

$$j_i = \frac{\text{the sum of all cycle diagrams containing cycle } i}{\Sigma/C} \tag{II.45}$$

The extreme currents $\mathbf{E}_i$ of unimolecular networks are "cycles," that is, cyclic sequences of reactions such as $X_1 \to X_2 \to X_3 \to X_k \to X_1$. A detailed proof appears in Section IV.C. The currents can be normalized so that all elements of E are either 0 or 1. Then the equation $\mathbf{v}^0 = \mathsf{E}\mathbf{j}$ says that v_k is the sum of j_i over all cycles $\mathbf{E}_i$ containing R_k. Summing (II.45) over such cycles gives the sum of all cycle diagrams containing R_k divided by Σ/C. Then subtracting a similar expression for R_{-k} gives (II.44). Hence the choice (II.45) for j_i gives the correct net reaction velocities. The cycle terms may be factored out to get

$$j_i = \frac{\text{cycle } i}{\Sigma / C} \times \text{the sum of all directional diagrams for cycle } i \qquad \text{(II.46)}$$

where "cycle i" is the product of the effective rate constants around the ith extreme current, and a "directional diagram for cycle i" is a product of effective rate constants, which is calculated as follows.

Take any partial diagram (Hill's "partial diagram" is called a "spanning tree" in graph theory) and delete all lines (but not the points) in cycle i. Then direct the remaining lines toward the points that used to be on the cycle. Take the product of effective rate constants of the reactions corresponding to the directed lines of the diagram. I have called such diagrams "directional diagrams for the cycle" because they are analogous to Hill's "directional diagrams for the species (states)." See Fig. 1.

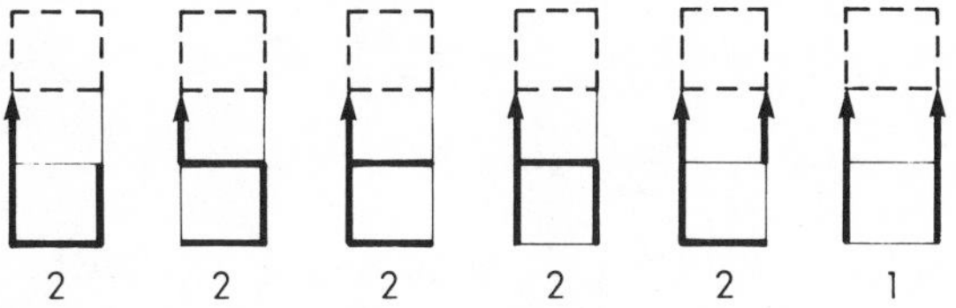

Fig. 1. The 11 directional diagrams for the dashed extreme current (cf. Fig. 2.8 of Hill[54]). The numbers below diagrams indicate the numbers of equivalent diagrams by symmetry.

F. The Equilibrium Simplex

Equilibrium steady states are defined by $\mathbf{J}_N = 0$. Hence from (II.38) $\mathbf{v}^0$ lies in the null space of F_N where the *detailed balance condition* $v_i^0 = v_j^0$ holds for every pair of forward and reverse reactions. This $r/2$-dimensional null space S_E is the intersection of $r/2$ hyperplanes with equations $v_i^0 = v_j^0$. Since hyperplanes are convex sets and the intersection of convex sets is convex, the set of equilibrium states is convex. Also, if $\mathbf{v}^0$ is an equilibrium state, so is $\lambda\mathbf{v}^0$ for any $\lambda \geqslant 0$. Hence the states form a convex cone $\mathcal{C}_E$. Its intersection with the hyperplanes $\mathbf{e}_v^t\mathbf{v} = 1$ is called Π_E. Only states in ri $\mathcal{C}_E$ and ri Π_E are equilibrium states of the network of interest; however the discussion is simpler if we consider the closed sets $\mathcal{C}_E$ or Π_E.

The extreme points of Π_E may be found as in Section II.C by deleting reactions until deletion of further reactions gives a matrix B such that (II.19) is inconsistent. Suppose the reaction R_i has not been deleted. Then deletion of its reverse reaction R_j would result in a steady state that does not satisfy the detailed balance condition $v_i^0 = v_j^0$. Hence we cannot delete

v_j^0. We must delete all other reactions to arrive at an *equilibrium extreme current* (EEC) consisting of equal flows in R_i and R_j. Thus the columns of the matrix of equilibrium extreme currents E_E are zero, except for two elements corresponding to v_i^0 and v_j^0, which are both equal. As before, the columns may be scaled arbitrarily. With suitable scaling $\mathsf{E}_E = \frac{1}{2}\mathsf{F}_E$; then the similarity between (II.18) and (II.39) implies $\mathbf{j} = \mathbf{J}_E$. Thus the projection of a nonequilibrium state $(\mathbf{J}_E, \mathbf{J}_N)$ on Π_E is $(\mathbf{J}_E, \mathbf{0})$. It is therefore appropriate to call $\mathbf{J}_E$ the equilibrium component of $\mathbf{J}$.

To prove that Π_E is a simplex, note that every vertex of Π_E is a scalar multiple of a column of E_E. This matrix has $r/2$ columns, one for each pair of forward and reverse reactions. The columns must be linearly independent, so Π_E has $r/2$ linearly independent vertices, hence is an $(r/2) - 1$-dimensional simplex. Therefore $\mathcal{C}_E$ is an $r/2$-dimensional simplicial cone.

The detailed balance condition is a restriction on $\mathcal{C}_v$. Hence

$$\mathcal{C}_E \subset \mathcal{C}_v \tag{II.47}$$

Their dimensions are $r/2$ and $r - d$, respectively. Also the boundary of $\mathcal{C}_E$ lies entirely in the coordinate hyperplanes $v_i = 0$ and is thus entirely contained in the boundary of $\mathcal{C}_v$. Since $\mathcal{C}_E$ has $r/2 - d$ fewer dimensions than $\mathcal{C}_v$, it lies in the intersection of $\mathcal{C}_v$ and an $(r - d) - (r/2 - d) = r/2$-dimensional linear subspace of S_v. Similarly Π_E is an $r/2$-dimensional slice through Π_v. Note that the vertices of Π_E are also obtained as vertices of Π_v by the procedure of Section II.C.

Choose the simplicial decomposition of $\mathcal{C}_v$ so that $\mathcal{C}_E$ is a facet of $\mathcal{C}_{v1}$, and let this facet be given by $j_i^* = 0$ for $i = r/2 + 1, \ldots, r - d$. Then the *equilibrium subset* of D_C is

$$D_{CE} \equiv \left\{ (\mathbf{h}, \mathbf{j}_E^*, \mathbf{0}, 1) \mid \mathbf{h} \in R_+^n, \mathbf{j}_E^* \in R_+^{r/2} \right\} \tag{II.48}$$

Note that the dimension of the equilibrium subset is $n + r/2$, which is $r/2 - d$ dimensions less than D_C (neglecting the dimension for the index i). The equilibrium subset of D_T is obtained by setting $\mathbf{J}_N = 0$ to get

$$D_{TE} = \left\{ \left(\beta\boldsymbol{\mu}, \mathbf{J}_E, \mathbf{0} \mid \beta\boldsymbol{\mu} \in R^n, \mathbf{J}_E \in R_+^{r/2} \right) \right\} \tag{II.49}$$

which also has dimension $n + r/2$. Finally, the equilibrium subset of D_K is a complicated hypersurface D_{KE} where $r/2 - d$ of the rate constants are determined by $\mathbf{C}$ and the remaining $r/2 + d$ rate constants.

At equilibrium, thermodynamics considers the reciprocal densities $\mathbf{h}$ to be equivalent to the parameters $\beta\boldsymbol{\mu}$. Since $\mathbf{J}_E = \mathbf{j}$, an equally good set of thermodynamic parameters is $(\mathbf{h}, \mathbf{j})$, the convex parameters. Far from equilibrium, the net fluxes that are usually used for near equilibrium thermodynamics become inelegant. The convex parameters are much better. If free energy is to be considered, an even better choice consists of the *convex thermodynamic parameters* $(\beta\boldsymbol{\mu}, \mathbf{j}^*, i)$. A reformulation of non-equilibrium thermodynamics and statistical mechanics into these parameters is under investigation but is beyond the scope of this chapter.

G. The Nonlinear Equations of Motion and Multiple Steady States

The nonlinear equations of motion take a simple form when expressed using convex parameters. From (II.35), (II.36), and (II.18), the mapping ϕ_{KC}^{-1} is given by

$$k_j = (\mathsf{E}_i^* \mathbf{j}^*)_j \prod_{i=1}^{n} h_i^{\kappa_{ij}} \tag{II.50}$$

Switching to redundant parameters and using (II.31) produces

$$\mathbf{v}_j = (\mathsf{E}\mathbf{j})_j \prod_{i=1}^{n} x_i^{\kappa_{ij}} \tag{II.51}$$

where the dimensionless concentration vector $\mathbf{x}$ is defined by

$$x_i = h_i \mathrm{X}_i \tag{II.52}$$

The nonlinear equations of motion (II.2) now become

$$\dot{\mathrm{x}}_i = h_i \sum_{j=1}^{r} \nu_{ij} (\mathsf{E}\mathbf{j})_j \prod_{i=1}^{n} x_i^{\kappa_{ij}}$$

which may be expressed in matrix form as

$$\dot{x} = (\operatorname{diag} \mathbf{h})\, \underline{\boldsymbol{\nu}} (\operatorname{diag} \mathsf{E}\mathbf{j}) \exp(\underline{\boldsymbol{\kappa}}^t \ln \mathbf{x}) \tag{II.53}$$

When $\mathbf{x} = \mathbf{e}_n = (1, 1, \ldots, 1)^t$ the right-hand side is $(\operatorname{diag} \mathbf{h})\underline{\boldsymbol{\nu}}\mathsf{E}\mathbf{j}$, which vanishes by (II.19). Hence each of the polynomials on the right-hand side is a linear combination of the factors $(x_i - 1)$, for $i = 1, \ldots, n$. Hence each term of these polynomials is divisible by one of the factors $(x_i - 1)$, for $i = 1, \ldots, n$.

Example II.7. The Oregonator was defined in Example II.1 and E was calculated in Example II.5. If x, y and z are the dimensionless concentrations,

$$(\operatorname{diag} \mathsf{E}\mathbf{j}) \exp(\underline{\boldsymbol{\kappa}}^t \ln \mathbf{x}) = (j_2 y, j_1 x y, (j_1 + j_2) x, j_2 x^2, (j_1 + j_2) z)^t$$

and the nonlinear equations are obtained by mutliplying on the left by $\underline{\boldsymbol{\nu}}$. We write the equations so that every term contains a factor of $(x-1)$, $(y-1)$, or $(z-1)$ to get

$$\frac{\dot{x}}{h_X} = -(2j_2x + j_1y + j_2 - j_1)(x-1) + (j_2 - j_1)(y-1)$$

$$\frac{\dot{y}}{h_Y} = -j_1y(x-1) - (j_1+j_2)(y-1) + (j_1+j_2)(z-1)$$

$$\frac{\dot{z}}{h_Z} = 2(j_1+j_2)[(x-1)-(z-1)]$$

In factorizing, we divided each polynomial by $x-1$ to obtain a quotient and a remainder. The remainder was divided by $y-1$ to obtain a quotient and a second remainder. The second remainder must be divisible by $z-1$.

We will now show that when a network has multiple steady states, a fixed set of points $\{\mathsf{Ej}, \mathsf{Ej}', \mathsf{Ej}'', \ldots\} \subset \mathcal{C}_v$ represents the set of associated steady states, for infinitely many different sets of parameters $\mathbf{h}, \mathbf{h}', \mathbf{h}'', \ldots$. When a network has multiple steady states there is a set of points $\{(\mathbf{h},\mathbf{j}), (\mathbf{h}',\mathbf{j}'), (\mathbf{h}'',\mathbf{j}''), \ldots,\} \subset D_R$ that are physically accessible with the same rate constants and constraints. That is, $\mathbf{k}$ obtained from (II.50) and $\mathbf{C}$ obtained from (II.5) satisfy

$$\mathbf{k}(\mathbf{h},\mathbf{j}) = \mathbf{k}(\mathbf{h}',\mathbf{j}') = \mathbf{k}(\mathbf{h}'',\mathbf{j}'') = \cdots$$
$$\mathbf{C}(\mathbf{h},\mathbf{j}) = \mathbf{C}(\mathbf{h}',\mathbf{j}') = \mathbf{C}(\mathbf{h}'',\mathbf{j}'') = \cdots \qquad \text{(II.54)}$$

The steady states must be different, that is, $\mathbf{h} \neq \mathbf{h}', \ldots$. Define $\mathbf{x}$ in (II.52) using the first steady state $\mathbf{h}$; then the steady-state concentration is $X_i^0 = h_i^{-1}$, so at steady state $\mathbf{x}^0 = \mathbf{h}/\mathbf{h} = \mathbf{e}$. Thus the solution $\mathbf{e}$ of the steady-state conditions

$$\sum_{j=1}^{r} \nu_{ij}(\mathsf{Ej})_j \prod_{i=1}^{n} x^{\kappa_{ij}} = 0 \qquad \text{(II.55)}$$

for $i = 1, \ldots, n$ corresponds to the parameter point $(\mathbf{h},\mathbf{j})$. The existence of multiple steady states for this $\mathbf{j}$ means that these equations have other solutions, which we call $\mathbf{a}'(\mathbf{j}), \mathbf{a}''(\mathbf{j}), \ldots$. If the solution $\mathbf{a}'(\mathbf{j})$ corresponds to the steady state $(\mathbf{h}',\mathbf{j}')$, whose concentration at steady state is $X_i^0 = h_i'^{-1}$, the dimensionless concentration at steady state is therefore $\mathbf{h}/\mathbf{h}'$ and this is $\mathbf{a}'(\mathbf{j})$. Hence

$$\mathbf{a}'(\mathbf{j}) = \frac{\mathbf{h}}{\mathbf{h}'} \qquad \text{(II.56)}$$

The velocity vector in the primed steady state is Ej'; alternatively, its jth component may be obtained from (II.51) using $\mathbf{x} = \mathbf{a}'(\mathbf{j})$. Hence for $j = 1, \ldots, r$,

$$(\mathsf{Ej}')_j = (E\mathbf{j})_j \prod_{i=1}^{n} a_i'(\mathbf{j})^{\kappa_{ij}} \qquad \text{(II.57)}$$

If this sytem of equations is solved for $\mathbf{j}'$ as a function of $\mathbf{j}$, the solution will be unique only if Π_v is a simplex because the redundant parameters have been used. Nevertheless, the point in $\mathcal{C}_v$ represented by $\mathbf{j}'$ is unique. Similarly, for each of the other distinct solutions $\mathbf{a}''(\mathbf{j}), \mathbf{a}'''(\mathbf{j}), \ldots$ of (II.55) corresponding unique points $\mathsf{E}\mathbf{j}'', \mathsf{E}\mathbf{j}''', \ldots$ exist in $\mathcal{C}_v$. If there are $\rho(\mathbf{k}, \mathbf{C})$ accessible steady states of the system for fixed rate constants, there are $\rho(\mathbf{k}, \mathbf{C})$ points in $\mathcal{C}_v$, called *isorate-constant points*, which represent these steady states. If the first steady state has parameters $(\mathbf{h}, \mathbf{j})$, from (II.56) the second steady state has parameters $(\mathbf{h}/\mathbf{a}'(\mathbf{j}), \mathbf{j}')$, and, in analogy with this result, the mth steady state has parameters $(\mathbf{h}/\mathbf{a}^{(m)}(\mathbf{j}), \mathbf{j}^{(m)})$. Such a set of steady states exists for every $\mathbf{h} \in R^n_+$, however all such sets correspond to the same set of isorate-constant points $\{\mathsf{E}\mathbf{j}, \mathsf{E}\mathbf{j}', \ldots, \mathsf{E}\mathbf{j}^{(m)}, \ldots\} \subset \mathcal{C}_v$. Thus one set of isorate-constant points corresponds to infinitely many sets of multiple steady states.

Note that (II.55) is unchanged if $\mathbf{j}$ is replaced by $\lambda\mathbf{j}$. Hence the set of sets of isorate-constant points that differ only by a scale factor λ may be represented by a single set of isorate-constant points in Π_v. Thus the complete picture of multiple steady states in a network may be expressed by giving all sets of isorate-constant points in Π_v.

If the solutions of $\underline{\nu}\mathbf{v}(\mathbf{X}, \mathbf{k}) = 0$ are plotted in $\mathbf{X} - \mathbf{k}$ space, multiple steady states can be seen to arise from the presence of a fold such that more than one value of $\mathbf{X}$ corresponds to a single $\mathbf{k}$. If the point $\mathbf{k} \in R^r_+$ is at the fold, the steady-state conditions must have a multiple root $\mathbf{x} = \mathbf{e}$, one root coming from each branch of the surface that joins at the fold. Such points $\mathbf{k} \in R^r_+$ are called *multiple steady-state bifurcation points*. For each such $\mathbf{k}$, there are $\rho(\mathbf{k}, \mathbf{C})$ points in D_K that are mapped by ϕ_{KC} into a set of $\rho(\mathbf{k}, \mathbf{C})$ corresponding multiple steady-state bifurcation points in D_C. The steady-state solutions of (II.53) do not depend on $\mathbf{h}$; hence these multiple steady-state bifurcation points depend only on $\mathbf{j}$, and after scaling $\mathbf{j}$, form a subset of Π_v. This subset may be divided into exit points and entry points. An *exit point* has a multiple root $\mathbf{x} = \mathbf{e}$, whereas an *entry point* has a multiple root $\mathbf{x} = \mathbf{a}$ in addition to the root $\mathbf{x} = \mathbf{e}$. The sets of entry and exit points may intersect if $\mathbf{a} = \mathbf{e}$; then $\mathbf{x} = \mathbf{e}$ is at least a triple root and the corresponding point in Π_v is a *critical point* or *tristate point*. Example II.8 illustrates how to obtain equations for entry, exit and critical points.

Example II.8. Consider the model (Schlög;[27, 28] Nicolis and Prigogine, Ref. 2, p. 170):

$$R_1, R_3: \quad \square \rightleftharpoons \mathrm{X}, \qquad v_1 = k_1, \qquad v_3 = k_3 X$$
$$R_2, R_4: \quad 2\mathrm{X} \rightleftharpoons 3\mathrm{X}, \qquad v_2 = k_2 X^2, \qquad v_4 = k_4 X^3$$

Then $\underline{\nu} = (1, 1, -1, -1)$, and in Example II.4 we obtained E and showed that Π_v is a square. The nonlinear equation of motion (II.53) becomes

$$\frac{\dot{x}}{h} = -(j_1 + j_2)x^3 + (j_2 + j_3)x^2 - (j_3 + j_4)x + (j_4 + j_1)$$

The edge of the fold has two roots $x = 1$ and is an exit point. Dividing the right-hand side by $(x-1)^2$ yields the factorization

$$\frac{\dot{x}}{h} = -(x-1)^2[(j_1+j_2)x + 2j_1 + j_2 - j_3] - (x-1)[3j_1 + j_2 - j_3 + j_4]$$

A double root occurs if and only if the second term vanishes, so that the equation of the exit points is

$$3j_1 + j_2 - j_3 + j_4 = 0$$

which is plotted in Fig. 2 as a line segment. When two roots are equal there is usually a third unequal root. When $\mathbf{Ej} \in \Pi_v$ is an entry point, the roots are $1, a, a$. Since $\mathbf{j}$ corresponds to the root $x = 1$, which is the third unequal root, we can find the entry points by writing the equation of motion as

$$\frac{\dot{x}}{h} \approx -(x-1)(x-a)^2 = -x^3 + (1+2a)x^2 - (a^2+2a)x + a^2$$

and comparing coefficients with the original equation to obtain

$$\frac{j_1+j_2}{1} = \frac{j_2+j_3}{1+2a} = \frac{j_3+j_4}{a^3+2a} = \frac{j_4+j_1}{a^2}$$

One of these equations is redundant; eliminating a from the other two gives the curve

$$(j_3 - j_1)^2 = 4(j_1+j_2)(j_1+j_4)$$

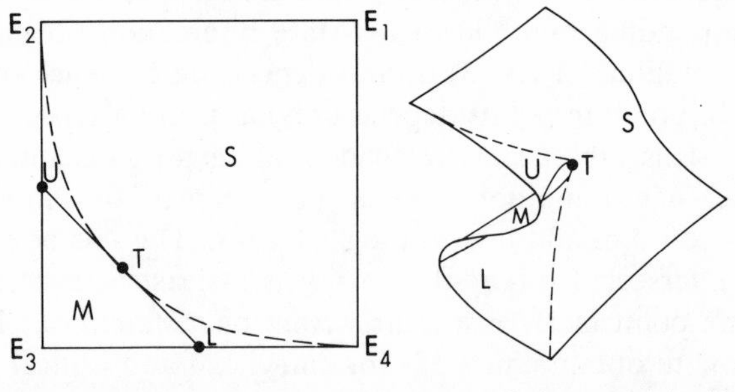

Fig. 2. Comparison between the sets of steady states of the model in Example II.8, as represented in stoichiometric network analysis (left) and in catastrophe theory (right). In the first case **h** and **j** may be scaled to eliminate two parameters. In catastrophe theory, X and t may be scaled to eliminate two parameters. Thus in both cases, only two parameters remain, and a homeomorphism exists between the sets of steady states in the two representations. The sets marked U, M, and L refer to the upper, middle, and lower steady states that occur for certain sets of fixed rate constants.

for the entry points. This curve appears as a dashed line in Figure 2. The entry and exit points intersect at the tristate point T in the figure. Corresponding regions of Π_v and the usual fold picture have been marked S (single root), U (upper root), M (middle root), and L (lower root).

If the steady state is near the common boundary of M and U, the rate constant may be changed so that $\rho(\mathbf{k}, \mathbf{C})$ changes from 3 to 1. This boundary is called an exit point because the system must then switch to a distant steady state. The new steady state is the corresponding entry point.

The situation becomes more complicated when $d \neq 1$. When $\mathbf{j}$ is a steady state on the fold, $\mathbf{x} = \mathbf{e}$ must be a double root of the system of equations; hence at least one of the polynomial equations (II.55) must have a double root. A polynomial has this property if and only if it can be put into the form

$$\sum_{ij} a_{ij}(x_i - 1)(x_j - 1)$$

where a_{ij} is a polynomial in $\mathbf{j}$ and $\mathbf{x}$. The relationships among the components of $\mathbf{j}$ that must be satisfied to give the polynomial this form are the equations of the fold. Since convex parameters are nonnegative, it is easy to recognize whether such conditions can be satisfied for physical steady states.

Example II.9. Consider the possibility of multiple steady states in the Oregonator (Example II.7). We inspect the coefficients of $x - 1, y - 1, z - 1$ to see if there is a polynomial where all coefficients can vanish for some $\mathbf{Ej} \in \Pi_v$. There is no such polynomial; hence the Oregonator does not have multiple steady states.

The fold at the right of Fig. 2 appears in $R^n \times D_K$, the product of the concentration and parameter spaces. It helps visualize the $\rho(\mathbf{k}, \mathbf{C})$ steady states at each point in D_K. Although D_K has dimension (neglecting the index i) $n + r - d$, usually dimensionless parameters are used to eliminate one rate constant per species and one additional rate constant by scaling time, to give a parameter space with the minimal possible dimension, which is $r - d - 1$. This is also the dimension of Π_v. Since there is only one steady state for each point in Π_v, since the minimum number of dimensionless rate constant parameters and $\dim \Pi_v$ are the same, and since nearby points on the fold correspond to nearby points on Π_v, we consider Π_v to be an unfolding of the surface in $R^n \times D_K$. In all cases I have examined, the surface in $R^n \times D_K$ is homeomorphic to Π_v.

The general form of the nonlinear equations (II.53) is precisely the form that is studied in *singular perturbation theory*.[55] If the dimensionless concentrations $\mathbf{x}$ are all roughly equidistant from the steady state $\mathbf{e}$, usually all the polynomials $\underline{\boldsymbol{\nu}}(\operatorname{diag} \mathbf{Ej})\exp(\underline{\boldsymbol{\kappa}}^t \ln \mathbf{x})$ are of roughly the same order of magnitude. Hence the factor diag **h** gives each equation a characteristic time scale for the motion. These time scales can be many orders of magnitude different from one another as a consequence of the wide range of possible steady-state concentrations. Singular perturbation theory treats the dynamics of systems like (II.53) when the dynamical variables have widely differing time scales.

H. Physical Limitations to the Accessible States

Molecules occupy a nonzero volume; hence one mole of particles of type i cannot be packed into a less than h_i^0 (say) liters. Thus

$$h_i > h_i^0 \tag{II.58}$$

is a physical constraint equivalent to an upper limit on the solubility of X_i. On the average, molecules in a dense medium at room temperature cannot collide more frequently than 10^{14} times per second. From this condition comes an upper limit to reaction rate constants and reaction velocities; hence each current component has an upper bound and

$$j_i < j_i^0 \tag{II.59}$$

The part of parameter space satisfying two such restrictions for arbitrary $\mathbf{h}^0 \in R_+^n$ and $\mathbf{j}^0 \in R_+^{r-d}$ is the *potentially accessible region corresponding to* $\mathbf{h}^0$ *and* $\mathbf{j}^0$.

Not all this region is accessible in practice because the number of external species is usually too small to allow all effective rate constants to be manipulated by varying the external concentrations. The *physically accessible region* (Section II.A) is a subset of the potentially accessible region for the physically appropriate $\mathbf{h}^0$ and $\mathbf{j}^0$. It is determined by the reaction mechanism from which the network was obtained by deleting the external species. Since the external species are not specified by the definition of a network, each network can be obtained from many mechanisms and can have many physically accessible regions. Thus we cannot discuss this region because it is not a network property. On the other hand, the restrictions (II.58) and (II.59) on **h** and **j**, which come from restrictions on solubilities and collision rates, can be discussed without reference to the physical context, that is, without reference to the particular reactions and choice of internal and external species that give rise to the network.

In this section we prove in general that these restrictions are insignificant in the following sense. Every dynamical phenomenon that occurs in the potentially inaccessible part of D_R may be observed as well in the potentially accessible part of D_R. Attempts have been made to reject certain oscillatory mechanisms as being a priori "unphysical" by arguing that oscillations cannot occur for parameters within the potentially accessible region. One cannot reject mechanisms using such arguments because we will soon prove that if a network oscillates for any rate constants, then it also oscillates *in the same way* for potentially accessible parameters. The only difference between the two oscillations is a scaling of time and concentration. The aforementioned attempts have also introduced the restriction that only a small fraction of the external reactants in the overall reaction should be consumed during a cycle of oscillation.[56] We prove that this restriction can always be satisfied within the potentially accessible region and thus cannot be used to reject networks a priori.

Choose any $(\mathbf{h},\mathbf{j}) \in D_R$ and any $\lambda > 0$, $\mu > 0$. Let $\mathbf{h}' \equiv \lambda\mathbf{h}$, $\mathbf{j}' \equiv \mu\mathbf{j}$, $t' = t/\lambda\mu$ (t is time). If the general form of the nonlinear equations (II.53) is converted to the new parameters, the equations take exactly the same form they had previously. Hence if the equations had a solution $\mathbf{x}(t)$ for parameters $\mathbf{h},\mathbf{j}$, the transformed equations have an identical solution $\mathbf{x}(t')$, which occurs for parameters $\mathbf{h}',\mathbf{j}'$. It is always possible to choose λ and μ such that $\lambda\mathbf{h} > \mathbf{h}^0$ and $\mu\mathbf{j} < \mathbf{j}^0$; then $\mathbf{h}',\mathbf{j}'$ lies in the potentially accessible region of D_R.

Note that all events take place $\lambda\mu$ times faster. For example, if the period of oscillation is originally τ, the new period is $\tau' = \tau/\lambda\mu$. If $\lambda\mu > 1$, $\tau' < \tau$, so events take place faster than before. Stability does not depend on the dynamical time scale, so $(\mathbf{h},\mathbf{j})$ and $(\lambda\mathbf{h}, \mu\mathbf{j})$ have exactly the same stability classification.

Suppose the equations have a limit cycle oscillation and $x_i(t)$ oscillates between $x_i^{\max}$ and $x_i^{\min}$. The range of $x_i(t)$ is $\Delta x_i \equiv x_i^{\max} - x_i^{\min} > 0$. If X_i is the concentration for the parameters $\mathbf{h},\mathbf{j}$, and X_i' is the concentration for the parameters $\mathbf{h}',\mathbf{j}'$, the ranges of X_i and X_i' may be related using (II.52) to get

$$h_i \Delta X_i = \Delta x_i = h_i' \Delta X_i' \tag{II.60}$$

We can always satisfy the condition $\mathbf{h}' > \mathbf{h}^0$ by choosing λ large enough. Assume $\lambda > 1$. From (II.60), $\Delta X_i' = \Delta X_i/\lambda$, so $\Delta X_i' < \Delta X_i$. Hence the amplitude of the oscillation decreases. A similar argument shows that $X_i^{\max}$, the maximum concentration of the ith species during the oscillation, scales in the same way, so that $X_i'^{\max} = X_i^{\max}/\lambda$. Hence we may choose λ so that $X_i'^{\max}$ is less than the limit set by the solubility of the species. Summarizing, any dynamical feature of the nonlinear equations may be made to appear in the potentially accessible region of D_R by choosing λ sufficiently large and μ sufficiently small. This scaling multiplies all frequencies by $\lambda\mu$ and all concentrations by λ^{-1}.

This scaling is equivalent to a change in the rate constants. Since stoichiometry is not affected, the amount of an external species consumed during a cycle of oscillation is also multiplied by λ^{-1} and becomes smaller. Thus the condition that a negligibly small fraction of each external reactant be consumed during a cycle of oscillation may be satisfied by making λ sufficiently large. Since condition (II.58) imposes a lower limit on λ but no upper limit, an arbitrarily small consumption during a limit cycle may be obtained within the potentially accessible region. Since this result holds for all networks, whatever the values of $\mathbf{h}_0$ and $\mathbf{j}_0$, these criteria can never be used to reject a network a priori.

It has also been argued (Ref. 2, p. 157) that the period of oscillation of a practical oscillator should be about 1 min. Since the period scales as $\tau' = \tau/\lambda\mu$, τ' might be extremely small or large when λ and μ are chosen to meet the criteria discussed above. Thus certain models cannot meet this criterion for certain $\mathbf{h}^0, \mathbf{j}^0$. The extremely fast oscillation with period τ' might not be observable until new measurement techniques are developed. The situation is analogous to the problem of extending visual astronomy to radio astronomy and X-ray astronomy. One cannot rule out such mechanisms, but they cannot be studied experimentally until new techniques are available.

Can thermodynamics limit the parameters of the network to only part of the domains D_K, D_C, or D_T? In Section II.E [see (II.40)] we showed that every point in D_T has the form $(\beta\boldsymbol{\mu}, \mathbf{J}_E, \mathbf{J}_N)$, where $\boldsymbol{\mu}$ is the vector of chemical potentials of the n internal species, $\mathbf{J}_E$ is the sum of forward and reverse reaction rates for each pair of forward and reverse reactions, and $\mathbf{J}_N$ is the net reaction rate for the same pairs of reactions. Clearly every $\beta\boldsymbol{\mu} \in R^n$ is consistent with thermodynamics. One may think of $\mathbf{J}_E \in R_+^{n/2}$ as being determined by the activation energies of the $r/2$ pairs of reactions, thus every $\mathbf{J}_L \in R_+^{n/2}$ is consistent with thermodynamics. Finally, the net rates $\mathbf{J}_N \in D^{r/2-d}(\mathbf{J}_E)$ may be nonzero when the system has external species that are considered to have fixed concentrations. When thermodynamics is applied to such a system, the free energy change $\Delta G°$ associated with each pair of reactions depends on both the internal and external species. Any set of chemical potentials for the standard states of the external species is consistent with thermodynamics. The external species give the network $r/2 - d$ more degrees of freedom in the parameters than the corresponding closed system. The only restriction on these extra degrees of freedom is that the reaction velocities remain positive, that is, $\mathbf{J}_N \in D^{r/2-d}(\mathbf{J}_E)$. Every such nonequilibrium flux must be consistent with some choice of chemical potentials for the external species. We conclude that every point in D_T is consistent with thermodynamics, therefore every point in D_K or D_C is also consistent with thermodynamics.

In this section we have proved that the full range of mathematically possible dynamical phenomena of the network can occur in the potentially accessible region. Neither thermodynamics, nor solubilities, nor upper limits to the reaction velocities can prevent any mathematically possible dynamical phenomenon from being physically possible. All that is needed is a mechanism that gives the network when the appropriate species are treated as external species, and sufficient flexibility in the ratios of concentrations of the external species to reach the appropriate part of the potentially accessible region of parameter space. Upper bounds to the solubilities of the external species are not restrictive for the same reason discussed for internal species. There should not be any lower bounds; thus adequate flexibility in the external species must exist whenever enough reactions involve a unique external species. Scaling the dynamical phenomenon to satisfy (II.58) and (II.59) frequently forces some species to have very low concentrations and some reactions to be very slow at steady state. Hence a mathematically possible dynamical pattern that might be useful to a biological species in controlling morphogenesis is physically possible whenever an appropriate mechanism can occur. Scaling may force certain morphogens to have extremely low concentrations, and certain reactions to be extremely slow at steady state. Consequently the time scale of the development might necessarily be very slow. Perhaps this is why morphogenesis is slow and the chemical species and reactions controlling it have not yet been found.

III. ELEMENTARY STABILITY ANALYSIS

A. The Linearized Dynamics

The dynamics close to steady state can be described by the linearized equation

$$\frac{d\zeta}{dt} = \mathsf{M}_{\zeta}\zeta \tag{III.1}$$

for the deviation from steady state $\zeta \equiv \mathbf{X} - \mathbf{X}^0$. The nonlinear equations (II.2) may be put in this form by making a Taylor series expansion of the ith component about $\mathbf{X}^0$.

$$(\underline{\nu}\mathbf{v})_i = \sum_{j=1}^{n} \nu_{ij} \left[v_j(\mathbf{X}^0, \mathbf{k}) + \sum_{m=1}^{n} \frac{\partial v_j}{\partial X_m^0}(\mathbf{X}^0, \mathbf{k})(X_m - X_m^0) + \cdots \right] \tag{III.2}$$

The leading term vanishes at steady state by (II.11). Define the *effective power function* in terms of the redundant parameters

$$\kappa_{ijR}(\mathbf{p}_R) \equiv \frac{\partial \log v_j(\mathbf{X}^0(\mathbf{p}_R), \mathbf{k}(\mathbf{p}_R))}{\partial \log X_i^0} \tag{III.3}$$

and then write the second term in (III.2) as

$$\sum_{j=1}^{n} \nu_{ij} \sum_{m=1}^{n} \frac{v_j(\mathbf{X}^0, \mathbf{k})}{X_m^0} \kappa_{mjR}(\mathbf{p}_R)(X_m - X_m^0)$$

Using (II.18) and (II.35), convert this to

$$\sum_{j=1}^{n} \nu_{ij} \sum_{q=1}^{f} E_{jq} j_q \sum_{m=1}^{n} \kappa_{mjR}(\mathbf{p}_R) h_m (X_m - X_m^0)$$

to obtain

$$\mathsf{M}_{\zeta R}(\mathbf{h},\mathbf{j}) = \underline{\boldsymbol{\nu}}(\operatorname{diag}\mathsf{E}\mathbf{j})\, \underline{\boldsymbol{\kappa}}_R^{t}(\mathbf{h},\mathbf{j})(\operatorname{diag}\mathbf{h}) \tag{III.4}$$

as the matrix of the linearized system as a function of the redundant convex parameters, where diag $\mathbf{h}$ is the diagonal matrix whose diagonal is $\mathbf{h}$.

By making a change of variable, this matrix can be replaced by any cyclic permutation of the matrices on the right of (III.4). For example, if $\zeta' = (\operatorname{diag}\mathbf{h}^{1/2})\zeta$ and $\zeta'' = (\operatorname{diag}\mathbf{h})\zeta$, the linearized system has the matrices

$$\mathsf{M}'_{\zeta R}(\mathbf{h},\mathbf{j}) = (\operatorname{diag}\mathbf{h}^{1/2})\, \underline{\boldsymbol{\nu}}(\operatorname{diag}\mathsf{E}\mathbf{j})\, \underline{\boldsymbol{\kappa}}_R^{t}(\mathbf{h},\mathbf{j})(\operatorname{diag}\mathbf{h}^{1/2}) \tag{III.5}$$

$$\mathsf{M}''_{\zeta R}(\mathbf{h},\mathbf{j}) = (\operatorname{diag}\mathbf{h})\, \underline{\boldsymbol{\nu}}(\operatorname{diag}\mathsf{E}\mathbf{j})\, \underline{\boldsymbol{\kappa}}_R^{t}(\mathbf{h},\mathbf{j}) \tag{III.6}$$

The parameter h_i multiplies the ith column of $\mathsf{M}_{\zeta R}$ and the ith row of $\mathsf{M}''_{\zeta R}$. The matrix $\mathsf{M}'_{\zeta R}$ is symmetrical in the parameters $\mathbf{h}$.

The linearized equation may be written in terms of extents of reactions instead of concentrations using the one-to-one mapping (II.28) from S_ζ to S_X. To each steady state $\mathbf{X}^0 \in S_X$ corresponds a unique $\boldsymbol{\xi}^0 \in S_\xi$. Let $\boldsymbol{\eta} \equiv \boldsymbol{\xi} - \boldsymbol{\xi}^0$. Then (II.28) implies $\boldsymbol{\zeta} = \underline{\boldsymbol{\nu}}\boldsymbol{\eta}$, and from (III.1) and (III.4)

$$\underline{\boldsymbol{\nu}}\dot{\boldsymbol{\eta}} = \underline{\boldsymbol{\nu}}(\operatorname{diag}\mathsf{E}\mathbf{j})\, \underline{\boldsymbol{\kappa}}_R^{t}(\mathbf{h},\mathbf{j})(\operatorname{diag}\mathbf{h})\, \underline{\boldsymbol{\nu}}\boldsymbol{\eta}$$

These d independent equations are sufficient to determine $\boldsymbol{\eta}$ if we make the convention that $\boldsymbol{\xi} \in S_\xi$, which implies $\boldsymbol{\eta} \in S_\xi$, a d-dimensional subspace. We may remove $\underline{\boldsymbol{\nu}}$ from the left of both sides, provided the projection operator P_ξ (see Section II.D) is used to ensure that the expression on the right lies in S_ξ. The resulting linearized equation for $\boldsymbol{\eta}$ has the same form as (III.1) and the matrix is

$$\mathsf{M}_{\eta R}(\mathbf{h},\mathbf{j}) = \mathsf{P}_\xi(\operatorname{diag}\mathsf{E}\mathbf{j})\, \underline{\boldsymbol{\kappa}}_R^{t}(\mathbf{h},\mathbf{j})(\operatorname{diag}\mathbf{h})\, \underline{\boldsymbol{\nu}} \tag{III.7}$$

Note that (ignoring P_ξ) the expressions for $\mathsf{M}_{\zeta R}$ (III.4) and $\mathsf{M}_{\eta R}$ differ only by a cyclic permutation of the matrices on the right-hand side. If $\boldsymbol{\eta}'$

$= (\operatorname{diag} \mathsf{Ej})^{-1}\boldsymbol{\eta}$, the linearized equation for the evolution of $\boldsymbol{\eta}'$ has the matrix

$$\mathsf{M}'_{\eta R}(\mathbf{h},\mathbf{j}) = \mathsf{P}_{\xi}\,\underline{\boldsymbol{\kappa}}^{t}_{R}(\mathbf{h},\mathbf{j})(\operatorname{diag}\mathbf{h})\,\underline{\boldsymbol{\nu}}(\operatorname{diag}\mathsf{Ej}) \tag{III.8}$$

which is the final cyclic permutation.

All these expressions for the matrix of the linearized system may be converted to convex parameters by replacing Ej with $\mathsf{E}^*_i\mathbf{j}^*$. The effective power function may also contain additional parameters beyond those shown. Two interesting special cases occur when $\underline{\boldsymbol{\kappa}}(\mathbf{p})$ is constant and when $\underline{\boldsymbol{\kappa}}(\mathbf{p})$ can vary over a wide range via the additional parameters.

If $\underline{\boldsymbol{\kappa}}(\mathbf{p}) = \underline{\boldsymbol{\kappa}}$, a constant matrix, it is convenient to define a purely numerical matrix containing all essential information for the ith extreme current. Let

$$\mathsf{S}^{(i)} \equiv -\,\underline{\boldsymbol{\nu}}(\operatorname{diag}\mathbf{E}_i)\,\underline{\boldsymbol{\kappa}}^{t} \tag{III.9}$$

where $\mathbf{E}_i$ is the corresponding column of E. It is also convenient to define

$$\mathsf{V}(\mathbf{j}) \equiv -\,\underline{\boldsymbol{\nu}}(\operatorname{diag}\mathsf{Ej})\,\underline{\boldsymbol{\kappa}}^{t} \tag{III.10}$$

Then the matrix of the linearized system may be decomposed into sums over extreme currents, or expressed using $\mathsf{V}(\mathbf{j})$ as follows

$$\mathsf{M}_{\zeta R}(\mathbf{h},\mathbf{j}) = -\sum_{i=1}^{f} j_i \mathsf{S}^{(i)}\operatorname{diag}\mathbf{h} = -\mathsf{V}(\mathbf{j})\operatorname{diag}\mathbf{h} \tag{III.11}$$

$$\mathsf{M}'_{\zeta R}(\mathbf{h},\mathbf{j}) = -\sum_{i=1}^{f} j_i(\operatorname{diag}\mathbf{h}^{1/2})\mathsf{S}^{(i)}\operatorname{diag}\mathbf{h}^{1/2} = -(\operatorname{diag}\mathbf{h}^{1/2})\mathsf{V}(\mathbf{j})\operatorname{diag}\mathbf{h}^{1/2} \tag{III.12}$$

$$\mathsf{M}''_{\zeta R}(\mathbf{h},\mathbf{j}) = -\sum_{i=1}^{f} j_i(\operatorname{diag}\mathbf{h})\mathsf{S}^{(i)} = -(\operatorname{diag}\mathbf{h})\mathsf{V}(\mathbf{j}) \tag{III.13}$$

Example III.1. The Oregonator of Example II.1 has the kinetic matrix

$$\underline{\boldsymbol{\kappa}} = \begin{pmatrix} 0 & 1 & 1 & 2 & 0 \\ 1 & 1 & 0 & 0 & 0 \\ 0 & 0 & 0 & 0 & 1 \end{pmatrix}$$

Using $\underline{\boldsymbol{\nu}}$ and E from Example II.5, we obtain from (III.9) and (III.11)

$$\mathsf{S}^{(1)} = \begin{pmatrix} 0 & 1 & 0 \\ 1 & 1 & -1 \\ -2 & 0 & 2 \end{pmatrix} \qquad \mathsf{S}^{(2)} = \begin{pmatrix} 3 & -1 & 0 \\ 0 & 1 & -1 \\ -2 & 0 & 2 \end{pmatrix}$$

$$\mathsf{M}_{\zeta R}(\mathbf{h},\mathbf{j}) = -\begin{bmatrix} 3h_1 j_2 & h_2(j_1 - j_2) & 0 \\ h_1 j_1 & h_2(j_1 + j_2) & -h_3(j_1 + j_2) \\ -2h_1(j_1 + j_2) & 0 & 2h_3(j_1 + j_2) \end{bmatrix}$$

The closely related matrix $\mathsf{M}''_{\zeta R}(\mathbf{h}, \mathbf{j})$ can be obtained by linearizing the nonlinear equation given in Example II.7 or by using (III.13).

The other interesting special case, when the effective power function can vary widely, is illustrated by modifying the Oregonator kinetics.

Example III.2. Introduce the parameters $(q_1, \ldots, q_6) \in R^6_+$ and let

$$\underline{\kappa} = \begin{bmatrix} 0 & q_1 & q_2 & q_3 & 0 \\ q_4 & q_5 & 0 & 0 & 0 \\ 0 & 0 & 0 & 0 & q_6 \end{bmatrix}$$

which is a generalization of $\underline{\kappa}$ in Example III.1. From (III.9) and (III.11) we obtain

$$\mathsf{M}_\zeta(\mathbf{h}, \mathbf{j}, \mathbf{q}) = -\begin{bmatrix} h_1(j_1q_1 - j_1q_2 - j_2q_2 + 2j_2q_3) & h_2(j_1q_5 - j_2q_4) & 0 \\ h_1j_1q_1 & h_2(j_1q_5 + j_2q_4) & -h_3j_2q_6 \\ -2h_1(j_1 + j_2)q_2 & 0 & h_3(j_1 + 2j_2)q_6 \end{bmatrix}$$

The increased flexibility coming from the new parameter $\mathbf{q}$ changes the network stability problem considerably, as we will see in Sections III.E and IV.D.

In the remainder of this subsection let us consider the consequences of reducing the r or n variables down to d variables, the minimum number possible. Without loss of generality, the first d variables may be assumed to be independent. All arrays having a subscript that indexes species are now divided into subarrays that are subscripted I (independent) and D (dependent), as for example: $\mathbf{X}_I, \mathbf{X}_D, \underline{\nu}_I, \underline{\nu}_D, \underline{\gamma}_I, \underline{\gamma}_D, \underline{\kappa}_I, \underline{\kappa}_D$. Then writing (II.5) as

$$\underline{\gamma}_I \mathbf{X}_I + \underline{\gamma}_D \mathbf{X}_D = \mathbf{C}$$

and solving gives

$$\mathbf{X}_D = \underline{\gamma}_D^{-1}\mathbf{C} - \underline{\gamma}_D^{-1}\underline{\gamma}_I \mathbf{X}_I \tag{III.14}$$

The matrix M_ζ in (III.1) splits into four submatrices. The independent components obey the equation of motion

$$\dot{\zeta}_I = \mathsf{M}_{II}\zeta_I + \mathsf{M}_{ID}\zeta_D$$

However $\zeta_D = \mathbf{X}_D - \mathbf{X}_D^0$, so by (III.14)

$$\dot{\zeta}_I = \mathsf{M}^*\zeta_I$$

$$\mathsf{M}^* = \mathsf{M}_{II} - \mathsf{M}_{ID}\underline{\gamma}_D^{-1}\underline{\gamma}_I \tag{III.15}$$

Expressing this in terms of the redundant parameters gives

$$\mathsf{M}^*_{\zeta R}(\mathbf{h}, \mathbf{j}) = \underline{\nu}_I(\operatorname{diag} \mathbf{Ej})\left[\,\underline{\kappa}_I^t \operatorname{diag} \mathbf{h}_I - \underline{\kappa}_D^t(\operatorname{diag} \mathbf{h}_D)\underline{\gamma}_D^{-1}\underline{\gamma}_I\right] \tag{III.16}$$

A similar calculation for $\boldsymbol{\zeta}''$ and $\mathsf{M}''_{\zeta R}$ gives

$$\mathsf{M}^{*\prime\prime}_{\zeta R}(\mathbf{h},\mathbf{j}) = (\operatorname{diag}\mathbf{h}_I)\,\underline{\boldsymbol{\nu}}_I(\operatorname{diag}\mathsf{E}\mathbf{j})\Big[\,\underline{\boldsymbol{\kappa}}_I^{t} - \underline{\boldsymbol{\kappa}}_D^{t}(\operatorname{diag}\mathbf{h}_D)\,\underline{\boldsymbol{\gamma}}_D^{-1}\,\underline{\boldsymbol{\gamma}}_I(\operatorname{diag}\mathbf{h}_I^{-1})\Big] \tag{III.17}$$

The complicated formula for $\mathsf{M}^{*}_{\zeta R}(\mathbf{h},\mathbf{j})$ is usually much more disadvantageous than the presence of the dependent variables. The most useful matrices are $\mathsf{M}_{\zeta R}(\mathbf{h},\mathbf{j})$ and $\mathsf{M}''_{\zeta R}(\mathbf{h},\mathbf{j})$.

B. Stability of Systems and Networks

Let $\mathbf{u}$ represent any of the perturbed variables $\boldsymbol{\zeta}$, $\boldsymbol{\zeta}'$, $\boldsymbol{\zeta}''$, $\boldsymbol{\eta}$, or $\boldsymbol{\eta}'$ and let the nonlinear equation of motion be

$$\dot{\mathbf{u}} = \mathbf{f}(\mathbf{u},\mathbf{p}) \tag{III.18}$$

Let the linearized system be

$$\dot{\mathbf{u}} = \mathsf{M}(\mathbf{p})\mathbf{u} \tag{III.19}$$

where $\mathsf{M}(\mathbf{p})$ is any of the matrices discussed in the previous section and the parameter vector $\mathbf{p}$ lies in the domain D.

The formal solution of (III.19) is

$$\mathbf{u}(t) = (\exp \mathsf{M}(\mathbf{p})t)\mathbf{u}(0) \tag{III.20}$$

where the exponential function is defined by a power series expansion. The series may be expressed as the product of two matrices,[57] the elements of one matrix being polynomials in t and those of the other being polynomials in $e^{\lambda_j t}$, where λ_j is an eigenvalue of $\mathsf{M}(\mathbf{p})$. Hence the general form of the ith component of $\mathbf{u}(t)$ is a sum over eigenvalues λ_j

$$u_i(t) = \sum_j \theta_{ij}(t)e^{\lambda_j t} \tag{III.21}$$

where $\theta_{ij}(t)$ is a polynomial in t. The imaginary part of λ_j ($\operatorname{Im}\lambda_j$) makes the corresponding term in (III.21) periodic and does not affect the stability. If the real parts of all the eigenvalues ($\operatorname{Re}\lambda_j$) are negative, $u_i(t)\to 0$ and the origin ($\mathbf{u}=\mathbf{0}$) is said to be *asymptotically stable*. If $\operatorname{Re}\lambda_j > 0$ for any j such that $\theta_{ij}(t) \neq 0$, then $u_i(t)$ increases without limit and the origin is said to be *unstable*. The remaining possibility is when $\operatorname{Re}\lambda_j = 0$ for some j and $\operatorname{Re}\lambda_j \leqslant 0$ for all j. This case can lead to stability or instability depending on the circumstances. We now examine this situation in detail by first considering how zero eigenvalues arise.

The motion of $\mathbf{u}$ is confined by (II.5) to a polyhedron $\Pi(\mathbf{C})$ of dimension d, such as $\Pi_X(\mathbf{C})$ or $\Pi_\xi(\mathbf{C})$. Hence $\dot{\mathbf{u}}$ must lie in a subspace of dimension d.

Since (III.19) states that $\dot{\mathbf{u}}$ is a linear combination of the columns of $\mathsf{M}(\mathbf{p})$, it follows that these columns must lie in this subspace. Hence

$$\operatorname{rank} \mathsf{M}(\mathbf{p}) \leqslant d \tag{III.22}$$

Now $\mathsf{M}(\mathbf{p})$ is either an $n \times n$ or an $r \times r$ matrix. Let $\rho = n - d$ in the first case, and $r - d$ in the second. Then (III.22) implies that $\mathsf{M}(\mathbf{p})$ has at least ρ zero eigenvalues. These eigenvalues contribute a constant term to (III.21) and a perturbation in the corresponding eigenspace (space spanned by the corresponding eigenvectors) takes the system into a new steady state. Thus there is a ρ-dimensional set of steady states near any particular steady state. Suppose the original steady state lies in $\Pi(\mathbf{C}^0)$. Recall that there must be a nearby steady state lying in $\Pi(\mathbf{C})$, for $\mathbf{C}$ close to $\mathbf{C}^0$. Since $\mathbf{C} \in R_+^{n-d}$ the nearby steady states form an $(n-d)$-dimensional set. If $\mathsf{M}(\mathbf{p})$ is $n \times n$, these steady states are the steady states associated with the generalized eigenspace of the $n - d$ vanishing eigenvalues. If $\mathsf{M}(\mathbf{p})$ is $r \times r$, the generalized eigenspace of the $r - d$ vanishing eigenvalues has $r - n$ more dimensions than the set of adjacent steady states (which is $n - d$ dimensional) in the set of nearby polyhedra $\Pi(\mathbf{C})$. These $r - n$ extra dimensions correspond to the $r - n$ physically meaningless reaction extents that have been removed by the projection operator P_ξ.

Perturbations outside of $\Pi(\mathbf{C}^0)$ are not physically possible because they violate the conservation condition (II.5). Since $\mathsf{M}(\mathbf{p})$ has a ρ-dimensional set of steady states that lie close to $\mathbf{X}^0$ and these steady states lie outside of $\Pi(\mathbf{C}^0)$, the perturbations from $\mathbf{X}^0$ to these steady states are irrelevant for stability and the ρ associated zero eigenvalues of $\mathsf{M}(\mathbf{p})$ are also irrelevant. We therefore remove a factor of λ^ρ from the characteristic polynomial

$$\det|\lambda \mathsf{I} - \mathsf{M}(\mathbf{p})| = \lambda^\rho \chi(\lambda, \mathbf{p}) \tag{III.23}$$

and define the *relevant eigenvalues* to be the solutions of

$$\chi(\lambda, \mathbf{p}) = 0 \tag{III.24}$$

The following precise definitions of stability are valid for the general (nonlinear) case. A system is *asymptotically stable* if there exists $\delta > 0$, such that every half-trajectory $\mathbf{u}(t)$ initially in the appropriate polyhedron with $\|\mathbf{u}(0)\| < \delta$, approaches the origin in the limit as t approaches infinity.

A system is *marginally stable* if, for every $\varepsilon > 0$, there exists $\delta > 0$ such that all half-trajectories $\mathbf{u}(t)$ initially in the appropriate polyhedron with $\|\mathbf{u}(0)\| < \delta$ satisfy $\|\mathbf{u}(t)\| < \varepsilon$ for all t.

A system is *stable* if it is asymptotically stable or marginally stable; otherwise it is *unstable*. Hence a system is unstable if and only if there exists $\varepsilon > 0$ such that for every $\delta > 0$, the set of half-trajectories $\mathbf{u}(t)$

initially in the appropriate polyhedron with $\|\mathbf{u}(0)\| < \delta$, contains a trajectory with $\|\mathbf{u}(t)\| \geqslant \varepsilon$ for some t.

For a linear system, the following conclusions follow easily from (III.21). The system is asymptotically stable if and only if $\text{Re}(\lambda_j) < 0$ for all relevant eigenvalues λ_j. A necessary (but not sufficient) condition for marginal stability is that $\text{Re}(\lambda_j) \leqslant 0$ for all relevant eigenvalues λ_j. A sufficient (but not necessary) condition for instability is that $\text{Re}(\lambda_j) > 0$ for some eigenvalue λ_j. These theorems determine the stability in all cases, except in the case when $\text{Re}(\lambda_j) \leqslant 0$ for all j and $\text{Re}(\lambda_j) = 0$ for some relevant eigenvalue λ_j. In this case, the system is marginally stable if $\theta_{ij}(t)$ in (III.21) contains no terms in t^k, $k \geqslant 1$, whenever $\text{Re}(\lambda_j) = 0$. It is otherwise unstable. This property of the functions $\theta_{ij}(t)$ may be ascertained by examining the *minimal polynomial* $\mu(\lambda, \mathbf{p})$, which is defined to be the polynomial of least degree that divides $\chi(\lambda, p)$ and has the property that $\mu(\mathsf{M}(\mathbf{p})) = 0$. If $\text{Re}(\lambda_j) = 0$, the origin is marginally stable if λ_j is a simple root of $\mu(\lambda, \mathbf{p})$ and unstable if λ_j is a multiple root. When instability occurs for this reason, the departure from steady state is as t^k rather than as an exponential; hence we call this case a *weak instabillty* to distinguish it from the exponential instability, which occurs when $\text{Re}(\lambda_j) > 0$.

Let the set of relevant eigenvalues of the linearized system with matrix $\mathsf{M}(\mathbf{p})$ be $\Lambda(\mathbf{p})$. Then for each $\mathbf{p} \in D$ the linearized system is asymptotically stable and exponentially unstable, respectively, in the domains

$$D_a^L \equiv \{\mathbf{p} \in D \,|\, \text{Re}(\lambda) < 0 \quad \text{for all} \quad \lambda \in \Lambda(\mathbf{p})\}$$
$$D_e^L \equiv \{\mathbf{p} \in D \,|\, \text{Re}(\lambda) > 0 \quad \text{for some} \quad \lambda \in \Lambda(\mathbf{p})\} \qquad \text{(III.25)}$$

The domain

$$D_{mw}^L \equiv \{\mathbf{p} \in D \,|\, \text{Re}(\lambda) \leqslant 0 \quad \text{for all } \lambda \in \Lambda(\mathbf{p})$$
$$\text{Re}(\lambda) = 0 \quad \text{for some } \lambda \in \Lambda(\mathbf{p})\} \qquad \text{(III.26)}$$

can be divided into two domains D_m^L and D_w^L, where the system is respectively marginally stable or weakly unstable. Then

$$D = D_a^L \cup D_m^L \cup D_w^L \cup D_e^L$$

and none of the four sets D_a^L, D_m^L, D_w^L, or D_e^L intersect. The linear system is stable on the set $D_s^L \equiv D_a^L \cup D_m^L$ and unstable on the set $D_u^L \equiv D_w^L \cup D_e^L$. It is *semistable* on the set $D_{\text{semi}}^L \equiv D_a^L \cup D_m^L \cup D_w^L$.

Further complications arise in nonlinear systems. One may write the nonlinear equation of motion as

$$\dot{\mathbf{u}} = \mathsf{M}(\mathbf{p})\mathbf{u} + \mathbf{g}(\mathbf{u}, \mathbf{p}) \qquad \text{(III.27)}$$

where $\mathbf{g}(\mathbf{u})$ contains the nonlinear corrections to the linearized system. If every eigenvalue of $\mathsf{M}(\mathbf{p})$ has a nonvanishing real part, the correction $\mathbf{g}(\mathbf{u})$ becomes negligible close to steady state and the motion near steady state is similar to the linearized systems. Hence a sufficient (but not necessary) condition for the nonlinear system to be asymptotically stable is that $\mathbf{p} \in D_a^L$. Also a sufficient (but not necessary) condition for the nonlinear system to be unstable is that $\mathbf{p} \in D_e^L$. These theorems treat all situations except the one where $\mathbf{p} \in D_{mw}^L = D_m^L \cup D_w^L$. Then the first term in (III.27) either does not give a radial component to the motion (marginally stable case) or gives a weakly divergent (t^k) radial dependence (unstable case). In these cases, the term $g(\mathbf{u})$ is not always negligible with respect to the linear terms and the nonlinear system may be asymptotically stable, marginally stable, or unstable, independently of the stability of the linearized system.

For each $\mathbf{p} \in D$, the nonlinear system is either asymptotically stable, marginally stable, or unstable. Corresponding to these cases we divide D into three mutually exclusive subsets D_a, D_m, and D_u respectively. The stable set is $D_s = D_a \cup D_m$. From the reasoning in the last paragraph,

$$D_a \supset D_a^L \tag{III.28}$$

$$D_u \supset D_e^L \tag{III.29}$$

and hence

$$D_m \subset D_{mw}^L \tag{III.30}$$

Let $\mathfrak{N}$ be the set of all chemical networks. The set of *asymptotically stable networks* is

$$\mathfrak{N}_a \equiv \{ N \in \mathfrak{N} \mid D_m = \varnothing, D_u = \varnothing \} \tag{III.31}$$

the set of *marginally stable networks* is

$$\mathfrak{N}_m \equiv \{ N \in \mathfrak{N} \mid D_m \neq \varnothing, D_u = \varnothing \} \tag{III.32}$$

and the set of *unstable networks* is

$$\mathfrak{N}_u \equiv \{ N \in \mathfrak{N} \mid D_u \neq \varnothing \} \tag{III.33}$$

The network stability problem is: "find $\mathfrak{N}_a$, $\mathfrak{N}_m$, and $\mathfrak{N}_u$."

This problem is extremely difficult because some networks in each set are networks whose linearizations are marginally stable or weakly unstable, with the nonlinear terms deciding the true stability. Hence we define sets of networks according to the stability of their linearization as follows. The set of *linearly asymptotically stable* networks is

$$\mathfrak{N}_a^L \equiv \{ N \in \mathfrak{N} \mid D_a^L = D \} \tag{III.34}$$

the set of *linearly marginally stable* networks is

$$\mathfrak{N}_m^L \equiv \{ N \in \mathfrak{N} \mid D_m^L \neq \emptyset, D_w^L = D_e^L = \emptyset \} \tag{III.35}$$

the set of *linearly weakly unstable* networks is

$$\mathfrak{N}_w^L \equiv \{ N \in \mathfrak{N} \mid D_w^L \neq \emptyset, D_e^L = \emptyset \} \tag{III.36}$$

and the set of *linearly exponentially unstable* networks is

$$\mathfrak{N}_e^L \equiv \{ N \in \mathfrak{N} \mid D_e^L \neq \emptyset \} \tag{III.37}$$

From these we define the set of *linearly semistable* networks to be $\mathfrak{N}_{\text{semi}}^L \equiv \mathfrak{N}_a^L \cup \mathfrak{N}_m^L \cup \mathfrak{N}_w^L$ and the set of *linearly stable networks* to be $\mathfrak{N}_s^L \equiv \mathfrak{N}_a^L \cup \mathfrak{N}_m^L$. From (III.28) and (III.29)

$$\mathfrak{N}_a \supset \mathfrak{N}_a^L, \qquad \mathfrak{N}_u \supset \mathfrak{N}_e^L \tag{III.38}$$

also, it follows from the definitions that

$$\mathfrak{N}_{\text{semi}}^L = \{ N \in \mathfrak{N} \mid D_e^L = \emptyset \} = \mathfrak{N} \setminus \mathfrak{N}_e^L \tag{III.39}$$

$$\mathfrak{N} = \mathfrak{N}_s^L \cup \mathfrak{N}_m^L \cup \mathfrak{N}_w^L \cup \mathfrak{N}_e^L \tag{III.40}$$

This terminology is summarized in Table II, which classifies the network according to its most unstable steady state, using the following ordering: asymptotically stable, marginally stable, weakly unstable, and exponentially unstable. (Because of a parallelism between the matrix sign stability problem and the network stability problem, I considered extending the sign stability terminology to the network problem. Then it would have been necessary to define a network to be "stable" if its steady states were all asymptotically stable, and "quasistable" if its steady states were all stable. Thus the term "stable" would have had a different meaning when referring to steady states than when referring to networks. I found this terminology so confusing that I rejected it. In this chapter, both networks and the sign stability problem are discussed using the parallel terminology of Table II, rather than the terminology of the sign stability literature.)

On the basis of the similarity with the sign stability problem, where considerable progress has already been made, it is probably easiest to find practical necessary and sufficient conditions for a network to be in $\mathfrak{N}_{\text{semi}}^L$, or equivalently, $\mathfrak{N}_e^L$. The next simplest set to characterize is probably $\mathfrak{N}_a^L$;

TABLE II
Summary of Stability Classifications for Networks

Network Stability Classification	Symbol	Classification of the Most Unstable Steady State	Sets That Cannot Be Empty	Sets That Must Be Empty	Conditions on the Relevant Eigenvalues of the Most Unstable Steady State
Linearly asymptotically stable	$\mathfrak{N}_a^L$	Linearly asymptotically stable	D_a^L	D_m^L, D_w^L, D_e^L	$\mathrm{Re}(\lambda) < 0$
Linearly marginally stable	$\mathfrak{N}_m^L$	Linearly marginally stable	D_m^L	D_w^L, D_e^L	$\mathrm{Re}(\lambda) = 0$[a]
Linearly stable (quasistable)	$\mathfrak{N}_s^L$	Linearly stable	D_s^L	D_w^L, D_e^L	$\mathrm{Re}(\lambda) \leqslant 0$[a]
Linearly weakly unstable	$\mathfrak{N}_w^L$	Linearly weakly unstable	D_w^L	D_e^L	$\mathrm{Re}(\lambda) = 0$[b]
Linearly semistable	$\mathfrak{N}_{\mathrm{semi}}^L$	Linearly semistable	D_{semi}^L	D_e^L	$\mathrm{Re}(\lambda) \leqslant 0$
Linearly exponentially unstable	$\mathfrak{N}_e^L$	Linearly exponentially unstable	D_e^L		$\mathrm{Re}(\lambda) > 0$
Asymptotically stable	$\mathfrak{N}_a$	Asymptotically stable	D_a	D_m, D_u	$\mathrm{Re}(\lambda) \leqslant 0$[c]
Marginally stable	$\mathfrak{N}_m$	Marginally stable	D_m	D_u	$\mathrm{Re}(\lambda) = 0$[c]
Stable	$\mathfrak{N}_s$	Stable	D_s	D_u	$\mathrm{Re}(\lambda) \leqslant 0$[c]
Unstable	$\mathfrak{N}_u$	Unstable	D_u		$\mathrm{Re}(\lambda) \geqslant 0$[c]

[a] Eigenvalues with $\mathrm{Re}(\lambda) = 0$ are all simple roots of the minimal polynomial.
[b] Some eigenvalues with $\mathrm{Re}(\lambda) = 0$ are degenerate roots of the minimal polynomial.
[c] Consideration of the nonlinear terms is necessary to determine the stability classification if $\mathrm{Re}(\lambda) = 0$.

then $\mathfrak{N}_m^L \cup \mathfrak{N}_w^L = \mathfrak{N}_{\text{semi}}^L \backslash \mathfrak{N}_a^L$. The next problem is to partition $\mathfrak{N}_{\text{semi}}^L \backslash \mathfrak{N}_a^L$ into $\mathfrak{N}_m^L$ and $\mathfrak{N}_w^L$; this would characterize $\mathfrak{N}_s^L$. Determining $\mathfrak{N}_a$, $\mathfrak{N}_m$, and $\mathfrak{N}_u$ is even more difficult because the nonlinear terms must be considered. At present not even $\mathfrak{N}_e^L$ has been fully characterized.

A network is an *extreme network* if it contains only one extreme current. The set of extreme networks is

$$\mathcal{E} \equiv \{ N \in \mathfrak{N} \mid \dim \Pi_v(N) = 0 \} \tag{III.41}$$

The extreme currents of the general network $N \in \mathfrak{N}$ are the columns of the matrix E that form the set $\{\mathbf{E}_i \mid i \in [1, f]\}$ which we also call E. To each vector $\mathbf{E}_i \in \mathsf{E}$, there corresponds an extreme network $E \in \mathcal{E}$ consisting of the reactions of N whose corresponding components of $\mathbf{E}_i$ do not vanish. Only the species in these reactions are species of E. We call E an *extreme subnetwork* of N. The correspondence between $\mathbf{E}_i$ and E can be represented by a mapping Ξ defined by $E = \Xi(N, \mathbf{E}_i)$. Then the set of extreme subnetworks of N is $\Xi(N, \mathsf{E})$. Without loss of generality, the reaction stoichiometries of an extreme subnetwork may be chosen so that its only extreme current is $\mathbf{e} = (1, 1, \ldots, 1)^t$. Then (III.9) becomes

$$\mathsf{S} = -\underline{\boldsymbol{\nu}}\,\underline{\boldsymbol{\kappa}}^t \tag{III.42}$$

Many of the stability properties of networks can be related to the stability properties of their extreme subnetworks. Hence we define

$$\mathcal{E}_i \equiv \mathfrak{N}_i \cap \mathcal{E}, \qquad \mathcal{E}_i^L \equiv \mathfrak{N}_i^L \cap \mathcal{E} \tag{III.43}$$

where i is any of the subscripts appearing in definitions (III.31) to (III.37).

C. Stability Proofs Using Hurwitz Determinants

The signs of the Hurwitz determinants determine the number of eigenvalues with positive, zero, and negative real parts. Here we summarize the important theorems; the details may be found in Gantmacher.[58]

The relevant part of the characteristic polynomial of $\mathsf{M}(\mathbf{p})$ is a polynomial in λ. The coefficient of λ^{d-i} will be called $\alpha_i(\mathbf{p})$. Then from (III.23)

$$\det|\lambda \mathsf{I} - \mathsf{M}| = \lambda^{\rho} \sum_{i=0}^{d} \lambda^{d-i} \alpha_i(\mathbf{p}) \tag{III.44}$$

where $\alpha_i(\mathbf{p})$ is a polynomial in $\mathbf{p}$. By writing $\lambda \mathsf{I} - \mathsf{M}$ rather than $\mathsf{M} - \lambda \mathsf{I}$, the signs of half the coefficients α_i have been reversed; this convention avoids factors of $(-1)^i$ that usually appear in certain equations. Note that the leading term of (III.44) is $\lambda^{\rho+d}$, so $\alpha_0(\mathbf{p}) = 1$. It is convenient to extend the sum to infinity and define $\alpha_i(\mathbf{p}) = 0$ for $i > d$.

Consider the infinite array A, which is defined to be

$$\begin{array}{ccccc} \alpha_1(\mathbf{p}) & \alpha_3(\mathbf{p}) & \alpha_5(\mathbf{p}) & \alpha_7(\mathbf{p}) & \cdots \\ \alpha_0(\mathbf{p}) & \alpha_2(\mathbf{p}) & \alpha_4(\mathbf{p}) & \alpha_6(\mathbf{p}) & \cdots \\ 0 & \alpha_1(\mathbf{p}) & \alpha_3(\mathbf{p}) & \alpha_5(\mathbf{p}) & \cdots \\ 0 & \alpha_0(\mathbf{p}) & \alpha_2(\mathbf{p}) & \alpha_4(\mathbf{p}) & \cdots \\ 0 & 0 & \alpha_1(\mathbf{p}) & \alpha_3(\mathbf{p}) & \cdots \\ 0 & 0 & \alpha_0(\mathbf{p}) & \alpha_2(\mathbf{p}) & \cdots \end{array} \tag{III.45}$$

The diagonal of A is $(\alpha_1, \alpha_2, \alpha_3, \ldots)$ and the columns of A contain the coefficients in descending order. The *Hurwitz determinant* $\Delta_i(\mathbf{p})$ is defined to be the determinant of the square matrix formed from the elements found in the first i rows and columns of A. Thus $\Delta_1(\mathbf{p}) = \alpha_1(\mathbf{p})$, $\Delta_2(\mathbf{p}) = \alpha_1(\mathbf{p})\alpha_2(\mathbf{p}) - \alpha_0(\mathbf{p})\alpha_3(\mathbf{p}), \ldots$. Such determinants are called the *principal minors* of A.

The Routh-Hurwitz Theorem (Gantmacher,[58] Theorem 4, p. 230). The number of eigenvalues λ_i with $\mathrm{Re}\,\lambda_i > 0$ equals the sum of the number of changes of sign in the sequences

$$1, \Delta_1(\mathbf{p}), \Delta_3(\mathbf{p}), \Delta_5(\mathbf{p}), \ldots$$

$$1, \Delta_2(\mathbf{p}), \Delta_4(\mathbf{p}), \Delta_6(\mathbf{p}), \ldots$$

Example III.3. Suppose the characteristic polynomial is

$$\lambda^4 + 10\lambda^3 - 20\lambda^2 + 6\lambda + 15 = 0$$

$$\mathsf{A} = \begin{array}{cccc} 10 & 6 & 0 & 0 \\ 1 & -20 & 15 & 0 \\ 0 & 10 & 6 & 0 \\ 0 & 1 & -20 & 15 \end{array}$$

$$\Delta_1 = \det(10) = 10, \qquad \Delta_2 = \det\begin{pmatrix} 10 & 6 \\ 1 & -20 \end{pmatrix} = -206$$

$$\Delta_3 = \det\begin{pmatrix} 10 & 6 & 0 \\ 1 & -20 & 15 \\ 0 & 10 & 6 \end{pmatrix} = -2736, \qquad \Delta_4 = \det \mathsf{A} = -41{,}040$$

The sequences are $1, 10, -2736$ and $1, -206, -41{,}040$. Each sequence has one sign change, so there are two eigenvalues with positive real parts.

If $\Delta_i(\mathbf{p}) > 0$ for $i = 1, \ldots, d$, the sequences have no sign changes, so $\mathrm{Re}\,\lambda_i < 0$ for all i, hence $\mathbf{p} \in D_a^L$. Conversely, if $\mathbf{p} \in D_a^L$, it follows that $\mathrm{Re}\,\lambda_i < 0$ for all i, so $\Delta_i(\mathbf{p}) > 0$ for $i = 1, \ldots, d$. Hence a necessary and sufficient condition for $\mathbf{p} \in D_a^L$ is that $\Delta_i(\mathbf{p}) > 0$ for $l = 1, \ldots, d$. Thus if

$[1,d] \equiv \{1,2,\ldots,d\}$

$$D_a^L = \{\mathbf{p} \in D \mid \Delta_i(\mathbf{p}) > 0 \qquad \text{for } i \in [1,d]\} \tag{III.46}$$

A similar line of argument yields

$$D_e^L = \{\mathbf{p} \in D \mid \Delta_j(\mathbf{p}) < 0 \qquad \text{for some } j \in [1,d]\} \tag{III.47}$$

$$D_{mw}^L = \{\mathbf{p} \in D \mid \Delta_i(\mathbf{p}) \geqslant 0 \qquad \text{for } i \in [1,d]$$

$$\text{for some } j \in [1,d], \Delta_j(\mathbf{p}) = 0\} \tag{III.48}$$

The polynomial $\chi(\lambda, \mathbf{p})$ may be factored into linear factors of the form $(\lambda - \lambda_i)$ corresponding to each real root λ_i, and into quadratic factors of the form $\lambda^2 - (\lambda_i + \bar{\lambda}_i) + \lambda_i\bar{\lambda}_i$ corresponding to each pair of complex conjugate roots λ_i and $\bar{\lambda}_i$. If $\Delta_i(\mathbf{p}) > 0$ for $i \in [1,d]$, $\mathrm{Re}(\lambda_i) < 0$, and the coefficients in the linear and quadratic factors are all real and positive. Hence if we multiply out these factors to obtain $\chi(\lambda, \mathbf{p})$, the coefficients we obtain must be positive. Thus the condition

$$\alpha_i(\mathbf{p}) > 0, \qquad i = 1, \ldots, d \tag{III.49}$$

must be satisfied for all $\mathbf{p} \in D_a^L$. Condition (III.49) may also be satisfied for some $\mathbf{p} \in D_{mw}^L \cup D_u^L$, so it is not a sufficient condition for asymptotic stability. However when (III.49) is satisfied, only about half the Hurwitz determinants need be examined, as stated in the next theorem.

Liénard-Chipart Theorem (Gantmacher,[58] Theorem 11, p. 263). If all functions in one of the four sets of functions

$$F_1 \equiv \{\Delta_i(\mathbf{p}), \alpha_d(\mathbf{p}), \alpha_j(\mathbf{p}) \mid i \text{ even}, j \text{ odd}\}$$

$$F_2 \equiv \{\Delta_i(\mathbf{p}), \alpha_d(\mathbf{p}), \alpha_j(\mathbf{p}) \mid i \text{ odd}, j \text{ even}\}$$

$$F_3 \equiv \{\Delta_i(\mathbf{p}), \alpha_d(\mathbf{p}), \alpha_j(\mathbf{p}) \mid i \text{ even}, j \text{ even}\}$$

$$F_4 \equiv \{\Delta_i(\mathbf{p}), \alpha_d(\mathbf{p}), \alpha_j(\mathbf{p}) \mid i \text{ odd}, j \text{ odd}\}$$

are positive, where i, $j \in [1,d]$, then $\mathbf{p} \in D_a^L$.

It follows from (III.46) and (III.49) that if $\mathbf{p} \in D_a^L$, all functions mentioned in the theorem are positive. Hence this theorem gives four sets of necessary and sufficient conditions for $\mathbf{p} \in D_a^L$. We may thus generalize

(III.46) to (III.48) to

$$D_a^L = \{\mathbf{p} \in D \mid f(\mathbf{p}) > 0 \quad \text{for all } f \in F_i\} \tag{III.50}$$

$$D_e^L = \{\mathbf{p} \in D \mid f(\mathbf{p}) < 0 \quad \text{for some } f \in F_i\} \tag{III.51}$$

$$D_{mw}^L = \{\mathbf{p} \in D \mid f(\mathbf{p}) \geqslant 0 \quad \text{for all } f \in F_i$$

$$f(\mathbf{p}) = 0 \quad \text{for some } f \in F_i\} \tag{III.52}$$

where F_i is any of the four sets of functions given in the Liénard-Chipart theorem or is $F_5 \equiv \{\Delta_i(\mathbf{p}) \mid i \in [1, d]\}$.

The sets of asymptotically stable, marginally stable, and unstable networks may now be reexpressed using one of the sets of functions F_i. Since F_i depends on the network N, we now write $F_i(N)$; similarly D becomes $D(N)$. From (III.34), (III.37), (III.50), and (III.51), we conclude that for any $i = 1, \ldots, 5$,

$$\mathfrak{N}_a^L = \{N \in \mathfrak{N} \mid f(\mathbf{p}) > 0 \quad \text{for all } \mathbf{p} \in D(N) \text{ and all } f \in F_i(N)\} \tag{III.53}$$

$$\mathfrak{N}_e^L = \{N \in \mathfrak{N} \mid f(\mathbf{p}) < 0 \quad \text{for some } \mathbf{p} \in D(N) \text{ and some } f \in F_i(N)\} \tag{III.54}$$

The networks that remain are

$$\mathfrak{N}_m^L \cup \mathfrak{N}_w^L = \{N \in \mathfrak{N} \mid f(\mathbf{p}) \geqslant 0 \quad \text{for all } \mathbf{p} \in D(N) \text{ and all } f \in F_i(N)$$

$$f(\mathbf{p}) = 0 \quad \text{for some } \mathbf{p} \in D(\mathfrak{N}) \text{ and some } f \in F_i(N)\} \tag{III.55}$$

The stability of these networks cannot be decided from the linear terms alone. Depending on the nonlinear terms, they may be stable, marginally stable, or unstable.

For all of the forms of $\mathbf{M}(\mathbf{p})$ given in Section III.A, the elements of $\mathbf{M}$ are first-order homogeneous polynomials in both $\mathbf{h}$ and $\mathbf{j}$. It follows from (III.44) that each polynomial $\alpha_i(\mathbf{h}, \mathbf{j})$ must be ith-order homogeneous in both $\mathbf{h}$ and $\mathbf{j}$. Then $\Delta_i(\mathbf{h}, \mathbf{j})$ must be homogeneous of order $i(i+1)/2$ in both $\mathbf{h}$ and $\mathbf{j}$.

Let $T(f, \mathbf{p})$ be the set of terms of a polynomial $f(\mathbf{p})$. Thus

$$f(\mathbf{p}) = \sum_{t \in T(f, \mathbf{p})} t \tag{III.56}$$

Every component of $\mathbf{p} \in D$ is positive; hence the sign of any element $t \in T(f, \mathbf{p})$ is independent of $\mathbf{p}$ and is never zero. The set of networks whose polynomials contain only positive terms in any one of the sets F_i of polynomials is

$$\mathfrak{N}_i^+ \equiv \{N \in \mathfrak{N} \mid t > 0 \quad \text{for all } t \in T(f,\mathbf{p})$$
$$T(f,\mathbf{p}) \neq \varnothing \quad \text{for all } f \in F_i(N)\} \tag{III.57}$$

If $t > 0$ for all $t \in T(f, p)$, then $f(\mathbf{p}) > 0$ for all $\mathbf{p} \in D(N)$. Hence from (III.53) and (III.38),

$$\mathfrak{N}_i^+ \subset \mathfrak{N}_a^L \subset \mathfrak{N}_a \tag{III.58}$$

A term $t \in T(f,\mathbf{p})$ is a *potentially dominant term* of f if it is possible to choose $\mathbf{p} \in D$ so that t is arbitrarily larger in magnitude than every other term in $T(f,\mathbf{p})$. The set of potential dominant terms of f is

$$T^D(f) \equiv \{t \in T(f,\mathbf{p}) \mid \forall \xi > 1, \exists \mathbf{p} \in D$$
$$\text{such that } |t| > \xi |t'| \ \forall t' \in T(f,\mathbf{p})\} \tag{III.59}$$

($\forall$ means "for all" and $\exists$ means "there exists.") If a network N has a polynomial $f \in F_i(N)$ containing a negative term that can dominate f, then f is negative for some $\mathbf{p} \in D$ and N is unstable by (III.54). The set of such networks is

$$\mathfrak{N}_i^{D-} \equiv \{N \in \mathfrak{N} \mid t < 0 \quad \text{for some } t \in T^D(f) \text{ for some } f \in F_i(N)\} \tag{III.60}$$

and is called the set of networks containing a *negative potentially dominant term*. Hence from (III.54) and (III.37) for $i \in [1,5]$,

$$\mathfrak{N}_i^{D-} \subset \mathfrak{N}_e^L \subset \mathfrak{N}_u \tag{III.61}$$

The networks that remain fall into two principal classes:

$$\mathfrak{N}_i^0 \equiv \{N \in \mathfrak{N} \mid T(f,\mathbf{p}) = \varnothing \quad \text{for some } f \in F_i(N)$$
$$t > 0 \quad \text{for all } t \in T(f,\mathbf{p}) \text{ for all } f \in F_i(N)\} \tag{III.62}$$

$$\mathfrak{N}_i^{I-} \equiv \{N \in \mathfrak{N} \mid t > 0 \quad \forall t \in T^D(f) \quad \forall f \in F_i(N)$$
$$t < 0 \quad \text{for some } t \in T(f,\mathbf{p}) \text{ for some } f \in F_i(N)\} \tag{III.63}$$

The set of networks with only positive terms is $\mathfrak{N}_i^+ \cup \mathfrak{N}_i^0$; hence the remaining networks have a negative term. In one of the remaining networks, if a negative term can be dominant, the network belongs to $\mathfrak{N}_i^{D-}$; otherwise there is a negative term, but all the dominant terms are positive, so the network belongs to $\mathfrak{N}_i^{I-}$. Hence for any $i \in [1,5]$,

$$\mathfrak{N}_i^{+} \cup \mathfrak{N}_i^{D-} \cup \mathfrak{N}_i^{0} \cup \mathfrak{N}_i^{I-} = \mathfrak{N} \tag{III.64}$$

For every network in $\mathfrak{N}_i^0$, either some Δ_i is identically zero or some α_i is identically zero. From (III.52), $D_{mw}^L = D$, so $D_e^L = \emptyset$, and $D_s^L = \emptyset$. Hence from (III.35) and (III.36)

$$\mathfrak{N}_i^0 \subset \mathfrak{N}_m^L \cup \mathfrak{N}_w^L \equiv \mathfrak{N}_{mw}^L \tag{III.65}$$

Of course, knowing that a network is in $\mathfrak{N}_{mw}^L$ does not resolve the stability problem because the stability of networks in $\mathfrak{N}_{mw}^L$ depends on the nonlinear terms, and such networks may be asymptotically stable, marginally stable, or unstable.

Knowledge of the signs and dominance of the terms in α_i and Δ_i is adequate to establish the stability classification of networks in $\mathfrak{N}_i^+$ and $\mathfrak{N}_i^{D-}$. The same information implies that for networks in $\mathfrak{N}_i^0$, the nonlinear terms decide the stability. Hence given the problem of determining the stability of an arbitrary network N, one strategy is to determine the signs and dominance of the terms of each $f \in F_i$ for some $i \in [1,5]$. If we are lucky, N will lie in $\mathfrak{N}_i^+$ or $\mathfrak{N}_i^{D-}$ and the problem will be solved. We will be less fortunate if N lies in $\mathfrak{N}_i^0$, for then the nonlinear terms must be considered; however we could not have avoided this complication. If N lies in $\mathfrak{N}_i^{I-}$ the stability problem is still unsolved because this fact alone does not determine whether a polynomial $f(\mathbf{p})$ is positive for all $\mathbf{p}$ or can be negative for some $\mathbf{p} \in D$. Only in this case do we have to look at the polynomial in greater detail, perhaps by using calculus to find where $f(\mathbf{p})$ attains its minimum, then by testing the sign of $f(\mathbf{p})$ at the minimum. If the set $\mathfrak{N}_i^{I-}$ were large, N might very well lie in $\mathfrak{N}_i^{I-}$ and this approach would usually require a good deal of calculation. If, on the other hand, the set $\mathfrak{N}_i^{I-}$ were small, this approach would quickly resolve the stability of most networks because few would lie in $\mathfrak{N}_i^0$ and most would lie in either $\mathfrak{N}_i^+$ or $\mathfrak{N}_i^{D-}$, where the stability is simple to determine.

I have a computer program that can construct the polynomials $\alpha_i(\mathbf{p})$ and $\Delta_i(p)$ from $\underline{\boldsymbol{\nu}}$ and $\underline{\boldsymbol{\kappa}}$. A very small fraction of the networks I have examined lie in $\mathfrak{N}_i^{I-}$, and all these networks possess small groups of terms whose sum can be negative and arbitrarily greater in magnitude than every other term in $T(f, \mathbf{p})$. Thus these networks are exponentially unstable. As a result of these computer experiments, we conjecture that $\mathfrak{N}_i^{I-} \subset \mathfrak{N}_e^L$, then from (III.58), (III.61), (III.64), (III.65), and (III.66),

$$\mathfrak{N}_a^L = \mathfrak{N}_i^+, \qquad \mathfrak{N}_{mw}^L = \mathfrak{N}_i^0, \qquad \mathfrak{N}_e^L = \mathfrak{N}_i^{D-} \cup \mathfrak{N}_i^{I-} \tag{III.66}$$

Then necessary and sufficient conditions for a network to be in $\mathfrak{N}_e^L$ would be identical to necessary and sufficient conditions for the existence of a negative coefficient in some polynomial $f \in F_i$, $i \in [1,5]$. Also, additional conditions for a polynomial's not vanishing identically would give conditions for a network to be in $\mathfrak{N}_a^L$. The ease of solving these problems appears to be in agreement with the discussion at the end of Section III.B and parallel developments in the sign stability problem.

The primary disadvantage to this approach is the large number of terms that can occur in the polynomials. The polynomials α_i have the largest number of terms when i is near the midpoint of the interval $[i, d]$ and the least number when $i = 1$ or $i = d$. The number in Δ_i increases extremely rapidly with i; thus one should never evaluate Δ_d, because by choosing F_i optimally, the largest Hurwitz determinant required is Δ_{d-1}. Hence F_5 is a poor choice because d rather than approximately $d/2$ Δ_i's must be constructed. Of the remaining possibilities, α_d must be constructed in all cases, so we minimize the number of α_j's required by avoiding α_{d-1}. If d is odd, the best set to use is then F_1; when d is even, the best set is F_2.

Example III.4. Let N be the network:

$$
\begin{aligned}
&R_1: X_1 + X_3 \to 2X_1 && v_1 = k_1 X_1 X_3 \\
&R_2: X_1 \to X_2 && v_2 = k_2 X_1 \\
&R_3: X_2 + X_3 \to X_4 && v_2 = k_3 X_2 X_3 \\
&R_4: X_4 \to 2X_3 && v_4 = k_4 X_4
\end{aligned}
$$

$$
\underline{\nu} = \begin{bmatrix} 1 & -1 & 0 & 0 \\ 0 & 1 & -1 & 0 \\ -1 & 0 & -1 & 2 \\ 0 & 0 & 1 & -1 \end{bmatrix} \qquad \underline{\kappa} = \begin{bmatrix} 1 & 1 & 0 & 0 \\ 0 & 0 & 1 & 0 \\ 1 & 0 & 1 & 0 \\ 0 & 0 & 0 & 1 \end{bmatrix} \qquad \mathsf{E} = \begin{bmatrix} 1 \\ 1 \\ 1 \\ 1 \end{bmatrix}
$$

The elements of each column of $\underline{\nu}$ sum to zero because of the conservation condition $X_1 + X_2 + X_3 + X_4 = C$. There are no other conservation conditions, so the rank of $\underline{\nu}$ is $d = n - 1 = 3$; hence $\alpha_i(\mathbf{p}) = 0$ for $i > 3$.

$$
\mathsf{M}_{\xi R}(\mathbf{h}, \mathbf{j}) = j_1 \begin{bmatrix} 0 & 0 & h_3 & 0 \\ h_1 & -h_2 & -h_3 & 0 \\ -h_1 & -h_2 & -2h_3 & 2h_4 \\ 0 & h_2 & h_3 & -h_4 \end{bmatrix}
$$

$$\alpha_1(\mathbf{h}, \mathbf{j}) = (h_2 + 2h_3 + h_4) j_1$$

$$\alpha_2(\mathbf{h}, \mathbf{j}) = (h_1 h_3 + h_2 h_3 + h_2 h_4) j_1^2$$

$$\alpha_3(\mathbf{h}, \mathbf{j}) = (2h_1 h_2 h_3 + h_1 h_3 h_4 + h_2 h_3 h_4) j_1^3$$

$$\Delta_1(\mathbf{h}, \mathbf{j}) = \alpha_1(\mathbf{h}, \mathbf{j})$$

$$\Delta_2(\mathbf{h}, \mathbf{j}) = (-h_1 h_2 h_3 + 2h_1 h_3^2 + h_2^2 h_3 + h_2^2 h_4 + h_2 h_3^2 + 2h_2 h_3 h_4 + 2h_2 h_4^2) j_1^3$$

$$\Delta_3(\mathbf{h}, \mathbf{j}) = \Delta_2(\mathbf{h}, \mathbf{j}) \alpha_3(\mathbf{h}, \mathbf{j})$$

Note the homogeneity of the polynomials. The factorization of Δ_d into $\Delta_{d-1}\alpha_d$ occurs in general because the right-hand column of the determinantal form of Δ_d contains zeros everywhere but in the last entry, which is α_d. Since d is odd, we consider only the polynomials $F_1 = \{\alpha_1, \alpha_3, \Delta_2\}$. The set $T(\Delta_2, \mathbf{p})$ contains the negative element $-h_1h_2h_3$, so $N \in \mathfrak{N}_1^{D-} \cup \mathfrak{N}_1^{I-}$. To show that $-h_1h_2h_3 \in T^D(\Delta_2)$, let $h_1 = 1$, $h_2 = 10^{-q}$, $h_3 = h_4 = 10^{-2q}$. Then

$$\Delta_2(\mathbf{h}, \mathbf{j}) = (-10^{-3q} + 4 \times 10^{-4q} + 5 \times 10^{-5q})$$

and the magnitude of the negative term becomes arbitrarily larger than the magnitudes of every other term as $q \to \infty$. Hence $N \in \mathfrak{N}_i^{D-} \subset \mathfrak{N}_u$. Since only one polynomial can be negative, from (III.50) and (III.51),

$$D_a^L = \{\mathbf{p} \in D \mid \Delta_2(\mathbf{p}) > 0\}$$

$$D_e^L = \{\mathbf{p} \in D \mid \Delta_2(\mathbf{p}) < 0\}$$

Since $\Delta_2(\mathbf{p})$ is continuous, the boundary of the unstable region is the surface

$$B = D_{mw}^L = \{\mathbf{p} \in D \mid \Delta_2(\mathbf{p}) = 0\}$$

and this is the bifurcation set. We have thus solved both the network stability problem and the stability diagram problem, although we do not yet have a clear picture of the surface D_{mw}^L. A method for approximating the zeros of large polynomials such as $\Delta_i(\mathbf{p})$ is discussed later. From the Routh-Hurwitz theorem one may conclude that two eigenvalues have $\text{Re}(\lambda_i) > 0$ everywhere in the unstable region. Since $\alpha_3(\mathbf{p})$ and $\alpha_2(\mathbf{p})$ never vanish, no zero relevant eigenvalues occur. Hence on the bifurcation set where $\text{Re}(\lambda_i) = 0$, λ_i must be purely imaginary. Thus a pair of complex conjugate eigenvalues crosses the imaginary axis as $\mathbf{p}$ traverses the bifurcation set. This situation is called a *Hopf bifurcation*.

D. Stability Proofs Using Lyapunov Functions

To begin with, we discuss systems with $\mathbf{p}$ fixed. If $\mathbf{p} \in D_a$, the origin $\mathbf{u} = \mathbf{0}$ is an asymptotically stable steady state of the nonlinear system (III.18). It is then always possible to find a differentiable function $L(\mathbf{u})$ with the following properties on a domain that contains the origin in its interior: (1) $L(\mathbf{u}) > 0$ if $\mathbf{u} \neq \mathbf{0}$, (2) $L(\mathbf{0}) = 0$, and (3) $dL(\mathbf{u})/dt < 0$ if $\mathbf{u} \neq \mathbf{0}$. $L(\mathbf{u})$ is called a *Lyapunov function* and is analogous to a potential energy that decreases as $\mathbf{u}$ approaches the steady state. Conversely, given any dynamical system for which a function with these properties exists, it follows that the origin must be asymptotically stable steady state. This is so because conditions 1 and 2 require that $L(\mathbf{u})$ have a local minimum at the origin, whereas condition 3 requires $\mathbf{u}$ to evolve so that $L(\mathbf{u})$ decreases toward the minimum. We have now established that the origin of the general system is asymptotically stable if and only if there exists a Lyapunov function $L(\mathbf{u})$.

Expand $L(\mathbf{u})$ in a Taylor series about $\mathbf{u} = \mathbf{0}$. The first two terms must vanish identically by conditions 1 and 2. Hence the first term that need not vanish is the quadratic term. It cannot be a consequence of the dynamics that the quadratic term must vanish for every Lyapunov function. Choose one whose quadratic term does not vanish; then for all sufficiently small $\mathbf{u}$, all the remaining terms are negligible, so there must exist a Lyapunov function of the form

$$L_Q(\mathbf{u}) \equiv \mathbf{u}'\mathsf{Q}\mathbf{u} \tag{III.67}$$

where Q may be taken to be a real symmetric matrix without any loss of generality. We say Q is *positive definite* if $\mathbf{u}'\mathsf{Q}\mathbf{u} > 0$ for all $\mathbf{u} \neq 0$. The set of positive definite real symmetric matrices is called $\mathbb{S}$. We conclude that the origin $\mathbf{u} = \mathbf{0}$ is asymptotically stable if and only if there exists $\mathsf{Q} \in \mathbb{S}$ such that $dL_Q(\mathbf{u})/dt < 0$ for all $\mathbf{u} \neq \mathbf{0}$.

To prove the asymptotic stability of a chemical system, it is sufficient to find a Lyapunov function for the linearized system (III.19) because the existence of such a function implies $\mathbf{p} \in D_a^L \subset D_a$ by (III.28). Hence the nonlinear system is asymptotically stable if there exists $\mathsf{Q} \in \mathbb{S}$ such that $dL_Q(u)/dt < 0$, where the time derivative is calculated using the *linearized* equation of motion, for all $\mathbf{u} \neq \mathbf{0}$. By (III.19)

$$\frac{dL_Q}{dt} = \dot{\mathbf{u}}'\mathsf{Q}\mathbf{u} + \mathbf{u}'\mathsf{Q}\dot{\mathbf{u}} = -\mathbf{u}'\mathsf{R}\mathbf{u}$$

where

$$\mathsf{R} = -\mathsf{M}'\mathsf{Q} - \mathsf{Q}\mathsf{M} \tag{III.68}$$

Hence asymptotic stability is proved if there exists $\mathsf{Q} \in \mathbb{S}$ such that $\mathsf{R} \in \mathbb{S}$.

Now let $\mathbf{p}$ vary and let Q, L_Q, and R be functions of $\mathbf{p}$. For future reference, the results we have proved can be expressed formally as follows:

$$D_a^L = \{\mathbf{p} \in D \mid \exists \mathsf{Q}(\mathbf{p}) \in \mathbb{S} \quad \text{such that } \mathsf{R}(\mathbf{p}) \in \mathbb{S}\} \tag{III.69}$$

Then from (III.28) and (III.34),

$$\mathfrak{N}_a^L = \{N \in \mathfrak{N} \mid \mathsf{Q}(\mathbf{p}) \in \mathbb{S} \quad \text{such that } \mathsf{R}(\mathbf{p}) \in \mathbb{S} \text{ for all } \mathbf{p} \in D\} \tag{III.70}$$

A matrix Q is positive semidefinite if $\mathbf{u}'\mathsf{Q}\mathbf{u} \geqslant 0$ for all $\mathbf{u} \neq 0$. Let the set of positive semidefinite real symmetric matrices be $\mathbb{S}_0$; then $\mathbb{S}_0 \supset \mathbb{S}$. If there exists $\mathsf{Q} \in \mathbb{S}$ such that $\mathsf{R} \in \mathbb{S}_0$, then $dL_Q/dt \leqslant 0$ for all $\mathbf{u} \neq 0$ and the linearized system cannot leave the steady state, although small oscillations around it are possible. Thus the steady state cannot be exponentially unstable or weakly unstable; therefore it must be linearly stable. Hence

$$D_s^L = \{\mathbf{p} \in D \mid \exists \mathsf{Q}(\mathbf{p}) \in \mathbb{S} \quad \text{such that } \mathsf{R}(\mathbf{p}) \in \mathbb{S}_0\} \tag{III.71}$$

$$\mathfrak{N}_s^L = \{N \in \mathfrak{N} \mid \exists \mathsf{Q}(\mathbf{p}) \in \mathbb{S} \quad \text{such that } \mathsf{R}(\mathbf{p}) \in \mathbb{S}_0 \quad \text{for all } \mathbf{p} \in D\} \tag{III.72}$$

When $\mathsf{R} \in \mathbb{S}_0$, the stability classification of the steady state can be upgraded from stability to asymptotic stability if it is possible to show that $dL_Q/dt < 0$ for all $\mathbf{u} \neq \mathbf{0}$ in the domain of $\mathbf{u}$ that is accessible to the steady state in question. For example, perturbations of $\mathbf{X}$ outside $\Pi_X(C_0)$ violate the conservation condition (II.5). Perturbations $\zeta \notin S_X$ are not considered in the stability classification as it has been defined in this chapter. Hence asymptotic stability is proved if $\zeta'\mathsf{R}\zeta > 0$ for all $\zeta \in S_X$, $\zeta \neq \mathbf{0}$, even if $\mathsf{R} \notin \mathbb{S}$.

Several tests for the positive definiteness of an $n \times n$ symmetric matrix S are summarized in Barnett and Storey[59] (Theorems 2-8-5 and 2-8-6). A *principal minor* s_i is the determinant of the $i \times i$ matrix constructed from the first i rows and columns of S.

1. $\mathsf{S} \in \mathbb{S}$ if and only if $s_i > 0$ for $i = 1, \ldots, n$.
2. $\mathsf{S} \in \mathbb{S}$ if and only if there exists a real nonsingular matrix T such that $\mathsf{S} = \mathsf{T}'\mathsf{T}$. When such a matrix exists, a triangular matrix must also exist with the same properties. The triangular matrix can be calculated using

$$T_{11} = S_{11}^{1/2}, \qquad T_{jj} = \left(S_{jj} - \sum_{i=1}^{j-1} T_{ij}^2 \right)^{1/2}, \qquad j = 2, \ldots, n$$

$$T_{jk} = \frac{\left(S_{jk} - \sum_{i=1}^{j-1} T_{ij} T_{ik} \right)}{T_{jj}}, \qquad j = 1, \ldots, n, \qquad k = j+1, \ldots, n$$

$$T_{jk} = 0, \qquad j > k \tag{III.73}$$

   The condition for the existence and nonsingularity of T is $\mathsf{T}_{jj} \neq 0$, for $j = 1, \ldots, n$. This condition is necessary and sufficient for $\mathsf{S} \in \mathbb{S}$. (Note that the condition $T_{jj} > 0$ in Barnett and Storey is in error.)
3. $\mathsf{S} \in \mathbb{S}$ if and only if all eigenvalues of S are positive.

Similar tests for semidefiniteness are as follows.

1. If it is possible to permute rows and columns such that $s_i > 0$ for $i = 1, \ldots,$ rank S then S is positive semidefinite.
2. Test 2 for positive definiteness fails if $T_{jj} = 0$ for some $1 \leqslant j \leqslant n$. If the algorithm for calculating T can be carried through to completion and if all components of T are real and finite, then S is positive semidefinite.
3. S is positive semidefinite if all eigenvalues are nonnegative.

E. Theorems on Matrix Stability

The chemical network stability problem has been converted into a matrix stability problem for the matrix functions $\mathsf{M}(\mathbf{p})$ defined in Section

III.A. Considerable work on more general matrix stability problems has already been done; it was motivated primarily by theoretical problems in economics. In this section we relate the chemical network problem to some widely used definitions in matrix stability theory and summarize some of the main theorems. A useful summary of matrix stability theory is given in the book by Barnett and Storey,[59] and in the papers by Quirk and Ruppert,[46] Maybee and Quirk,[60] and Jefferies et al.[44]

In matrix stability theory, the stability properties of the steady state are often ascribed to the matrix M. Thus we say M is *asymptotically stable* (*marginally stable*, *stable*, *weakly unstable*, *semistable*, *exponentially unstable*) if the steady state $\mathbf{u}=\mathbf{0}$ of the linearized system (III.1) is asymptotically stable (marginally stable, stable, weakly unstable, semistable, exponentially unstable). When M is asymptotically stable it is sometimes[59] called a *stability matrix*. The matrix M is diagonal-stable (D-stable) if DM is asymptotically stable for every positive diagonal matrix D. M is *totally stable* if every principal submatrix of M (i.e., every submatrix whose determinant is a principal minor of M) is D-stable. From these definitions it follows that every totally stable matrix is D-stable and every D-stable matrix is asymptotically stable.

Consider any $\mathbf{j}$ such that $\mathsf{E}\mathbf{j} \in \Pi_v$, and suppose the matrix $-\mathsf{V}(\mathbf{j})$ defined in (III.10) is D-stable. Then for all $\mathbf{h} \in R^n_+$, $-(\operatorname{diag}\mathbf{h})V(\mathbf{j})$ is asymptotically stable, and from (III.13), $M''_{\xi R}(\mathbf{h},\mathbf{j})$ is stable. We then say that $\mathsf{E}\mathbf{j}$ is a *linearly asymptotically stable current*. A current $\mathsf{E}\mathbf{j}$ is linearly asymptotically stable if and only if $-\mathsf{V}(j)$ is D-stable. Then from (III.34)

$$\mathfrak{N}^L_a = \{ N \in \mathfrak{N} \mid -\mathsf{V}(\mathbf{j}) \text{ is D-stable for all } \mathbf{j} \in R^f_+ \} \qquad \text{(III.74)}$$

The network stability problem is therefore partly the problem of finding necessary and sufficient conditions for $-\mathsf{V}(\mathbf{j})$ to be D-stable for all $\mathbf{j} \in R^f_+$, when $\mathsf{V}(\mathbf{j})$ has the structure given in (III.10). Of course this only determines what we call *linear* asymptotic stability; the nonlinear terms must sometimes be considered to determine whether the current is asymptotically stable.

The *sign stability problem* is closely related to the network stability problem. Just as each chemical system is associated with a network N that represents a set of systems, each $n \times n$ real matrix U will be associated with a set of matrices $\Sigma(\mathsf{U}) \equiv \{\mathsf{M}(\mathbf{p}) \mid \mathbf{p} \in D\}$, called a *sign matrix set*, where $\mathbf{p}$ is an $n \times n$ parameter matrix, $D \equiv R^{n^2}_+$, and

$$M_{ij}(\mathbf{p}) \equiv U_{ij} p_{ij} \qquad \text{(III.75)}$$

The sign matrix set is analogous to a network. Since the sign matrix set depends only on whether each element of U is positive, negative, or zero, we may assume $U_{ij} \in \{-1,0,1\}$.

Each sign matrix set has domains on which $\mathbf{M}(\mathbf{p})$ has various stability properties. These domains are called D_a, D_m, D_w, D_e, D_{mw}, D_u, D_{semi}, D_s and are defined, as for linearized networks, by Table II. Continuing the analogy, each sign matrix set can be classified according to the most unstable matrix in it. We say $\Sigma(\mathsf{U})$ is *asymptotically stable*, *marginally stable*, *weakly unstable*, *exponentially unstable*, *marginally* or *weakly unstable*, *unstable*, *semistable*, or *stable*, in analogy with the network classifications of Table II. Then we may define sets of sign matrix sets having various stability classifications as follows. Let $\mathcal{U}$ be the set of all sign matrix sets. The set of *asymptotically stable* sign matrix sets is

$$\mathcal{U}_a \equiv \{\Sigma \in \mathcal{U} \mid D_a = D\} \tag{III.76}$$

The set of *marginally stable* sign matrix sets is

$$\mathcal{U}_m \equiv \{\Sigma \in \mathcal{U} \mid D_m \neq \emptyset, \qquad D_w = D_e = \emptyset\} \tag{III.77}$$

the set of *weakly unstable* sign matrix sets is

$$\mathcal{U}_w \equiv \{\Sigma \in \mathcal{U} \mid D_w \neq \emptyset, \qquad D_e = \emptyset\} \tag{III.78}$$

and the set of *exponentially unstable* sign matrix sets is

$$\mathcal{U}_e \equiv \{\Sigma \in \mathcal{U} \mid D_e \neq \emptyset\} \tag{III.79}$$

Then $\mathcal{U}_s \equiv \mathcal{U}_a \cup \mathcal{U}_m$, $\mathcal{U}_{\text{semi}} \equiv \mathcal{U}_a \cup \mathcal{U}_m \cup \mathcal{U}_w$, $\mathcal{U}_u \equiv \mathcal{U}_w \cup \mathcal{U}_e$.

If $\Sigma(\mathsf{U}) \in \mathcal{U}_a$, we will say U is *sign asymptotically stable*. Similarly, other stability classifications of $\Sigma(\mathsf{U})$ may be ascribed to U by prefixing the term "sign." (Our terminology differs from the terminology in the literature in two places. We say that U is "sign asymptotically stable" where the literature[44] says that U is "sign stable"; we say that U is "sign stable" where the literature says that U is "sign quasistable.")

Necessary and sufficient conditions for both sign asymptotic stability and sign semistability are known. These problems involve only the signs of $\text{Re}(\lambda)$ for $\lambda \in \Lambda$ and are therefore much easier than the sign stability problem, where one must distinguish between marginal stability and weak instability in the case where $\text{Re}(\lambda) \leqslant 0$ for all $\lambda \in \Lambda$, and $\text{Re}(\lambda) = 0$ for some $\lambda \in \Lambda$. Recent work on the sign asymptotic stability and sign semistability problems is discussed by Jefferies et al.[44] We now state the two main theorems.

Quirk-Ruppert-Maybee Theorem. An $n \times n$ matrix U is sign semistable if and only if it satisfies the following three conditions:

(α). $U_{ii} \leqslant 0$ for all i.
(β). $U_{ij}U_{ji} \leqslant 0$, for all $i \neq j$.

(γ). $U_{i(1)i(2)} \cdots U_{i(k-1)i(k)} U_{i(k)i(1)} = 0$, for each sequence of $k \geqslant 3$ distinct indices $i(1), \ldots, i(k)$.

Jefferies[45] gave examples of matrices U that satisfy the conditions of this theorem, where U is marginally stable or weakly unstable. His subsequent work led to

Jefferies' Theorem. An $n \times n$ matrix U is sign asymptotically stable if and only if it satisfies conditions α, β, and γ of the Quirk-Ruppert-Maybee theorem plus the following conditions:

(δ). In every R_A-coloring of the undirected graph G_A, all vertices are black.

(ε). The undirected graph G_A admits a $(V \sim R_A)$-complete matching.

For an explanation of the terminology in δ and ε see Ref. 44.

Sign asymptotically stable matrices are proved to be totally stable in Ref. 59, hence are D-stable and stable. The necessity of using graph theory to state conditions δ and ε in Jefferies' theorem suggests that graph theory should play a major role in any necessary and sufficient conditions for D-stability, and in the network stability problem. Diagrammatic methods have been introduced by several authors and are discussed in the next two sections.

The stoichiometric network stability problem degenerates into the sign stability problem when the effective power function $\underline{\boldsymbol{\kappa}}(\mathbf{p})$ can vary over a suitable range. All the techniques that we apply to the network problem may be applied to the sign stability problem.

Example III.5. The matrix $\mathsf{M}_\zeta(\mathbf{h},\mathbf{j},\mathbf{q})$ for the Oregonator was given in Example III.2 when $\underline{\boldsymbol{\kappa}}(\mathbf{p})$ is a function of the parameters $\mathbf{q} \in R^6_+$. From the forms of the matrix elements and the parameter ranges, an equivalent parametrization is

$$\begin{bmatrix} -r_1 & \pm r_2 & 0 \\ -r_3 & -r_4 & r_5 \\ r_6 & 0 & -r_7 \end{bmatrix}$$

where the new ranges are $r_1 > r_3 - \frac{1}{2} r_6$, $r_2 > 0$, $r_3 > 0$, $r_4 > 0$, $r_5 > 0$, $r_6 > 0$, $r_7 > 2r_5$. With a slight modification in the ranges of r_1 and r_7, this network stability problem becomes identical to the sign stability problem for the matrices

$$\begin{pmatrix} -1 & -1 & 0 \\ -1 & -1 & 1 \\ 1 & 0 & -1 \end{pmatrix} \quad \text{and} \quad \begin{pmatrix} -1 & 1 & 0 \\ -1 & -1 & 1 \\ 1 & 0 & -1 \end{pmatrix}$$

Both matrices violate condition γ, so they are not sign semistable. The same conclusion applies to the network because the modifications to the parameter ranges did not affect the matrix elements that violate γ. Hence the network is exponentially unstable.

IV. NETWORKS HAVING SIMPLE LYAPUNOV FUNCTIONS

A. Mixing Stability

In this subsection we show that there is a function $L_M(\zeta, \mathbf{h})$ that is a Lyapunov function for a large number of extreme networks. Because L_M does not depend on $\mathbf{j}$, it is also a Lyapunov function for all mixtures of such networks.

Choose the redundant parameters and let $\mathsf{Q}(\mathbf{h},\mathbf{j}) \equiv \operatorname{diag}\mathbf{h}$. Then for all $(\mathbf{h},\mathbf{j}) \in D_R$, $\mathsf{Q}(\mathbf{h},\mathbf{j}) \in \mathbb{S}$, the set of positive definite real symmetric matrices, so the function

$$L_M(\zeta,\mathbf{h}) \equiv \zeta^t(\operatorname{diag}\mathbf{h})\zeta \tag{IV.1}$$

could be a Lyapunov function for some networks. A sufficient condition for this is that $\zeta^t\mathsf{R}(\mathbf{h},\mathbf{j})\zeta > 0$ for accessible deviations ζ from steady state. Evaluating $\mathsf{R}(\mathbf{h},\mathbf{j})$ using (III.68) and (III.11) for $\mathsf{M}_{\zeta R}(\mathbf{h},\mathbf{j})$, gives

$$\zeta^t\mathsf{R}(\mathbf{h},\mathbf{j})\zeta = 2\sum_{i=1}^{f} j_i\zeta^t(\operatorname{diag}\mathbf{h})\mathsf{S}^{(i)}_{\text{sym}}(\operatorname{diag}\mathbf{h})\zeta \tag{IV.2}$$

where

$$\mathsf{S}^{(i)}_{\text{sym}} \equiv \tfrac{1}{2}(\mathsf{S}^{(i)t} + \mathsf{S}^{(i)}) \tag{IV.3}$$

A sufficient condition for this expression to be nonnegative is that for all $i \in [1, f]$, $\mathsf{S}^{(i)}_{\text{sym}} \in \mathbb{S}_0$. This will be so if every extreme subnetwork satisfies

$$\mathsf{S}_{\text{sym}} \in \mathbb{S}_0 \tag{IV.4}$$

Extreme networks with this property are called *mixing stable*. A network is *mixing stable* if its extreme subnetworks are mixing stable. If $\mathcal{E}_{ms}$ is the set of mixing stable extreme networks and $\mathfrak{N}^L_{ms}$ is the set of mixing stable networks, then from (III.72)

$$\mathfrak{N}^L_s \supset \mathfrak{N}^L_{ms} \equiv \{N \in \mathfrak{N} \mid \Xi(N,\mathsf{E}) \subset \mathcal{E}_{ms}\} \tag{IV.5}$$

A similar statement may be made for domains. The steady states of a stable network are linearly stable, so we define the *cone of mixing stable currents* to be

$$\mathcal{C}_{ms} \equiv \left\{\Sigma_i j_i\mathbf{E}_i \,|\, \mathbf{j} \in R^f_+,\ j_i = 0 \text{ for all } i \text{ such that } \Xi(N,\mathbf{E}_i) \notin \mathcal{E}_{ms}\right\} \tag{IV.6}$$

If $\mathsf{E}\mathbf{j} \in \mathcal{C}_{ms}$, the terms of (IV.2) are all nonnegative and departure from the steady state is impossible. Thus

$$D^L_s \supset \{(\mathbf{h},\mathbf{j}) \in D_R \,|\, \mathsf{E}\mathbf{j} \in \mathcal{C}_{ms}\} \tag{IV.7}$$

An extreme network is *mixing asymptotically stable* if it is mixing stable and

$$\mathsf{S}\underline{\gamma}^t = 0 \tag{IV.8}$$

$$\operatorname{rank}\mathsf{S} = d \tag{IV.9}$$

We now prove that a network is asymptotically stable if all its extreme subnetworks are mixing asymptotically stable. If $\mathcal{E}_{ma}$ is the set of mixing asymptotically stable extreme networks and $\mathfrak{N}^L_{ma}$ is the set of mixing asymptotically stable networks, we will prove

$$\mathfrak{N}^L_a \supset \mathfrak{N}^L_{ma} \tag{IV.10}$$

where $\mathfrak{N}^L_{ma} \equiv \{N \in \mathfrak{N} \,|\, \Xi(N, \mathsf{E}) \subset \mathcal{E}_{ma}\}$.

For any network N, consider any extreme current $\mathbf{E}_i$ and its corresponding extreme subnetwork E. Let E^i be the network with the same species (in the same order) as N, and only those reactions of N whose components of $\mathbf{E}_i$ do not vanish. Thus the stoichiometries of the reactions of E^i are multiples of the stoichiometries of the corresponding reactions of E. The matrix $\mathsf{S}^{(i)}$ of N, given by (III.9), also applies to E^i. It may also be obtained from S of E by extending the rows by zeros for each nonparticipating species (i.e., species of N that does not appear in E), by adding rows of zeros for each nonparticipating species, and by permuting rows and columns to put the species in the proper order. Let $\underline{\gamma}^{(i)}$ be the conservation matrix for E^i. This matrix may be obtained from $\underline{\gamma}$ of E by extending the rows by zeros for each nonparticipating species, by adding for each nonparticipating species a row of the form $(0, \ldots, 0, 1, 0, \ldots, 0)$ with 1 in the column associated with the species, and by permuting columns to put the species in the proper order. If $E \in \mathcal{E}_{ma}$, it then follows from (IV.8) that

$$\mathsf{S}^{(i)}\underline{\gamma}^{(i)t} = 0 \tag{IV.11}$$

Let d_i be the dimension of $\Pi_X(\mathbf{C}^0)$, the concentration polyhedron of E^i. Then d_i is the rank of the stoichiometric matrix of E^i, which must have the same rank as the stoichiometric matrix of E. Since $\mathsf{S}^{(i)}$ and S clearly have the same rank, (IV.9) yields

$$\operatorname{rank}\mathsf{S}^{(i)} = d_i \tag{IV.12}$$

Thus E^i is also mixing asymptotically stable. We also need the relation

$$\underline{\gamma}^{(i)}\mathsf{S}^{(i)} = 0 \tag{IV.13}$$

which is a consequence of (III.42) and (II.6). Conditions IV.11 to IV.13 hold for every extreme current whose corresponding subnetwork is in $\mathcal{E}_{ma}$. Also of great importance are the subspaces S^i_X and S^i_C that are spanned by the columns of the stoichiometric matrix of E^i and the rows of $\underline{\gamma}^{(i)}$, respectively.

We now return to the network E and consider the conjecture that for some $\zeta \in S_X$, $\zeta \neq 0$, the projection of $(\operatorname{diag}\mathbf{h})\zeta$ on S_X^i vanishes for all $i \in [1, f]$. Since S_X is the direct sum of these subspaces, this conjecture would be true only if $(\operatorname{diag}\mathbf{h})\zeta$ were orthogonal to S_X; that is, orthogonal to ζ, which implies $\zeta^t(\operatorname{diag}\mathbf{h})\zeta = 0$. Since $h_i > 0$ for $i \in [1, n]$, this condition cannot hold when $\zeta \neq 0$ and the conjecture is never true. Now choose any $\zeta \in S_X$ and keep ζ fixed. There must be some subspace S_X^i on which the projection of $(\operatorname{diag}\mathbf{h})\zeta$ does not vanish. Fix i so that E^i is an extreme network with this property. Then we may write $(\operatorname{diag}\mathbf{h})\zeta = \mathbf{w} + \mathbf{z}$ where $\mathbf{w} \in S_X^i$, $\mathbf{w} \neq 0$, $\mathbf{z} \in S_C^i$. Since S_C^i is spanned by the $n - d_i$ rows of $\underline{\gamma}^{(i)}$, there exists $\mathbf{u} \in R^{n-d_i}$ such that $\mathbf{z} = \underline{\gamma}^{(i)t}\mathbf{u}$. Then

$$\zeta^t(\operatorname{diag}\mathbf{h})\mathsf{S}^{(i)}(\operatorname{diag}\mathbf{h})\zeta = \mathbf{w}^t\mathsf{S}^{(i)}\mathbf{w} + \mathbf{u}^t\underline{\gamma}^{(i)}\mathsf{S}^{(i)}\left(\mathbf{w} + \underline{\gamma}^{(i)t}\mathbf{u}\right)\mathbf{w}^t\mathsf{S}^{(i)}\underline{\gamma}^{(i)t}\mathbf{u}. \tag{IV.14}$$

The second and third terms vanish by (IV.13) and (IV.11). Since $\mathsf{S}^{(i)}$ is a positive semidefinite matrix of rank d_i, there must exist a subspace of dimension d_i such that $\mathbf{w}^t\mathsf{S}^{(i)}\mathbf{w} > 0$ for all $\mathbf{w} \neq 0$ in this subspace. Condition (IV.13) implies that this subspace cannot be contained in the subspace S_C^i spanned by the rows of $\underline{\gamma}^{(i)}$. Hence the subspace lies in the complementary subspace S_X^i which has dimension d_i, the same dimension as the required subspace. Hence $\mathbf{w}^t\mathsf{S}^{(i)}\mathbf{w} > 0$ for all $\mathbf{w} \in S_X^i$, $\mathbf{w} \neq 0$, and expression (IV.14) is strictly positive for the chosen $\zeta \in S_X$, $\zeta \neq 0$. The term in (IV.2) coming from E^i is thus positive and the remaining terms are nonnegative by (IV.4). Hence $\zeta^t\mathsf{R}\zeta > 0$ for the chosen ζ. Since this argument is valid for every $\zeta \in S_X$, $\zeta \neq 0$, we conclude that $\zeta^t\mathsf{R}\zeta > 0$ for all such ζ and then the steady state is linearly asymptotically stable. Since this argument holds for any steady state, N is stable and (IV.10) is proved.

Of the two conditions (IV.8) and (IV.9) in the definition of mixing asymptotic stability, the latter is essential and the former is not. If $\operatorname{rank}\mathsf{S} < d$, $\mathbf{w}^t\mathsf{S}^{(i)}\mathbf{w}$ can vanish for $\mathbf{w} \in S_X^i$ and linear asymptotic stability does not hold for the network E^i. Thus (IV.9) is necessary. Note that $\mathbf{w}$ and $\mathbf{z}$ are not independent (both being determined by ζ). If they were independent, the last term in (IV.14) could be made negative and dominant as $\mathbf{w} \to 0$ and (IV.8) would be a necessary condition. However it is not. In a later example we will show that the left-hand side of (IV.14) can be positive definite for $\zeta \in S_X$, even when (IV.8) is violated.

Example IV.1. For the second extreme network E^2 of the Oregonator, given in Example III.1,

$$2\mathsf{S}_{\text{sym}}^{(2)} = \begin{pmatrix} 6 & -1 & -2 \\ -1 & 2 & -1 \\ -2 & -1 & 4 \end{pmatrix}$$

and the principal minors are $s_1 = 6$, $s_2 = 11$, and $s_3 = 30$. Hence $\mathsf{S}^{(2)}_{\mathrm{sym}} \in \mathcal{S}$, condition (IV.4) is satisfied, and E^2 is mixing stable. The reactions of E^2 do not have a conservation condition, so $\underline{\gamma}$ has no rows and (IV.8) is satisfied. Since rank $\mathsf{S}^{(2)} = 3$, all the conditions for mixing asymptotic stability are satisfied. Hence $E^2 \in \mathcal{E}_{ma}$.

We now strengthen the preceding result and obtain sufficient conditions for network asymptotic stability that apply when some extreme networks are mixing stable but not mixing asymptotically stable. Consider any current $\mathbf{v}^0 \in \mathrm{ri}\,\mathcal{C}_{ms}$, the relative interior of the cone of mixing stable currents. Since $\mathcal{C}_{ms}$ is convex, it is possible to write $\mathbf{v}^0 = \mathsf{E}\mathbf{j}$, where $j_i > 0$ for all i such that $E^i \in \mathcal{C}_{ms}$. Hence the term in the expansion (IV.2) of $\zeta^t\mathsf{R}\zeta$ that corresponds to any mixing stable extreme network E^i has the sign of $\zeta^t(\mathrm{diag}\,\mathbf{h})\mathsf{S}^{(i)}_{\mathrm{sym}}(\mathrm{diag}\,\mathbf{h})\zeta$, which must be nonnegative by (IV.4). Hence every term in $\zeta^t\mathsf{R}\zeta$ is nonnegative. Let S_{Xma} be the direct sum of the subspaces S^i_X of all the mixing asymptotically stable extreme networks. Formally,

$$S_{Xma} \equiv \bigoplus_i S^i_X \qquad \text{for } E^i \in \Xi(N, \mathsf{E}) \cap \mathcal{E}_{ma} \tag{IV.15}$$

Now consider the special case where $\dim S_{Xma} = d$. Then $S_X = S_{Xma}$ and it may be shown that there exists a mixing asymptotically stable extreme network E^i such that the projection of $(\mathrm{diag}\,\mathbf{h})\zeta$ on S^i_X does not vanish. The proof is almost identical to the one given in the second paragraph before Example IV.1. The same argument may then be used to conclude that the term in expansion (IV.2) corresponding to E^i is positive. Hence $\zeta^t\mathsf{R}\zeta > 0$ and L_M is a Lyapunov function. This proof applies to any current $\mathbf{v}^0 \in \mathrm{ri}\,\mathcal{C}_{ms}$, so for any network such that $\dim S_{Xma} = d$,

$$D^L_a \supset \{(\mathbf{h}, \mathbf{j}) \in D_R \mid \mathsf{E}\mathbf{j} \in \mathrm{ri}\,\mathcal{C}_{ms}\} \tag{IV.16}$$

If every extreme network is mixing stable $\mathcal{C}_{ms} = \mathcal{C}_v$ and $\mathsf{E}\mathbf{j}$ must lie in the interior of $\mathcal{C}_{ms}$ because the domain of $\mathbf{j}$ is the open orthant R^f_+. Hence

$$\mathfrak{N}^L_a \supset \{N \in \mathfrak{N} \mid \Xi(N, \mathsf{E}) \subset \mathcal{E}_{ms}, \qquad \dim S_{Xma} = d\} \tag{IV.17}$$

The mixing stability property allows a network N^1 to combine with any other mixing stable network N^2 and retain its stability. Note that N^1 may have a conservation condition $\underline{\gamma}^{(1)}\mathbf{X} = \mathbf{C}$, while the combined network N may not ($\underline{\gamma}$ has no rows). The stability proof for the combined network is completely unaffected by whether or not N has any of the conservation conditions of N^1. However in many applications N has a conservation matrix $\underline{\gamma}$ (i.e., $\underline{\gamma}$ has some rows). This conservation condition must also apply to all subnetworks of N, including N^1. (These subnetworks could also have additional conservation conditions that do not apply to N.) To prove the stability of N, it is sufficient to know that the subnetworks have the mixing stability property whenever the conservation law $\underline{\gamma}\mathbf{X} = \mathbf{C}$ of N is not violated. We say such subnetworks have *constrained mixing stability* or *mixing stability with respect to the conservation matrix* $\underline{\gamma}$.

Given $\underline{\gamma}$ and a network N^1 with conservation matrix $\underline{\gamma}^{(1)}$ (whose rows span S_C^1), we are interested in whether N^1 can mix to form a stable network with conservation matrix $\underline{\gamma}$. Since the full network cannot conserve what N^1 does not conserve, the rows of $\underline{\gamma}$ involving only the species of N^1 must span a subspace of S_C^1. Since the constraint $\underline{\gamma}\mathbf{X} = \mathbf{C}$ can be expressed in many equivalent forms by taking linear combinations of the rows, constrained mixing stability depends only on the subspace spanned by the rows of $\underline{\gamma}$, not on $\underline{\gamma}$ directly. Hence a particular case ($\underline{\gamma}$) of constrained mixing stability is identifiable with a subspace T of S_C^1. Let ψ be the set of subspaces of S_C^1 for which N^1 has constrained mixing stability. If $T \in \psi$ and new constraints are added, we get a new matrix $\underline{\gamma}^*$ having more rows. The rows now span a possibly higher dimensional subspace $U \in \psi$, which must contain T. Thus when $U \supset T \in \psi$, then $U \in \psi$. Let us delete all subspaces of ψ that contain other subspaces of ψ, and call the resulting set $\psi_{\min}$. This set then has the property that if $T \in \psi_{\min}$ and V is any subspace of T, then $V \notin \psi_{\min}$. One may easily reconstruct ψ from $\psi_{\min}$. Mixing stability (unconstrained) occurs when $\psi_{\min}$ contains only the subspace $\{\mathbf{0}\}$. It is not known whether $\psi_{\min}$ always contains a single subspace or whether it can contain more than one subspace.

Given any extreme network E^i with the conservation matrix $\underline{\gamma}^{(i)}$, and any matrix $\underline{\gamma}$ whose rows span a subspace of the subspace spanned by the rows of $\underline{\gamma}^{(i)}$, we may use $\underline{\gamma}$ to eliminate some of the dependent concentrations. From (III.14)

$$\zeta_D = -\underline{\gamma}_D^{-1}\underline{\gamma}_I\zeta_I \tag{IV.18}$$

As in Section III.A, the subscripts I and D refer to the independent and dependent species. Define

$$\mathsf{L} = (\operatorname{diag}\mathbf{h}_D)\,\underline{\gamma}_D^{-1}\underline{\gamma}_I(\operatorname{diag}\mathbf{h}_I)^{-1} \tag{IV.19}$$

then

$$\zeta^t(\operatorname{diag}\mathbf{h})\mathsf{S}^{(i)}(\operatorname{diag}\mathbf{h})\zeta = \zeta_I^t(\operatorname{diag}\mathbf{h}_I)\mathsf{S}^{*(i)}(\operatorname{diag}\mathbf{h}_I)\zeta_I \tag{IV.20}$$

where

$$\mathsf{S}^{*(i)} \equiv \mathsf{S}_{II}^{(i)} - \mathsf{S}_{ID}^{(i)}\mathsf{L} - \mathsf{L}^t\mathsf{S}_{DI}^{(i)} + \mathsf{L}^t\mathsf{S}_{DD}^{(i)}\mathsf{L} \tag{IV.21}$$

The same equation applies to the corresponding symmetric parts $\mathsf{S}_{\text{sym}}^{(i)}$ and $\mathsf{S}^{*(i)}_{\text{sym}}$. Hence (IV.2) becomes

$$\zeta^t\mathsf{R}(\mathbf{h},\mathbf{j})\zeta = 2\sum_{i=1}^{f} j_i\zeta_I^t(\operatorname{diag}\mathbf{h}_I)\mathsf{S}^{*(i)}_{\text{sym}}(\operatorname{diag}\mathbf{h}_I)\zeta_I \tag{IV.22}$$

and the earlier discussion may be applied to $\mathsf{S}^{*(i)}_{\text{sym}}$. E^i is (*constrained*) *mixing stable with respect to* $\underline{\gamma}$ if

$$\mathsf{S}^{*(i)}_{\text{sym}} \in \mathcal{S}_0 \tag{IV.23}$$

for all $\mathbf{h} \in R^n_+$. Let $\mathcal{E}_{cms}(N)$ be the set of extreme networks that are mixing stable with respect to the constraint matrix of the network N. Then (IV.5) generalizes to

$$\mathfrak{N}^L_s \supset \mathfrak{N}^L_{cms} \equiv \{N \in \mathfrak{N} \mid \Xi(N, \mathsf{E}) \subset \mathcal{E}_{cms}(N)\} \tag{IV.24}$$

If we let $\mathcal{C}_{cms}(N)$ be the cone of currents that are mixing stable relative to the constraint matrix of the network N, (IV.7) generalizes to

$$D^L_s \supset \{(\mathbf{h}, \mathbf{j}) \in D_R \mid \mathsf{E}_\mathbf{j} \in \mathcal{C}_{cms}(N)\} \tag{IV.25}$$

That is, when the dynamics of the network N are linearized about steady state, they are stable on the domain given.

Let us consider the possible asymptotic stability of a network with the constraint matrix $\underline{\gamma}$, such that each extreme network E^i is constrained mixing stable with respect to $\underline{\gamma}$. Asymptotic stability can be proved by showing that there is a positive term in (IV.2), or equivalently (IV.22). Hence we look for sufficient conditions for the existence of such a term. Let δ_i be the rank of $\mathsf{S}^{(i)}$. By (IV.13), $\delta_i \leqslant d_i$. Since E^i is constrained mixing stable, there must be a subspace S^i_δ of dimension δ_i such that $\mathbf{w}^t\mathsf{S}^{(i)}\mathbf{w} > 0$ when the projection of $\mathbf{w}$ on S^i_δ does not vanish. Hence the term in (IV.2) coming from E^i will be positive whenever the projection of $(\text{diag}\,\mathbf{h})\boldsymbol{\zeta}$ on S^i_δ does not vanish. Suppose S_X of N is the direct sum of the subspaces S^i_δ for all the extreme subnetworks E^i. Then if $(\text{diag}\,\mathbf{h})\boldsymbol{\zeta}$ has a nonvanishing projection on S_X, it must have a nonvanishing projection on some S^i_δ, and the corresponding term in (IV.2) must be positive. We proved earlier that $(\text{diag}\,\mathbf{h})\boldsymbol{\zeta}$ always has a nonvanishing projection on S_X if $\boldsymbol{\zeta} \neq 0$, by showing that otherwise $\boldsymbol{\zeta}^t(\text{diag}\,\mathbf{h})\boldsymbol{\zeta}$ must vanish. Hence if we define

$$S_{Xcma} \equiv \bigoplus_i S^i_\delta \tag{IV.26}$$

then N is linearly asymptotically stable if $S_X = S_{Xcma}$. Since $S_{Xcma} \subset S_X$, we only need to check that $\dim S_{Xcma} = d$. Thus (IV.17) generalizes to

$$\mathfrak{N}^L_a \supset \mathfrak{N}^L_{cma} \equiv \{N \in \mathfrak{N} \mid \Xi(N, \mathsf{E}) \subset \mathcal{E}_{cms}(N), \dim S_{Xcma} = d\} \tag{IV.27}$$

If $N \in \mathfrak{N}^l_{cma}$, we say that N is *constrained mixing asymptotically stable.* When $\dim S_{Xcma} = d$, the corresponding statement for domains is

$$D^L_a \supset \{(\mathbf{h}, \mathbf{j}) \in D_R \mid \mathsf{E}\mathbf{j} \in \text{ri}\,\mathcal{C}_{cms}(N)\} \tag{IV.28}$$

This is a generalization of (IV.16).

Example IV.2. Consider the irreversible mass action Michaelis-Menten mechanism for the conversion of a substrate S to a product P using an enzyme E and the enzyme-substrate complex F. $R_1 : \square \to S$, $R_2 : E + S \to F$, $R_3 : F \to E + P$, $R_4 : P \to \square$. This network is extreme and $\mathsf{E} = \mathbf{e}$. Take $\mathbf{X} = (S, E, P, F)^t$. Then

$$\underline{\nu} = \begin{bmatrix} 1 & -1 & 0 & 0 \\ 0 & -1 & 1 & 0 \\ 0 & 0 & 1 & -1 \\ 0 & 1 & -1 & 0 \end{bmatrix} \qquad \mathsf{S} = \begin{bmatrix} 1 & 1 & 0 & 0 \\ 1 & 1 & 0 & -1 \\ 0 & 0 & 1 & -1 \\ -1 & -1 & 0 & 1 \end{bmatrix}$$

$\mathsf{S}_{\mathrm{sym}} \notin \mathbb{S}_0$, so the network is not mixing stable.

To see whether the network is constrained mixing stable relative to its own conservation matrix $\underline{\gamma} = (0 \quad 1 \quad 0 \quad 1)$, we consider E independent and F dependent. Then if $\mathbf{h} \equiv (s, e, p, f)$,

$$\mathsf{L} = f(0 \quad 1 \quad 0)(\mathrm{diag}(s \quad e \quad p))^{-1} = (0 \quad \alpha \quad 0)$$

where $\alpha \equiv f/e$. From (IV.2)

$$\mathsf{S}^* = \begin{pmatrix} 1 & 1 & 0 \\ 1+\alpha & (1+\alpha)^2 & 0 \\ 0 & \alpha & 1 \end{pmatrix}$$

The three principal minors of $2\mathsf{S}^*_{\mathrm{sym}}$ are 2, $4\alpha + 3\alpha^2$, and $8\alpha + 4\alpha^2$. Hence $\mathsf{S}^*_{\mathrm{sym}}$ is positive definite for all $\alpha = f/e > 0$. Since (IV.23) is satisfied, this extreme network is constrained mixing stable with respect to $\underline{\gamma}$. Also, $\delta = 3 = d$, so $\dim S_{Xcma} = d$. By (IV.27) the network is constrained mixing asymptotically stable, and therefore linearly asymptotically stable.

We now give a test for constrained mixing asymptotic stability in a whole network. A sufficient condition for the asymptotic stability of the steady state is that $\zeta^t \mathsf{R}(\mathbf{h},\mathbf{j})\zeta > 0$ for all $\zeta \neq 0$. Note that from (IV.18), $\zeta_I = 0$ if and only if $\zeta \neq 0$. Now write (IV.22) as

$$\zeta^t \mathsf{R}(\mathbf{h},\mathbf{j})\zeta = \zeta_I^t \mathsf{R}_I(\mathbf{h},\mathbf{j})\zeta_I \tag{IV.29}$$

where

$$\mathsf{R}_I(\mathbf{h},\mathbf{j}) \equiv 2(\mathrm{diag}\,\mathbf{h}_I)\left(\sum_{i=1}^{f} j_i \mathsf{S}^{*(i)}_{\mathrm{sym}}\right)(\mathrm{diag}\,\mathbf{h}_I) \tag{IV.30}$$

Then the steady state is asymptotically stable if $\mathsf{R}_I(\mathbf{h},\mathbf{j}) \in \mathbb{S}$, or, from (IV.30) if

$$\sum_{i=1}^{f} j_i \mathsf{S}^{*(i)}_{\mathrm{sym}} \in \mathbb{S} \tag{IV.31}$$

The network is therefore linearly asymptotically stable if this condition is satisfied for all $(\mathbf{h},\mathbf{j}) \in D_R$. Hence

$$\mathfrak{N}_a^L \supset \left\{ N \in \mathfrak{N} \mid \sum j_i \mathsf{S}^{*(i)}_{\mathrm{sym}} \in \mathbb{S}, \qquad \text{for all } (\mathbf{h},\mathbf{j}) \in D_R \right\} \tag{IV.32}$$

Condition (IV.31) can be established by constructing the polynomials corresponding to the principal minors. These polynomials frequently have no negative terms, as in Example IV.2. The polynomials contain far fewer terms than the polynomials representing the Hurwitz determinants; therefore this test should be practical for very large networks if the polynomials are constructed on a computer. If the polynomials contain some negative terms, asymptotic stability can still usually be established on a domain of D_R be using the techniques for treating polynomials that are discussed later in connection with the Hurwitz determinants.

B. Some Fundamental Topological Concepts for Reaction Networks

Networks having the Lyapunov functions we will discuss can be characterized by what I call *topological properties*. Topology is a branch of mathematics in which the connectiveness of sets is examined without regard to the size or shape of the sets. For example, all plane ellipses have the same topological properties because the size and shape are unimportant. An ellipse and a line have different topological properties because one can trace around an ellipse but one cannot trace around a line. By "topological properties of reaction networks" I mean the connectiveness properties of various types of diagrams that can be used to represent the networks. For example, on some types of diagrams, feedback cycles appear as closed curves. Networks without feedback cycles are stable, hence stability is a consequence of a topological property of a network's diagram.

Topological concepts have been introduced by a number of people, notably Delattre,[5] Hyver,[6] Horn, Feinberg, and Jackson,[17–19] Solimano and Beretta,[7–9] and Clarke.[36, 38, 39] The most important of these concepts will now be developed in a unified approach that will lead to a deeper understanding of the role of extreme networks.

The expressions that appear on each side of a chemical pseudoreaction are called *complexes* (Horn). Each complex has a corresponding *complex vector* $\mathbf{y} \in \overline{R}^n_+$ whose ith component is the stoichiometric coefficient of X_i in the complex. For example, if only species X_1 and X_2 are present, the complexes X_1, $2X_1 + X_2$, and $\square$ correspond to the complex vectors $(1,0)^t$, $(2,1)^t$ and $(0,0)^t$, respectively. The complex vector corresponding to a complex is denoted by parentheses around the complex thus: $(1,0)^t = (X_1)$, $\mathbf{0} = (\square)$. Denote the set of complex vectors of a network by Y; we also interpret Y as a matrix whose columns are the vectors $\mathbf{y}_i$. Let the number of complexes be x. The jth reaction now can be written as a *chemical reaction* R_j or as a *complex vector reaction* R_j^c whose general form is

$$\mathbf{y}_{r(j)} \rightarrow \mathbf{y}_{p(j)} \qquad \text{(IV.33)}$$

where $r(j)$ and $p(j)$ are the column indices in Y of the complex vectors corresponding to the *reactant complex* and the *product complex* of R_j^c, respectively. For example, the chemical reaction $X + Y \rightarrow Z$ corresponds to the complex vector reaction $(X + Y) \rightarrow (Z)$.

The set of species in a complex is called an *interactant* and is denoted "species **y**," where **y** is the corresponding complex vector. For example, $\{X_1\}$ = species $(1,0)^t$ = species (X_1), $\emptyset$ = species $\mathbf{0}$ = species $(\square)$. Solimano formerly used the term "interaction subset" and is now using "interactant." For $\mathbf{y}_i, \mathbf{y}_j \in Y$, species $\mathbf{y}_i \cap$ species $\mathbf{y}_j = \emptyset$ if and only if $\mathbf{y}_i^t\mathbf{y}_j = 0$. Two interactants *overlap* if (species $\mathbf{y}_i$) $\cap$ species $y_j \neq \emptyset$. Let $\mathcal{X}$ be the set of species. An *interactant overlap path* from $X_1 \in \mathcal{X}$ to $X_2 \in \mathcal{X}$ is a sequence (which may contain only one interactant) of interactants, species $\mathbf{y}_{\gamma(1)}$, ..., species $\mathbf{y}_{\gamma(k)}$, such that successive pairs overlap—that is, $\mathbf{y}_{\gamma(i)}^t \mathbf{y}_{\gamma(i+1)} \neq 0$—and $X_1 \in$ species $\mathbf{y}_{\gamma(1)}$, $X_2 \in$ species $\mathbf{y}_{\gamma(k)}$. For any $X \in \mathcal{X}$, the *knot* containing X is the subset $K \subset \mathcal{X}$ consisting of all species that can be reached from X via an interactant overlap path. Let the set of knots be $\mathcal{K}$. If K_i and K_j are knots and $K_i \cap K_j \neq \emptyset$, they contain a common species X. Hence there is an interactant overlap path from any $X_i \in K_i$ to any $X_j \in K_j$ via X. Therefore X_i and X_j lie in the same knot, and hence $K_i = K_j$. If $K_i \neq K_j$ it follows that $K_i \cap K_j = \emptyset$; thus $\mathcal{K}$ divides $\mathcal{X}$ into equivalence classes.

To obtain a less abstract definition of a knot, write down the chemical symbols for each species, and for each interactant, draw a closed curve around only the species in the interactant. The curves must be drawn so that they intersect as little as possible. If the interior of every closed curve is colored red, there will be a number of connected red regions that correspond to the knots. A knot is the set of all species whose symbols are contained in one connected red region. Examples appear in Hyver[6].

Example IV.3. The Oregonator of Example II.1 has the following complexes: Y, X, Y + Y, $\square$, 2X + 2Z, 2X, 2Z, and *f*Y. The corresponding complex vectors are $\mathbf{y}_1 \equiv (0,1,0)^t$, $\mathbf{y}_2 \equiv (1,0,0)^t$, $\mathbf{y}_3 \equiv (1,1,0)^t$, $\mathbf{y}_4 \equiv \mathbf{0}$, $\mathbf{y}_5 \equiv (2,0,2)^t$, $\mathbf{y}_6 \equiv (2,0,0)^t$, $\mathbf{y}_7 \equiv (0,0,2)^t$. The complex vector reactions are $R_1^c: \mathbf{y}_1 \rightarrow \mathbf{y}_2$, $R_2^c: \mathbf{y}_3 \rightarrow \mathbf{y}_4$, $R_3^c: \mathbf{y}_2 \rightarrow \mathbf{y}_5$, $R_4^c: \mathbf{y}_6 \rightarrow \mathbf{y}_4$, and $R_5^c: \mathbf{y}_7 \rightarrow \mathbf{y}_8$. The interactants are $\emptyset$, $\{X\}$, $\{Y\}$, $\{Z\}$, $\{X,Y\}$, $\{X,Z\}$. The interactants $\{X,Y\}$ and $\{X,Z\}$ overlap, so every pair of species can be joined by an interactant overlap path. Hence there is only one knot $K_1 \equiv \{X,Y,Z\}$.

When the complex $\square$ appears, species $(\square) = \emptyset$, so $\emptyset$ is an interactant. Note that for every other interactant "species **y**," $\emptyset \cap$ species $\mathbf{y} = \emptyset$. Hence the interactant $\emptyset$ cannot be part of an interactant overlap path between any two species; therefore the presence of $\square$ does not affect the set of knots. Note also that it is incorrect to write $\square \subset \emptyset$ because $\square$ is a complex, not a set; however, one may write $\emptyset \subset \emptyset$. This argument shows that the symbol $\emptyset$ cannot be used for $\square$ as many authors have done.

A *graph* G is defined to be a set P of distinct points π_i, $i \in [1,n]$, and a set L of unordered pairs of distinct points π_i–π_j, $i \neq j$, π_i, $\pi_j \in P$, called *lines*. The "line" π_i–π_i is called a *loop* and is not permitted in true graphs. If two "lines" $(\pi_i–\pi_j)_1$ and $(\pi_i–\pi_j)_2$ between the same pair of points can be considered different from each other, we are dealing with *multigraphs*, not graphs. Multigraphs that contain loops are called *pseudographs*.[61]

A *k-path* from π_i to π_j is a sequence of k lines $\pi_{\gamma(1)} - \pi_{\gamma(2)}$, $\pi_{\gamma(2)} - \pi_{\gamma(3)}, \ldots, \pi_{\gamma(k)} - \pi_{\gamma(k+1)}$, where $i = \gamma(1)$ and $j = \gamma(k+1)$. We define the *components* of G by stating that each component is a *subgraph*, with the property that two points are in the same component if and only if there exists a path between them. A k-path is a *k-cycle* if $\gamma(1) = \gamma(k+1)$ and the path does not cross itself ($\gamma(p) \neq \gamma(q)$ for all other $p \neq q$). Loops in a pseudograph are called *l-cycles*. A graph without cycles is called a *tree*.

Four types of graph can be constructed for a reaction network. In each case we have the option of allowing loops or multiple lines, and of considering reversible reactions once or twice. The definitions are given for pseudographs, with reversible reactions counted only once. The *complex graph* G_C has a point π_i for each complex or complex vector $\mathbf{y}_i$, and a line π_i–π_j if either of the reactions $\mathbf{y}_i \to \mathbf{y}_j$ or $\mathbf{y}_j \to \mathbf{y}_i$ occur. For the Oregonator, which was defined in Example II.1, we get a tree graph with three components.

$$G_C\colon\ \mathrm{Y} \overset{1}{—} \mathrm{X} \overset{3}{—} 2\mathrm{X} + 2\mathrm{Z} \qquad 2\mathrm{Z} \overset{5}{—} f\mathrm{Y}$$
$$\mathrm{X} + \mathrm{Y} \overset{2}{—} \square \overset{4}{—} 2\mathrm{X}$$

Each line has been labeled with the reaction number. The *interactant pseudograph* G_I has a point π_i for each interactant I_i and a line π_i–π_j for each reaction that takes a species in I_i into a species in I_j, or vice versa. Reversible reactions are only counted once. For the Oregonator we get

$$G_I\colon\ \{\mathrm{X},\mathrm{Y}\} \overset{2}{—} \varnothing \overset{4}{—} \{\mathrm{X}\} \overset{1}{—} \{\mathrm{Y}\} \overset{5}{—} \{\mathrm{Z}\}$$
$$\big|\, 3$$
$$\{\mathrm{X},\mathrm{Z}\}$$

This is a tree with one component. The *species pseudograph* G_S has a point π_i for each species X_i and a line π_i–π_j whenever a reaction converts X_i into X_j, or vice versa (the other species in the reaction are ignored). For the Oregonator we get

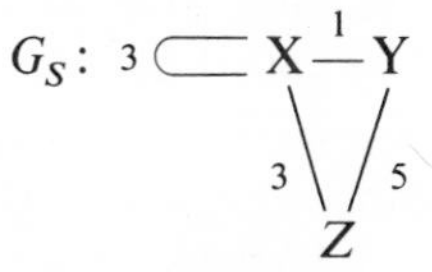

This one-component pseudograph contains a 1-cycle and a 3-cycle. Finally, the *knot pseudograph* G_K contains a point π_i for each knot K_i, and a line π_i–π_j for each reaction that takes some species in K_i into some species in K_j, or vice versa. Reversible reactions are only counted once. For the Oregonator we get a pseudograph with three loops on one point.

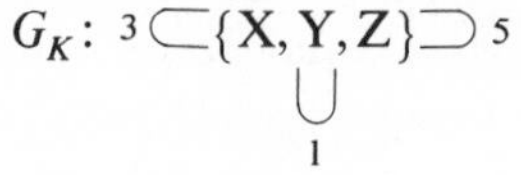

The example shows a tendency for the graph to become progressively more connected in the sequence G_C, G_I, G_S, G_K. The number of components decreases; the number of cycles (including loops) increases. Knot graphs were discovered by Hyver,[6] rediscovered by Sinonaglu,[62] and used extensively by Solimano, Beretta, and co-workers. Species graphs are closely related to the directed pseudographs developed by Clarke (see Section V). Interactant graphs are new but are related to the "interaction subset diagram" used by Solimano *et al.* The components of the complex graph were defined by Horn to be *linkage classes*; however he has used the term "complex graph" with a meaning different (Ref. 19, see Fig. 2) from ours.

A path in each of these graphs roughly corresponds to a complex path, an interactant path, a species path, and a knot path of the network. A *complex path* is a sequence of complex vectors $y_{\gamma(1)}, \ldots, y_{\gamma(k+1)}$ such that there is a sequence of k complex vector reactions $\mathbf{y}_{\gamma(1)} \rightarrow y_{\gamma(2)} \rightarrow \cdots \rightarrow y_{\gamma(k+1)}$. If either $\mathbf{y}_i \rightarrow \mathbf{y}_j$ or $y_i \leftarrow y_j$ occurs, we write $y_i — y_j$. A *weak complex path* is a sequence of k complex vectors such that we may write $\mathbf{y}_{\gamma(1)} — y_{\gamma(2)} — \cdots — y_{\gamma(k+1)}$. A path in G_C corresponds to a weak complex path. An *interactant path* is a sequence of interactants $I_{\gamma(1)}, \ldots, I_{\gamma(k+1)}$ such that some reaction consumes a species in $I_{\gamma(i)}$ and produces a species in $I_{\gamma(i+1)}$, for $i \in [1,k]$. In a *weak interactant path* we do not consider the direction of the reaction, and this path corresponds to a path on G_I. A *species path* is a sequence of species $X_{\gamma(1)}, \ldots, X_{\gamma(k+1)}$, such that some reaction has $X_{\gamma(i)}$ as a reactant and $X_{\gamma(i+1)}$ as a product, for $i \in [1,k]$. In a *weak species path* the direction of the reaction is ignored, and this path corresponds to a path on G_S. A *knot path* is a sequence of knots $K_{\gamma(1)}, \ldots, K_{\gamma(k+1)}$ such that a reaction consumes species in $K_{\gamma(i)}$ and produces species in $K_{\gamma(i+1)}$, for all $i \in [1,k]$. In a *weak knot path* we do not consider the direction of the reaction, and this path corresponds to a path in G_K. A (weak) complex path is called *internal* if it does not involve **0** and a (weak) interactant path is internal if it does not involve Ø.

We now define an (*internal*) (*weak*) *complex k-cycle*, an (*internal*) (*weak*) *interactant k-cycle*, an (*internal*) (*weak*) (*species k-cycle*, and an (*internal*) (*weak*) *knot k-cycle* to be the path of the same type, where $\gamma(1) = \gamma(k+1)$. A network having no (internal)(weak) complex cycles is called an (*internal*)(*weak*) *complex tree*. Note that our notation for sets of networks having particular stability properties has always involved the use of lowercase subscripts on $\mathfrak{N}$. For sets of networks that are defined by their topological properties, we use uppercase subscripts. The set of (internal)(weak) complex tree networks is denoted ($\mathfrak{N}_{IWCT}$, $\mathfrak{N}_{ICT}$, $\mathfrak{N}_{WCT}$) $\mathfrak{N}_{CT}$ or simply $\mathfrak{N}_{(I)(W)CT}$, where various options are possible. Similarly $\mathfrak{N}_{IT}$, $\mathfrak{N}_{ST}$, and $\mathfrak{N}_{KT}$ are the sets of interactant-tree, species-tree, and knot-tree networks.

Suppose every network with a cycle of type Y has a cyle of type Z. Here Y and Z can be "(internal)(weak) complex," "(internal)(weak) interactant," "(internal)(weak) knot," or "(internal)(weak) species." If a network is a Z-tree ($N \in \mathfrak{N}_{ZT}$), it cannot have a Z-type cycle; so by our supposition it cannot have a Y-type cycle either and must be a Y-tree ($N \in \mathfrak{N}_{YT}$). We have now proved that *if a Y-type cycle implies the existence of a Z-type cycle, then* $\mathfrak{N}_{ZT} \subset \mathfrak{N}_{YT}$.

If a network has a (weak)(internal) complex cycle $\mathbf{y}_{\gamma(1)}, \ldots, \mathbf{y}_{\gamma(k)}, \mathbf{y}_{\gamma(1)}$, it then has the (weak)(internal) interactant cycle $\mathrm{I}_{\gamma(1)}, \ldots, \mathrm{I}_{\gamma(k)}, \mathrm{I}_{\gamma(1)}$, where $\mathrm{I}_{\gamma(i)} =$ species $y_{\gamma(i)}$. Hence $\mathfrak{N}_{(W)(I)IT} \subset \mathfrak{N}_{(W)(I)CT}$. Two different complexes can have the same interactant (e.g., 2X + Y and X + Y), so an interactant cycle does not imply a complex cycle; thus $\mathfrak{N}_{(W)(I)IT}$ is a proper subset of $\mathfrak{N}_{(W)(I)CT}$. If a network has the (weak) internal interactant cycle above, it must have a (weak) species cycle, $\mathrm{X}_{\gamma(1)}, \ldots, \mathrm{X}_{\gamma(k)}$, where $\mathrm{X}_{\gamma(i)} \in \mathrm{I}_{\gamma(i)}$. Hence $\mathfrak{N}_{(W)ST} \subset \mathfrak{N}_{(W)IIT}$. Two different interactants can contain the same species, so $\mathfrak{N}_{(W)ST}$ is a proper subset. If a network has the (weak) species cycle above, it must also have a (weak) knot cycle $\mathrm{K}_{\gamma(1)}, \ldots, \mathrm{K}_{\gamma(k)}$, where $\mathrm{X}_{\gamma(i)} \in \mathrm{K}_{\gamma(i)}$. Hence $\mathfrak{N}_{(W)KT} \subset \mathfrak{N}_{(W)ST}$. Two different species can appear in the same knot, so a knot cycle does not imply a species cycle; hence $\mathfrak{N}_{(W)KT}$ is a proper subset of $\mathfrak{N}_{(W)ST}$. These results can be stated

$$\mathfrak{N}_{(W)KT} \subset \mathfrak{N}_{(W)ST} \subset \mathfrak{N}_{(W)IIT} \subset \mathfrak{N}_{(W)ICT} \qquad \text{(IV.34)}$$

where all subsets are proper subsets. These relationships explain the increasing connectivity of the graphs G_C, G_I, G_S, and G_K of the Oregonator. The Oregonator is unstable; hence it cannot be that all complex tree networks or all interactant tree networks are stable. On the other hand, we will prove the asymptotic stability of all reversible knot-tree networks with mass action kinetics.

A network is *reversible* if for every complex vector reaction $\mathbf{y}_i \rightarrow \mathbf{y}_j$, there is a *reverse reaction* $\mathbf{y}_j \rightarrow \mathbf{y}_i$. This definition of reversibility depends only on stoichiometry, not on the rate laws. Let the set of reversible networks be

$\mathfrak{N}_R$. In a reversible network, a species path exists between species of the same interactant, provided the other interactant of the reaction is not Ø. For example, a species path between X and Y in a network containing $X + Y \rightleftarrows Z$ is X, Z, Y. However no species path between X and Y involves the reaction $X + Y \rightleftarrows \square$ because neither $\square$ nor this reaction affects G_S. Let $\mathfrak{N}_{RME}$ be the set of reversible networks with monomolecular exits (i.e., every reaction having the interactant Ø must have a single species as the other interactant). In these networks there is a species path between every pair of species in a knot. If the network has a knot cycle, it must also have a species cycle. Hence

$$\mathfrak{N}_{RME} \cap \mathfrak{N}_{ST} \subset \mathfrak{N}_{KT} \tag{IV.35}$$

and from (IV.34)

$$\mathfrak{N}_{RME} \cap \mathfrak{N}_{ST} = \mathfrak{N}_{RME} \cap \mathfrak{N}_{KT} \tag{IV.36}$$

The asymptotic stability proof we will later give for reversible mass action knot-tree networks therefore implies the asymptotic stability of all reversible mass action species tree networks with monomolecular exits.

C. Complex Balanced Networks

Complex balanced networks were first studied by Horn, Feinberg, and Jackson.[13–19] In this section we show that the extreme subnetworks of such networks are complex cycles and that complex cycles are mixing asymptotically stable.

From the point of view of stoichiometric network analysis, a network is defined to be a matrix $\underline{\boldsymbol{\nu}}$ and a reaction velocity function $\mathbf{v}(\mathbf{X},\mathbf{p})$. For power law networks, $\mathbf{v}(\mathbf{X},\mathbf{p})$ is determined by $\underline{\boldsymbol{\kappa}}$ according to (I.1), thus these networks are completely specified by a pair of matrices $\underline{\boldsymbol{\nu}}$ and $\underline{\boldsymbol{\kappa}}$. The matrices determine the general dynamics by (II.53). Any such network may be represented by many possible sets of reactions. The set of reactions determines the set of complexes. Hence a power law stoichiometric network does not have an unique set of complexes. In the theory that follows, any set of complexes may be used. Later on it will be necessary to use only sets of complexes with certain properties.

The set of complex vector reactions (IV.33) may be represented by a matrix W having elements W_{ij} defined by

$$W_{ij} = \begin{cases} 1 & \text{if} \quad i = p(j) \\ -1 & \text{if} \quad i = r(j) \\ 0 & \text{otherwise} \end{cases} \tag{IV.37}$$

It then follows from the definitions of $\underline{\nu}$ and Y that

$$\underline{\nu} = \mathsf{YW} \tag{IV.38}$$

The net rate of formation of the species is $\underline{\nu}\mathbf{v} = \mathsf{YW}\mathbf{v}$, and the net rate of formation of the complexes is $\mathsf{W}\mathbf{v}$. Note that a sufficient but not necessary condition for the species steady-state condition $\underline{\nu}\mathbf{v}^0 = 0$ to be satisfied is that

$$\mathsf{W}\mathbf{v}^0 = 0 \tag{IV.39}$$

This equation defines a linear subspace that intersects the current cone $\mathcal{C}_v$ in a subcone $\mathcal{C}_{CB}$ of *complex balanced* currents, where the complexes are produced and consumed at the same rate. A network is called *complex balanced* if $\mathcal{C}_v = \mathcal{C}_{CB}$. This relation will hold if the dimension of the solution spaces of (IV.39) and $\underline{\nu}\mathbf{v} = 0$ are the same; that is, if $\delta = 0$, where

$$\delta \equiv \dim \mathcal{C}_v - \dim \mathcal{C}_{CB} \tag{IV.40}$$

Note that in general $\delta \geqslant 0$. The hyperplane $\mathbf{e}'\mathbf{v} = 1$ intersects these cones in the polytopes Π_v and Π_{CB}. The *complex balanced current polytope* Π_{CB} has δ fewer dimensions than Π_v and lies in the intersection of Π_v with a linear subspace. Hence δ is the number of independent hyperplane constraints required to restrict Π_v to Π_{CB}.

Every equilibrium extreme subnetwork (EEN) consists of a reaction and its reverse. The corresponding complex vector reactions are $\mathbf{y}_i \rightleftharpoons \mathbf{y}_j$; hence the subnetwork produces and consumes both complexes at the same rate. Every EEN is complex balanced. It follows that Π_E must lie in the intersection of Π_{CB} and a linear subspace.

We now obtain an alternate formula for δ and thereby prove that δ is the quantity HJF called the *deficiency*. The complexes in each component of the complex graph G_C form a set called a *linkage class*. Let there be l linkage classes and denote the set of linkage classes by $\mathcal{L} \equiv \{L_i \mid i \in [1,l]\}$. From the properties of G_C it follows that $\mathsf{Y} = \bigcup_i L_i$, and $L_i \cap L_j = \emptyset$ for all $i, j \in [1,l]$, $i \neq j$.

Let $\mathbf{W}_i^\dagger$ be the ith row of W, and let the most general linear dependence among the x rows of W be

$$\sum_{i=1}^{x} \mu_i \mathbf{W}_i^\dagger = \mathbf{0} \tag{IV.41}$$

From (IV.37), the jth component of this vector equality reads

$$\mu_{p(j)} = \mu_{r(j)} \tag{IV.42}$$

thus $\mu_i = \mu_j$ whenever $\mathbf{y}_i — \mathbf{y}_j$. It then follows that all complexes in a linkage class must have the same coefficient μ_i, so we may define l independent coefficients $\mu_1^*, \ldots, \mu_l^*$ and write (IV.41) as

$$\sum_{i=1}^{l} \mu_i^* \sum_{\mathbf{y}_j \in L_i} \mathbf{W}_j^\dagger = 0$$

This is the most general linear relation among the rows of W; thus there are l independent linear relations, for $i \in [1, l]$

$$\sum_{\mathbf{y}_j \in L_i} W_j^\dagger = 0$$

Since W has x rows, the number of linearly independent rows must be $x - l$. Each $\mathbf{W}_j^\dagger \in R^r$, so the rows span a subspace of dimension $x - l$ of R^r. The complex balancing condition (IV.39) states that $\mathbf{v}^0 \in \bar{R}_+^r$ must be orthogonal to all the rows of W. Hence $\mathbf{v}^0$ lies in the orthogonal complementary subspace whose dimension is $r - (x - l)$; thus

$$\dim \mathcal{C}_{CB} = r - x + l \qquad \text{(IV.43)}$$

Obtaining $\dim \mathcal{C}_v$ from Table I, we derive

$$\delta = x - l - d \qquad \text{(IV.44)}$$

from (IV.40). HJF defined the deficiency using this equation. It is often used to calculate δ because x, l, and d are easily determined.

For the general network (not necessarily complex balanced) there will be extreme solutions of (IV.39) subject to the constraints $\mathbf{v}^0 \in \bar{R}_+^r$ and $e^t\mathbf{v}^0 = 1$. These extreme solutions may be found, as for $\underline{\nu}\mathbf{v}^0 = 0$, by deleting columns of W until the remaining equation has a unique solution. Each solution of (IV.39) corresponds to a flow pattern among the complexes in which the total inflow equals the total outflow at each complex. Deleting the jth column of W implies that the jth reaction is not in the flow pattern; thus the inflows and outflows can be equal only if the flows can be balanced by the remaining reactions. When the maximum number of reactions has been deleted, subject to the condition that a complex balanced steady state exists, the corresponding flow pattern must look like

$$y_{\gamma(1)} \to y_{\gamma(2)} \to \cdots \to y_{\gamma(k)} \to y_{\gamma(1)} \qquad \text{(IV.45)}$$

which is a complex k-cycle. Flows containing branching are not extreme because the elements of W are 0, 1, and -1, and this permits all subcycles to be suppressed. The extreme complex balanced flow (IV.45) is the *extreme complex balanced current* $\mathbf{v}^0 \in \overline{R}^r_+$, defined by

$$v_i^0 = \begin{cases} \dfrac{1}{k} & \text{if } i \in \{\gamma(1), \ldots, \gamma(k)\} \\ 0 & \text{otherwise} \end{cases} \tag{IV.46}$$

The extreme CB currents are extreme points of Π_{CB}; however they are often not extreme currents (i.e., extreme points of Π_v), as illustrated in the example. Thus complex cycles are not always extreme networks.

Example IV.4. Suppose that $r = 5$, $n = 4$, $x = 5$, $l = 1$:

$$\mathrm{W + X \xrightarrow{1} X + Y \xrightarrow{2} Y + Z \xrightarrow{3} Z + W \xrightarrow{4} W + Y \xrightarrow{5} W + X}$$

$$\mathsf{Y} = \begin{pmatrix} 1 & 0 & 0 & 1 & 1 \\ 1 & 1 & 0 & 0 & 0 \\ 0 & 1 & 1 & 0 & 1 \\ 0 & 0 & 1 & 1 & 0 \end{pmatrix} \qquad \mathsf{W} = \begin{pmatrix} -1 & 0 & 0 & 0 & 1 \\ 1 & -1 & 0 & 0 & 0 \\ 0 & 1 & -1 & 0 & 0 \\ 0 & 0 & 1 & -1 & 0 \\ 0 & 0 & 0 & i & -1 \end{pmatrix}$$

From (IV.38) we obtain $\underline{\nu}$ and then calculate the extreme currents:

$$\underline{\nu} = \begin{pmatrix} -1 & 0 & 1 & 0 & 0 \\ 0 & -1 & 0 & 0 & 1 \\ 1 & 0 & -1 & 1 & -1 \\ 0 & 1 & 0 & -1 & 0 \end{pmatrix} \qquad \mathsf{E} = \begin{pmatrix} \frac{1}{2} & 0 \\ 0 & \frac{1}{3} \\ \frac{1}{2} & 0 \\ 0 & \frac{1}{3} \\ 0 & \frac{1}{3} \end{pmatrix}$$

There is one conservation condition, $\underline{\gamma} = (1 \quad 1 \quad 1 \quad 1)$ and $d = \operatorname{rank} \underline{\nu} = 3$. $\operatorname{Dim} \mathcal{C}_v = r - d = 2$; however $\delta = x - l - d = 1$, so $\mathcal{C}_{CB}$ has one fewer dimension than $\mathcal{C}_v$. The extreme CB current is $\mathbf{E}_{CB} = (1, 1, 1, 1, 1)^t/5 = \frac{2}{5}\mathbf{E}_1 + \frac{3}{5}\mathbf{E}_2$. Thus it lies in the relative interior of $\mathcal{C}_v$.

If a network is complex balanced ($\mathcal{C}_v = \mathcal{C}_{CB}$), every extreme current is an extreme complex balanced current, or equivalently, every extreme network is complex balanced. If the network is not complex balanced, the extreme complex balanced currents need not be extreme currents. Equivalently, the corresponding complex cycles need not be extreme networks.

The discussion so far is valid for any choice of complexes for the network. The deficiency and the cone $\mathcal{C}_{CB}$ depend on the complexes chosen. Now we restrict the discussion to networks of particular kinds and to complexes that have certain properties. Some general definitions are required.

A power law stoichiometric network is *pathological* if there is a species X_i and a reaction R_j such that $\nu_{ij} < 0$ and $\kappa_{ij} = 0$. In a pathological network, X_i is consumed by a reaction at a rate that is independent of X_i. Hence even when X_i has zero concentration, the reaction will continue to consume X_i, thereby driving the concentration negative. Pathological networks are not of physical interest.

A power law stoichiometric network has *mass action kinetics* if it is not pathological, and if for every pair of corresponding and reverse reactions R_i and R_j,

$$\boldsymbol{\nu}_i = -\boldsymbol{\kappa}_i + \boldsymbol{\kappa}_j \tag{IV.47}$$

where $\boldsymbol{\nu}_i$, $\boldsymbol{\kappa}_i$, and $\boldsymbol{\kappa}_j$ are columns of the matrices $\underline{\boldsymbol{\nu}}$ and $\underline{\boldsymbol{\kappa}}$. A *mass action network* is a power law stoichiometric network that exhibits mass action kinetics.

We now show that the complexes may be chosen for a mass action network such that

$$\underline{\boldsymbol{\kappa}} = -\mathsf{YW}^- \tag{IV.48}$$

where W^- contains only the negative elements of W (all ones have been replaced by zeros). Let the complex vector for the left (reactant) side of the reaction R_i be $\boldsymbol{\kappa}_i$. Let the complex vector for the right (product) side be

$$\mathbf{y} = \boldsymbol{\kappa}_i + \varepsilon \boldsymbol{\nu}_i$$

where ε is chosen as follows. If R_i is not reversible, ε may be any positive number that makes $\mathbf{y}$ nonnegative. Many choices are possible because mass action networks are not pathological. If R_i is reversible, let $\varepsilon = 1$. Then from the definition of a mass action network, $\mathbf{y} = \boldsymbol{\kappa}_j$. For every reaction, the complex vector on the left of the reaction is identical to the corresponding column of $\underline{\boldsymbol{\kappa}}$. Hence (IV.48) holds. Complexes chosen in this manner are called *mass action complexes*. Reversibile mass action networks have a unique set of mass action complexes, but all other mass action networks have infinitely many sets of mass action complexes. For these networks one should choose ε for the irreversible reactions to make $\mathcal{C}_{CB}$ as large as possible. For mass action networks, $\mathcal{C}_{CB}$ always is defined using a set of mass action complexes. If $\mathcal{C}_{CB} = \mathcal{C}_v$, the network is a *complex balanced mass action network*.

We next prove that every extreme complex balanced subnetwork (i.e., every complex cycle) is mixing asymptotically stable. Let us assume that the complexes and reactions of the extreme complex balanced network N are numbered as follows:

$$\mathbf{y}_1 \xrightarrow{1} \mathbf{y}_2 \xrightarrow{2} \mathbf{y}_3 \rightarrow \cdots \rightarrow \mathbf{y}_r \xrightarrow{r} \mathbf{y}_1$$

Then $W_{ij} = \delta_{ij+1} - \delta_{ij}$, where δ_{ij} is the Kronecker delta function ($\delta_{ij} = 1$ if $i = j, \delta_{ij} = 0$ otherwise) and all subscripts are calculated modulo r (i.e., $\delta_{1r+1} = \delta_{11} = 1$), and $\mathsf{W}^- = -\mathsf{I}$. From (III.42), (IV.38), and (IV.48),

$$\mathsf{S} = -\mathsf{YWY}^t \tag{IV.49}$$

The symmetric part of S is $\mathsf{S}_{\text{sym}} = \frac{1}{2}\mathsf{YHY}'$, where $\mathsf{H} \equiv -\mathsf{W} - \mathsf{W}'$. One may verify that $\mathsf{WW}' = \mathsf{H}$, so that $\mathsf{S}_{\text{sym}} = \frac{1}{2}(\mathsf{YW})(\mathsf{YW})'$. Since any matrix that can be put into this form is positive semidefinite, extreme complex balanced networks are mixing stable.

To prove mixing asymptotic stability, we must verify (IV.8) and (IV.9). Since (IV.13) holds in general,

$$\mathsf{S}\underline{\gamma}^t = (2\mathsf{S}_{\text{sym}} - \mathsf{S}')\underline{\gamma}^t = 2\mathsf{S}_{\text{sym}}\underline{\gamma}^t - (\underline{\gamma}\mathsf{S})^t$$

$$= 2\mathsf{S}_{\text{sym}}\underline{\gamma}^t = \mathsf{YW}(\mathsf{YW})^t\underline{\gamma}^t = \mathsf{YW}(\underline{\gamma}\mathsf{YW})^t$$

This vanishes because (IV.38) and (II.6) imply $\underline{\gamma}\mathsf{YW} = 0$; thus (IV.8) is satisfied.

To verify (IV.9), we first show that the most general solution of $\mathbf{u}'\mathsf{S} = 0$ is a subspace of dimension $n - d$. Since S is an $n \times n$ matrix, it will then follow that $\operatorname{rank}\mathsf{S} = d$. From (IV.49), we therefore consider the most general solution of $\mathbf{u}'\mathsf{YWY}' = 0$, which is equivalent to the system of equations

$$\sum_{ik} u_i(Y_{ik+1} - iY_{ik})Y_{lk} = 0 \qquad \text{(IV.50)}$$

for $l \in [1,n]$. Multiplying by u_l and summing over l yields the algebraic equation

$$f(\mathbf{z}) \equiv \sum_k (z_{k+1} - z_k)z_k = 0 \qquad \text{(IV.51)}$$

where $z_k = \sum_i u_i \mathrm{Y}_{ik}$. Since $f(\lambda\mathbf{z}) = \lambda^2 f(\mathbf{z})$, the most general solution to $f(\mathbf{z}) = 0$ can easily be obtained from the solution when $\mathbf{z}$ is restricted to the hyperplanes $\mathbf{e}'\mathbf{z} = n$. On this hyperplane the set of conditions $\partial f(\mathbf{z})/\partial z_k = 0$, for all $k \in [1,n]$, has a unique solution, which is $\mathbf{z} = \mathbf{e}$. (The reader should check this.) At this point, the matrix whose ijth element is $\partial^2 f(\mathbf{z})/\partial z_i \partial z_j$, is negative semidefinite. Hence $f(\mathbf{z})$ attains its absolute maximum on the hyperplane at the point $\mathbf{z} = \mathbf{e}$. Since $f(\mathbf{e}) = 0$, the only solution to (IV.51) on the hyperplane is $\mathbf{z} = \mathbf{e}$; the general solution is $\mathbf{z} = \lambda\mathbf{e}$, $\lambda \in R$. Hence the general solution satisfies $z_{k+1} = z_k$, which implies $\sum u_i(\mathrm{Y}_{ik+1} - \mathrm{Y}_{ik}) = 0$, which implies $\mathbf{u}'\mathsf{YW} = 0$, which implies $\mathbf{u}'\underline{\boldsymbol{\nu}} = 0$. Hence every solution of $\mathbf{u}'\mathsf{S} = 0$ is also a solution of $\mathbf{u}'\underline{\boldsymbol{\nu}} = 0$. Since $\operatorname{rank}\underline{\boldsymbol{\nu}} = d$, the solution space of $\mathbf{u}'\underline{\boldsymbol{\nu}}$ has dimension $n - d$; thus the most general solution of $\mathbf{u}'\mathsf{S} = 0$ lies in a space whose maximum dimension is $n - d$. Hence the minimum rank of S is d. Note that $\operatorname{rank}\mathsf{S} > d$ contradicts (IV.13) because $\underline{\gamma}$ has $n - d$ independent rows. Hence $\operatorname{rank}\mathsf{S} = d$. Since (IV.8) and (IV.9) are satisfied, and since extreme complex balanced networks are mixing stable, it follows that they are also mixing asymptotically stable. This conclusion may be expressed formally by defining $\mathcal{E}_{CBMA}$ to be the set of extreme complex balanced mass action networks. Then we have proven

$$\mathcal{N}_{ma} \supset \mathcal{E}_{CBMA} \tag{IV.52}$$

Let $\mathcal{N}_{RMA}$ be the set of reversible mass action networks. From Table I, every accessible perturbation $\Delta\mathbf{X} \in S_X$ can be represented as a linear combination $\sum_i \xi_i \boldsymbol{\nu}_i$ of the columns of $\underline{\boldsymbol{\nu}}$, because the columns span S_X. The column $\boldsymbol{\nu}_i$ contains the stoichiometry of reaction R_i. If $N \in \mathcal{N}_{RMA}$, then corresponding to R_i is an equilibrium extreme network, which we call E^i, consisting of R_i and its reverse reaction. The stoichiometric matrix of E^i has two columns which are $\boldsymbol{\nu}_i$ and $-\boldsymbol{\nu}_i$. Hence the accessible subspace of E^i, which we call S_X^i, is spanned by $\boldsymbol{\nu}_i$ only, and is one-dimensional. Since S_X is spanned by all the columns of $\underline{\boldsymbol{\nu}}$, it is therefore the direct sum of all the subspaces S_X^i, where i takes only those values such that E^i is an equilibrium extreme network. Every such E^i is mixing asymptotically stable (because it is a complex balanced mass action extreme network), so the union of the subspaces S_X^i is contained in S_{Xma}. But this union has already been shown to be S_X, so $S_X \subset S_{Xma}$. Also $S_{Xma} \subset S_X$, so $\dim S_{Xma} = \dim S_X = d$. Hence (IV.16) holds for any $N \in \mathcal{N}_{RMA}$; and from (IV.17)

$$\mathcal{N}_a^L \supset \{N \in \mathcal{N}_{RMA} \mid \Xi(N, \mathsf{E}) \subset \mathcal{E}_{ms}\} \tag{IV.53}$$

Example IV.5. Let us apply (IV.16) to the Oregonator, defined in Example II.1, with $f = 1$. To give R_5 mass action kinetics, we will replace 2Z by Z wherever this expression occurs (R_3 and R_5). When the reverse reactions are included, the stoichiometric matrix is

$$\underline{\boldsymbol{\nu}} = \begin{pmatrix} 1 & -1 & 1 & -2 & 0 & -1 & 1 & -1 & 2 & 0 \\ -1 & -1 & 0 & 0 & 1 & 1 & 1 & 0 & 0 & -1 \\ 0 & 0 & 1 & 0 & -1 & 0 & 0 & -1 & 0 & 1 \end{pmatrix}$$

where R_i is the reverse of R_{i-5}, for $i > 5$ (i.e., $\boldsymbol{\nu}_i + \boldsymbol{\nu}_{i-5} = 0$). A computation as in Example II.5 now gives the complete set of extreme currents

$$\mathsf{E} = \begin{bmatrix} 0 & 1 & 1 & 0 & 0 & 0 & 1 & 0 & 0 & 0 & 0 \\ 1 & 0 & 0 & 0 & 0 & 1 & 0 & 1 & 0 & 0 & 0 \\ 1 & 1 & 0 & 0 & 0 & 0 & 0 & 0 & 1 & 0 & 0 \\ 0 & 1 & 1 & 0 & 0 & 0 & 0 & 0 & 0 & 1 & 0 \\ 1 & 1 & 0 & 0 & 0 & 0 & 0 & 0 & 0 & 0 & 1 \\ 0 & 0 & 0 & 0 & 1 & 1 & 1 & 0 & 0 & 0 & 0 \\ 0 & 0 & 1 & 1 & 0 & 0 & 0 & 1 & 0 & 0 & 0 \\ 0 & 0 & 0 & 1 & 1 & 0 & 0 & 0 & 1 & 0 & 0 \\ 0 & 0 & 0 & 0 & 1 & 1 & 0 & 0 & 0 & 1 & 0 \\ 0 & 0 & 0 & 1 & 1 & 0 & 0 & 0 & 0 & 0 & 1 \end{bmatrix}$$

$\mathbf{E}_1$ and $\mathbf{E}_2$ were found previously; $\mathbf{E}_3$ is new; $\mathbf{E}_4$, $\mathbf{E}_5$, and $\mathbf{E}_6$ are the reverse currents of $\mathbf{E}_1$, $\mathbf{E}_2$, and $\mathbf{E}_3$, respectively; $\mathbf{E}_7, \ldots, \mathbf{E}_{11}$ are the equilibrium extreme currents of $R_1, \ldots, R_5$, respectively. Thus $r = 10$, $d = 3$, $n = 3$, $\dim \mathcal{C}_v = r - d = 7$, and $\dim \mathcal{C}_E = r/2 = 5$.

The forward reactions of the Oregonator (given in Example II.1) have mass action kinetics when the stoichiometries are changed by replacing 2Z by Z in R_3 and R_5. The reversible network will have mass action kinetics only if the kinetics of the reverse reactions are determined by (IV.47). Then

$$\underline{\kappa} = \begin{pmatrix} 0 & 1 & 1 & 2 & 0 & 1 & 0 & 2 & 0 & 0 \\ 1 & 1 & 0 & 0 & 0 & 0 & 0 & 0 & 0 & 1 \\ 0 & 0 & 0 & 0 & 1 & 0 & 0 & 1 & 0 & 0 \end{pmatrix}$$

From constructing $S^{(i)}$, for $i = 1, \ldots, 11$, and calculating the eigenvalues of $S^{(i)}_{\text{sym}}$, we discover that all extreme subnetworks are mixing stable except E^1 and E^4. A calculation shows that the mixing stable extreme currents span 60% of the content (volume) of Π_v. The relative interior of this region is asymptotically stable by (IV.16).

The two nonmixing subnetworks (E^1 and E^4) are involved in producing regions of instability. We will later show that E^1 is unstable. A careful analysis of the reversible Oregonator by Gibbs[63] using the Hurwitz determinant approach (Section III.C), has located an unstable region of Π_v that includes E^1, and a second isolated unstable region that appears near E^4 for $f < 1$. Thus the mixing stability test has succeeded wherever possible.

D. Trees and Sign Stability

This subsection introduces several general concepts that are useful for reaction networks. We give a proof of sign semistability for matrices whose corresponding graph is a tree. This proof uses a "tree function" that is used again in the next section to prove the asymptotic stability of reversible knot-tree networks. We define the feedback cycles of a matrix, a concept that plays an important role in Sections V and VI. The bare bones of the proof in this subsection can be found in Jefferies;[45] however he did not mention the concepts we consider important.

Consider a tree graph G, of points $\pi_i \in P$, and lines $\pi_i\text{–}\pi_j \in L$, as defined in Section IV.B. With each line $\pi_i\text{–}\pi_j$, we associate two *directed lines*, which are ordered pairs of points, written $\pi_i \to \pi_j$ and $\pi_j \to \pi_i$. If the lines of G are replaced by pairs of directed lines, we obtain the *directed graph* or *digraph* G^* whose set of directed lines is called L^*. Consider any mapping $f: L^* \to R$. Since G is a tree, we can choose one point π in each component, and define a related mapping $g: P \to R$ as follows. Let π_{ij} be the ith point in the jth component, let π_{1j} be the chosen point, and let $g(\pi_{1j}) \equiv \zeta_j$, a parameter to be determined. For each π_{ij}, $i \neq 1$,

$$g(\pi_{ij}) \equiv \zeta_j f(\pi_{1j} \to \pi_{\gamma(2)j}) f(\pi_{\gamma(2)j} \to \pi_{\gamma(3)j}) \cdots f(\pi_{\gamma(k)j} \to \pi_{ij}) \qquad \text{(IV.54)}$$

that is, the product of the directed line function along the path from π_{1j} to π_{ij}. Choose ζ_j so that $\sum_i g(\pi_{ij}) = 1$. We call g the *tree function* derived from f and the particular choice of one point per component.

The tree function helps construct Lyapunov functions. A nice illustration is the proof of the "if" part of the Quirk-Ruppert-Maybee theorem (Section III.E).

Given a matrix U, a cyclic product of k matrix elements $U_{\gamma(1)\gamma(2)} U_{\gamma(2)\gamma(3)} \cdots U_{\gamma(k)\gamma(1)}$ is called a *matrix feedback k-cycle*, or simply a *k-cycle*. The feedback is *positive* (*negative*) (the cycle *does not occur*) if the product is *positive* (*negative*) (*zero*). We now prove that if all cycles are negative 1-cycles or negative 2-cycles, then U is sign semistable. Associated with the $n \times n$ matrix U, is a graph G consisting of n points $\pi \in P$ and a set L of lines, one line π_i–π_j for each negative 2-cycle $U_{ij}U_{ij}$ that occurs (and no other lines). G cannot contain a cycle because a cycle of 2-cycles is forbidden by the absence of all k-cycles, $k > 2$. Hence G is a tree.

The matrix U is sign semistable if for all $\mathbf{p} \in R_+^{n^2}$, all eigenvalues λ of $\mathsf{M}(\mathbf{p})$, defined by (III.75), satisfy $\mathrm{Re}(\lambda) \leqslant 0$. Without loss of generality, we assume $U_{ij} \in \{-1, 0, 1\}$. Section V.D proves that each coefficient α_i of the characteristic equation is a sum of products of feedback cycles. Hence λ, thus $\mathrm{Re}(\lambda)$, are independent of the magnitude of the matrix elements that do not lie on feedback cycles. Therefore the sign semistability problem is unaffected if we assume that these elements are zero; then all elements of $\mathsf{M}(\mathbf{p})$ belong to 1-cycles or 2-cycles. Define $f: L^* \to R$ by

$$f(\pi_i \to \pi_j) = \frac{p_{ij}}{p_{ji}} \tag{IV.55}$$

Choose one point in each component of G and let g be the tree function derived from f and this choice of points. Let $g_i \equiv g(\pi_i)$, $i \in [1, n]$ define the vector $\mathbf{g}$. Then for each i, j such that $\pi_i \to \pi_j \in L^*$,

$$\frac{g_j}{g_i} = f(\pi_i \to \pi_j) = \frac{p_{ij}}{p_{ji}} \tag{IV.56}$$

Since $\mathbf{p} \in R_+^{n^2}$, $g_i > 0$ for all i; hence the function

$$L_S(\mathbf{u}, \mathbf{p}) \equiv \mathbf{u}^t(\mathrm{diag}\,\mathbf{g})\mathbf{u} \tag{IV.57}$$

is always positive. From (III.19) and (III.75),

$$\begin{aligned} \frac{dL_S}{dt} &= \sum_{ij} u_i(U_{ji}p_{ji}g_j + g_i U_{ij}p_{ij})u_j \\ &= \sum_i 2U_{ii}p_{ii}g_i u_i^2 + \sum_{i \neq j} U_{ij}(g_i p_{ij} - g_j p_{ji})u_i u_j \end{aligned}$$

where the identity $U_{ij} = -U_{ji}$, for all $i \neq j$, has been used. The first sum is nonpositive because $U_{ii} \leqslant 0$ and $p_{ii}, g_i, u_i^2 \geqslant 0$; the second sum vanishes by (IV.56). Hence L_S cannot increase and the system cannot evolve away from $\mathbf{u} = 0$. This implies $\mathrm{Re}(\lambda) \leqslant 0$ for all $\lambda \in \Lambda$, so U is proved to be sign semistable.

E. Mass Action Knot-Tree Skeleton Networks and Semidirected Species-Tree Networks

Instability requires feedback cycles. Hence certain kinds of tree networks (which do not have certain kinds of feedback cycles) ought to be stable. Interactant tree networks can be unstable—the Oregonator is an example. From (IV.34), the next smallest class of trees are the species trees. A proof of the stability of these networks appears at the end of this section. Next smallest are the knot trees. This section first proves the asymptotic stability of a set of networks having mass action kinetics, which is contained in the set of species-tree networks, but contains and is much larger than the set of knot-tree networks. Initially we treat only reversible networks. Then we extend the results to irreversible networks. Finally stability (but not asymptotic stability) is proved for species trees.

The theorems proved in this section complement the theory that appears in Section V. Here we prove that the absence of species cycles implies stability. Section V shows that the presence of certain types of species cycle implies instability.

A complex vector reaction $\mathbf{y}_i \rightarrow \mathbf{y}_i$ is *internal* if neither $\mathbf{y}_i$ nor $\mathbf{y}_j$ is $\mathbf{0}$; otherwise the reaction is *external*. Every network has a corresponding *skeleton network* consisting of all the internal reactions and none of the external reactions. The set of *knot-tree skeleton networks* ($\mathfrak{N}_{KTS}$) is the set of all networks whose skeleton networks are knot-tree networks. Clearly $\mathfrak{N}_{KTS} \supset \mathfrak{N}_{KT}$. To see that $\mathfrak{N}_{KTS}$ is much larger, consider any $N \in \mathfrak{N}_{KT}$ having two knots that are directly connected by a knot-graph line. Let X be a species in one of these knots, and let Y be a species in the other. Now add the external reaction $\mathrm{X} + \mathrm{Y} \rightleftarrows \square$ to obtain a new network N'. Since $N \in \mathfrak{N}_{KT}$, the skeleton of N' is in $\mathfrak{N}_{KT}$; hence $N' \in \mathfrak{N}_{KTS}$. However $N' \notin \mathfrak{N}_{KT}$ because the knot containing X and Y (which is the union of the corresponding knots of N) has a loop. The coalescence of two knots has changed a line into this loop; thus the knot graph is no longer a tree.

Let us first prove the asymptotic stability of all reversible mass action knot-tree skeleton networks. That is, we prove

$$\mathfrak{N}_a \supset \mathfrak{N}_R \cap \mathfrak{N}_{MA} \cap \mathfrak{N}_{KTS} \tag{IV.58}$$

where $\mathfrak{N}_{MA}$ is the set of networks with mass action kinetics. Toward the end of this section this result is partly generalized to the irreversible case. Stability then can be proved for all mass action networks (irreversible or reversible) whose semidirected species graphs are trees.

In the proof we replace $\mathcal{C}_v$ by a larger cone $\mathcal{C}_v^* \supset \mathcal{C}_v$. Doing this does not affect the formalism because $\mathbf{h}$ and $\mathbf{v}$ can be considered to be independent parameters equivalent to the kinetic parameters. The original reason for using $\mathbf{j}$ rather than $\mathbf{v}$ was to confine $\mathbf{v}$ to the part of R_+^r that corresponds to steady states, without having to consider complicated inequalities that might arise from the condition $v_i \geqslant 0$. Allowing $\mathbf{v}$ to take values outside $\mathcal{C}_v$ can result in a meaningless indication of "instability" in the nonsteady part of velocity space $\mathcal{C}^* \backslash \mathcal{C}_v$. Otherwise, the formalism remains valid on any cone $\mathcal{C}_v^* \supset \mathcal{C}_v$. Here we choose $\mathcal{C}_v^* \equiv \overline{R}_+^r$. The frames of $\mathcal{C}_v^*$ are the coordinate axes, and the basis vectors of R^r are *pseudoextreme currents*. Then $\mathsf{E} = \mathsf{I}$ and each reaction R_i becomes a *pseudoextreme network* E^i. Equation II.18 now reads $\mathbf{v} = \mathbf{j}$.

The knot graph of the skeleton network G_K has a point π_p for each knot K_p, $p \in [1,k]$, and a line $\pi_p - \pi_q$ for each pair of reactions $\mathbf{y}_s \rightleftarrows \mathbf{y}_t$, such that species $\mathbf{y}_s \in \mathrm{K}_p$ and species $\mathbf{y}_t \in \mathrm{K}_q$. Let $\kappa(i)$ be the index of the knot containing the products of reaction R_i. Let the reverse of this reaction be R_{-i}. These reactions can also be referred to as R_{-j} and R_j, respectively, because neither R_i nor R_j is "forward" in an absolute sense. Then $\kappa(-i)$ will mean the index of the knot containing the products of R_i, or equivalently, the reactants of R_i. The same may be said of the pseudoextreme networks E^i and E^{-i}, which correspond to the line $\pi_{\kappa(-i)} - \pi_{\kappa(i)}$ of G_K. The directed line $\pi_{\kappa(-i)} \rightarrow \pi_{\kappa(i)}$ will correspond to $E^i(\mathrm{R}_i)$ and the directed line $\pi_{\kappa(i)} \rightarrow \pi_{\kappa(-i)}$ will correspond to $E^{-i}(\mathrm{R}_{-i})$. Define the mapping $f: L^* \rightarrow R$ by

$$f(\pi_{\kappa(-i)} \rightarrow \pi_{\kappa(i)}) = \frac{j_{-i}}{j_i} \tag{IV.59}$$

where j_i is the weighting of the subnetwork E^i in $\mathbf{v}$ (i.e., $j_i = v_i$). Since G_K is a tree, we may choose one point in each graph component and let g be the tree function derived from f. Since the network is reversible and no reactions are permitted to vanish ($j_i \neq 0$), it follows that $g(\pi_p) > 0$ for all $p \in [1,k]$. For each line $\pi_{\kappa(-i)} - \pi_{\kappa(i)}$, from (IV.54), $g(\pi_{\kappa(i)})/g(\pi_{\kappa(-i)})$ is either $f(\pi_{\kappa(-i)} \rightarrow \pi_{\kappa(i)})$ or $f(\kappa_{\pi(-i)} \rightarrow \kappa_{\pi(-i)})$, depending on whether $\pi(-i)$ or $\pi(i)$ is closer to the special point chosen in this component of G_K. In the first (second) case E^i and R_i (E^{-i} and R_{-i}) are considered *forward* in the absolute sense, and E^{-i} and R_{-i} (E^i and R_i) are considered to be *reverse*. Henceforth the notation E^{-i} and R_{-i} always refers to reverse subnetworks and reactions. For each forward subnetwork E^i

$$\frac{g(\pi_{\kappa(i)})}{g(\pi_{\kappa(-i)})} = f(\pi_{\kappa(-i)} \rightarrow \pi_{\kappa(i)}) = \frac{j_{-i}}{j_i} \tag{IV.60}$$

Define the n-tuple $\boldsymbol{\rho}$ by $\rho_i \equiv g(\pi_p)$, where p is determined by $\mathrm{X}_i \in \mathrm{K}_p$. Then $\rho_i > 0$ for all $i \in [1,n]$. For $\mathbf{h} \in R_+^n$ the function

$$L_D(\zeta, \mathbf{h}, \mathbf{j}) \equiv \zeta^t(\operatorname{diag}\boldsymbol{\rho})(\operatorname{diag}\mathbf{h})\zeta \tag{IV.61}$$

is always positive and would be a Lyapunov function if $dL_D/dt < 0$ for all physically accessible $\boldsymbol{\zeta} \neq 0$. This trial Lyapunov function resembles (IV.1); so (IV.2) can be replaced with

$$\boldsymbol{\zeta}^t \mathsf{R} \boldsymbol{\zeta} = \sum_{i=1}^{r} j_i \boldsymbol{\zeta}^t (\operatorname{diag} \mathbf{h}) \left[\mathsf{S}^{(i)t} (\operatorname{diag} \boldsymbol{\rho}) + (\operatorname{diag} \boldsymbol{\rho}) \mathsf{S}^{(i)} \right] (\operatorname{diag} \mathbf{h}) \boldsymbol{\zeta} \quad \text{(IV.62)}$$

The steady state is linearly asymptotically stable if $\boldsymbol{\zeta}^t \mathsf{R} \boldsymbol{\zeta} > 0$ for all physically accessible $\boldsymbol{\zeta} \neq 0$. Note that the two terms inside the summation are equal because the matrices are the transpose of each other. Henceforth we treat only the second term.

The terms corresponding to an internal forward reaction R_i and its reverse R_{-i} can be paired to become

$$2\boldsymbol{\zeta}^t (\operatorname{diag} \mathbf{h})(\operatorname{diag} \boldsymbol{\rho}) \left[j_i \mathsf{S}^{(i)} (\operatorname{diag} \boldsymbol{\rho})^{-1} + j_{-i} \mathsf{S}^{(-i)} (\operatorname{diag} \boldsymbol{\rho})^{-1} \right] (\operatorname{diag} \boldsymbol{\rho})(\operatorname{diag} \mathbf{h}) \boldsymbol{\zeta} \quad \text{(IV.63)}$$

From (III.42) $\mathbf{S}^{(i)} = \boldsymbol{\nu}_i \boldsymbol{\kappa}_i^t$. Since all components of $\boldsymbol{\kappa}_i$ vanish except those whose corresponding species are in the knot $\mathrm{K}_{\kappa(-i)}$, and since the corresponding components of $\boldsymbol{\rho}$ all equal $g(\pi_{\kappa(-i)})$, then $\boldsymbol{\kappa}_i^t (\operatorname{diag} \boldsymbol{\rho})^{-1} = \boldsymbol{\kappa}_i^t / g(\pi_{\kappa(-i)})$. With this simplification, the inner brackets of (IV.63) become

$$-\frac{j_i}{g(\pi_{\kappa(-i)})} \boldsymbol{\nu}_i \boldsymbol{\kappa}_i^t - \frac{j_{-i}}{g(\pi_{\kappa(i)})} \boldsymbol{\nu}_{-i} \boldsymbol{\kappa}_{-i}^t$$

Since the coefficients are equal by (IV.60), this expression may be factored as

$$-\frac{j_i}{g(\pi_{\kappa(-i)})} \boldsymbol{\nu}_i (\boldsymbol{\kappa}_i^t - \boldsymbol{\kappa}_{-i}^t)$$

where we have used $\boldsymbol{\nu}_{-i} = -\boldsymbol{\nu}_i$. The last factor is simply $-\boldsymbol{\nu}_i^t$; hence (IV.63) becomes $2 j_i \beta_i^2 / g(\pi_{\kappa(-i)})$, where the scalar β_i is defined by

$$\beta_i \equiv \boldsymbol{\nu}_i^t (\operatorname{diag} \boldsymbol{\rho})(\operatorname{diag} \mathbf{h}) \boldsymbol{\zeta} \quad \text{(IV.64)}$$

We now show that the terms corresponding to any pair of external reactions have the same form. If the forward reaction R_i is $\mathbf{y}_j \to \mathbf{0}$, then $\boldsymbol{\kappa}_i = -\boldsymbol{\nu}_i$ so $\mathsf{S}^{(i)} = \boldsymbol{\nu}_i \boldsymbol{\nu}_i^t$. The reverse reaction $\mathbf{0} \to \mathbf{y}_j$ has $\boldsymbol{\kappa}_{-i} = \mathbf{0}$, so $\mathsf{S}^{(-i)} = \mathsf{0}$. Then (IV.63) also becomes $2 j_i \beta_i^2 / g(\pi_{\kappa(-i)})$. Combining these results gives

$$\boldsymbol{\zeta}^t \mathsf{R} \boldsymbol{\zeta} = 2 \sum_{i=1}^{r/2} j_i \beta_i^2 / g(\pi_{\kappa(-i)}) \quad \text{(IV.65)}$$

where the sum is taken over forward reactions only. The terms are all nonnegative, so the reversible networks whose skeletons are represented by knot-tree graphs are stable.

To prove asymptotic stability consider (IV.65), and note that because $j_i > 0$ and $g(\pi_{\kappa(-i)}) > 0$, the only way that $\boldsymbol{\zeta}'\mathsf{R}\boldsymbol{\zeta}$ could vanish is if $\beta_i = 0$ for all i. Then from (IV.64) $(\operatorname{diag}\boldsymbol{\rho})(\operatorname{diag}\mathbf{h})\boldsymbol{\zeta}$ must be orthogonal to every vector $\boldsymbol{\nu}_i$. From Table I we see that these vectors span S_X, so $(\operatorname{diag}\boldsymbol{\rho})$ $(\operatorname{diag}\mathbf{h})\boldsymbol{\zeta}$ would have to be orthogonal to S_X. Recall that every physically accessible perturbation $\boldsymbol{\zeta} \in S_X$. Hence $\boldsymbol{\zeta}'\mathsf{R}\boldsymbol{\zeta}$ can only vanish if $(\operatorname{diag}\boldsymbol{\rho})$ $(\operatorname{diag}\mathbf{h})\boldsymbol{\zeta}$ is orthogonal to $\boldsymbol{\zeta}$. This implies $\boldsymbol{\zeta}'(\operatorname{diag}\boldsymbol{\rho})(\operatorname{diag}\mathbf{h})\boldsymbol{\zeta} = 0$. No solution of this equation exists other than $\boldsymbol{\zeta} = 0$ because $(\operatorname{diag}\boldsymbol{\rho})(\operatorname{diag}\mathbf{h})$ is a positive diagonal matrix. Hence $\boldsymbol{\zeta}'\mathsf{R}\boldsymbol{\zeta} > 0$ for all perturbations $\boldsymbol{\zeta} \neq 0$ that are compatible with the conservation matrix $\underline{\boldsymbol{\gamma}}$; thus these networks are asymptotically stable. We have now proved (IV.58).

Next we extend this theorem to networks with irreversible reactions. The external reactions can be divided into two types. Those like $\mathbf{0} \rightarrow \mathbf{y}_j$ are called *entrance reactions*, and those like $\mathbf{y}_i \rightarrow \mathbf{0}$ are called *exit reactions*. First we consider networks $N \in \mathfrak{N}_{MA} \cap \mathfrak{N}_{KTS}$ whose only irreversible reactions are exit reactions. We examine how dropping some entrance reactions affects the proof of (IV.58). The paragraph containing (IV.65) explains that the entrance reactions have $\mathsf{S}^{(-i)} = 0$, thus contribute nothing to $\boldsymbol{\zeta}'\mathsf{R}\boldsymbol{\zeta}$. Hence these networks are stable. Note that the column of $\underline{\boldsymbol{\nu}}$ corresponding to an entrance reaction is minus one times the column for the corresponding exit reaction. Hence dropping the entrance reaction does not change the subspace S_X spanned by the column of $\underline{\boldsymbol{\nu}}$. The argument of the preceding paragraph can still be used to conclude that for every $\boldsymbol{\zeta} \in S_X$, one of the scalars $\beta_i \neq 0$. Hence (IV.65) implies asymptotic stability. We have now proved that *every mass action knot-tree skeleton network whose only irreversible reactions are exit reactions is asymptotically stable.*

Irreversible entrance reactions cause terms to drop out of (IV.65). Hence the network is still stable. The asymptotic stability proof will fail if and only if it is possible to choose $\boldsymbol{\zeta} \in S_X$, $\boldsymbol{\zeta} \neq 0$, so that every β_i appearing in the remaining terms of $\boldsymbol{\zeta}'\mathsf{R}\boldsymbol{\zeta}$ vanishes. From (IV.64) $(\operatorname{diag}\boldsymbol{\rho})(\operatorname{diag}\mathbf{h})\boldsymbol{\zeta}$ must be orthogonal to all columns of $\underline{\boldsymbol{\nu}}$ that do not correspond to irreversible entrance reactions. Let this set of columns form the matrix $\underline{\boldsymbol{\nu}}^*$, and let $\boldsymbol{\zeta} = \underline{\boldsymbol{\nu}}\boldsymbol{\xi}$. The asymptotic stability proof breaks down if and only if the equation

$$\underline{\boldsymbol{\nu}}^{*\prime}(\operatorname{diag}\boldsymbol{\rho})(\operatorname{diag}\mathbf{h})\,\underline{\boldsymbol{\nu}}\boldsymbol{\xi} = 0$$

has a solution $\boldsymbol{\xi} \neq 0$. We have now proved that *every mass action knot-tree skeleton network whose skeleton reactions are all reversible is stable. It is asymptotically stable if*

$$\text{rank } \underline{\nu}^* = d \qquad \text{(IV.66)}$$

where $\underline{\nu}^*$ *is the matrix of the columns of* $\underline{\nu}$ *that do not correspond to irreversible entrance reactions.*

When the skeleton network of N has irreversible reactions, a different approach is needed. Since the skeleton is a tree, we may delete all irreversible reactions from the skeleton to obtain a reversible knot-tree network N_{RS}. If any reactions were deleted, the knot graph G_{RS} of N_{RS} would have more than one component. All components are trees. Construct a directed graph G_D whose points represent these reversible tree components, and whose directed lines represent the irreversible reactions of the original network N between those components. Since G_D is a tree, it is possible to number the points of G_D so that every line is directed from a point with a lower number to a point with a higher number. The point of G_D with the lowest number corresponds to a reversible tree component of G_{RS} with the property that no irreversible internal reactions of the original network are "entrance" reactions to this component. Hence the dynamics of the species in this component cannot be influenced by any events outside this component. Since this component is a reversible mass action knot-tree skeleton network whose only irreversible reactions are internal "exit" reactions and the true irreversible external reactions of N, the species in this component have a stable steady state. Whatever occurs elsewhere, they cannot evolve away from this state. Hence the "exit" reactions cannot evolve away from their constant steady-state velocities.

Now consider the point of G_D with the next lowest assigned number. This point corresponds to a component of G_{RS}, which is in $\mathfrak{N}_R \cap \mathfrak{N}_{KTS}$. It may be connected to other components of G_{RS} by irreversible internal reactions of N. These may be subdivided into those that are "entrances" and "exits" from this component. Now an "entrance" corresponds to an arrow on G_D that must be directed from a lower numbered component to a higher numbered component. The only lower component is the one discussed in the previous paragraph and it has a stable steady state. Hence the velocity of this internal "entrance" reaction cannot deviate from its constant steady-state values. Thus this internal "entrance" reaction behaves like an external entrance reaction. We now use the earlier result to conclude that the species in this component of G_{RS} have a stable steady state. As a consequence, all "exit" reactions from this component of G_{RS} to other components of G_{RS} have velocities that cannot spontaneously deviate from their constant steady-state values.

The argument in the preceding paragraph should now be repeated until it has been applied to every component of G_{RS}. We then arrive at the conclusion that the steady state of N is stable. Hence *every mass action knot-tree skeleton network is stable.*

The preceding proof may be generalized to a proof of asymptotic stability. For each component of G_{RS}, (IV.66) must hold when all the irreversible entrance reactions are taken into consideration. These entrance reactions must include the irreversible entrance reactions of N as well as the "entrance" reactions that are actually internal irreversible reactions between the components of G_{RS}.

The following example proves that these knot-graph tests for stability sometimes work where the mixing stability tests do not.

Example IV.6. The extreme network $N \in \mathfrak{N}_{MA}$, defined by $\square \to W \to X + Z$; $\square \to Y$; $X + Y \to \square$; $Y + Z \to \square$ has

$$S = \begin{pmatrix} 1 & 0 & 0 & 0 \\ -1 & 1 & 1 & 0 \\ 0 & 1 & 2 & 1 \\ -1 & 0 & 1 & 1 \end{pmatrix}$$

and then $\det(S^t + S) < 0$. Hence N is not mixing stable. Rank $\underline{\nu} = 4 = d = n$. Hence the conservation matrix $\underline{\gamma}$ has no rows; thus the network cannot be constrained mixing stable. The mixing stability tests have not helped decide the stability. The skeleton network has only one reaction $W \to X + Z$. Since the corresponding knot-graph is a tree, this network is stable. The knot graph of the reversible part of the skeleton G_{RS} has no reactions. Hence (IV.65) reads $\zeta^t R \zeta = 0$, for all $\zeta \in S_X$. This network is not asymptotically stable.

Example IV.7. The mass action network defined by $\square \to X + Z; Z \to Y$; $X + Y \to \square$ is a knot-tree skeleton network; hence is stable. If the reaction $\square \to X + Z$ is replaced by the autocatalytic reaction $X \to 2X + Z$, the network becomes an extreme current of the Oregonator, which is proved to be exponentially unstable in Example V.6.

The stability proof in the irreversible case used the condition that each component of G_{RS} is a reversible knot-tree skeleton network. It also used the tree property of G_D to deduce that the points of G_D could be numbered so that every arrow went from a point with a lower number to a point with a higher number. Sometimes G_D has this property when it is not a tree. If the points of G_D represent components of G_{RS} that are reversible knot-tree skeleton networks, we can still conclude that N is stable.

Example IV.8. $\square \to X_1$; $\square \to X_2$; $X_1 + X_2 \to Y_1$; $\square \to Y_2$; $Y_1 + Y_2 \to Z_1 + Z_2$; $Z_1 \to \square$; $Z_2 \to \square$; $X_1 + X_2 \to Z_1$. This network has three knots $X \equiv \{X_1, X_2\}$, $Y \equiv \{Y_1, Y_2\}$, and $Z \equiv \{Z_1, Z_2\}$. The knot graph of the skeleton is

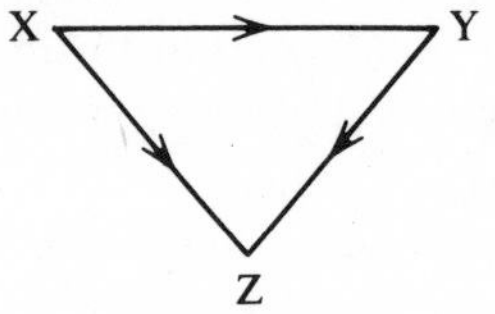

G_{RS} has three points and is completely disconnected; G_D resembles the accompanying diagram. If we number the vertices $X = 1$, $Y = 2$, and $Z = 3$, we can conclude that this network is stable.

This result can now be expressed as a theorem for "species-tree graph networks," provided we define the graph in the proper way. For each network N we define a *semidirected species graph* G_S, which has one point for each species (no point represents Ø), and lines that may or may not be directed according to the following rules. Lines will be directed when the corresponding reactions are irreversible; reversible reactions will correspond to undirected lines. Lines will connect every species on one side of a reaction with every species on the other side of the reaction. For example, the reversible reaction W + X⇄Y + Z will produce undirected lines between the pairs WY, WZ, XY, and XZ. The irreversible reaction W + X →Y + Z will produce four directed lines, one from the first species to the second species for each of the four previously mentioned pairs of species. Reactions containing □ on either side do not produce any lines on the species graph, whether the reaction is reversible or irreversible.

A *directed species cycle* is a closed non-self-intersecting path on the species graph that passes from species to species either along undirected arrows or along directed arrows with the proper orientation. No arrow can occur more than once in the cycle. Hence reversible reactions do not produce a species cycle because the single undirected arrow representing the reaction cannot be traversed in both directions in a cycle. A semidirected species graph is a *tree* if it has no directed species cycles.

Let N be a network whose semidirected species graph is a tree. We now prove that N is linearly stable. Let G_{RS} be the knot graph of the reversible reactions of the skeleton of N. If G_{RS} is not a tree, there must be cycles in G_{RS}. These cycles imply the existence of cycles around sequences of undirected lines of the species graph G_S contrary to hypothesis. Hence G_{RS} is a tree. Each component of G_{RS} must be a tree. We now set up the graph G_D as before. Each point of G_D represents all the knots in a tree component of G_{RS}. Each line of G_D represents an irreversible reaction between two knots. We now argue that if there is a directed cycle on G_D there must be a directed cycle in the species graph G_S. The argument is as follows.

Assume that we have a directed cycle along the irreversible reactions of G_D. There are corresponding directed arrows of G_S. When two arrows approach and leave a point ($\rightarrow \pi_1 \rightarrow$) on G_D, the corresponding directed arrows of G_S approach one species and leave a possibly different species (→X · · · Y→). To have a cycle on G_S, we must show that when these species are different, it is possible to pass between them along a path of undirected arrows (i.e., X⇄ · · · ⇄Y). Now these species may or may not lie in the same knot of G_{RS}. Consider the case when they do lie in the same knot. We use an argument similar to that at the end of Section IV.B to conclude that a species path must exist between the species. We know that there is an interactant overlap path between the species because, by definition, any two species in the same knot are linked by an interactant overlap path. We can also show there is a species-graph path between any pair of species in an interactant. Consider any two species U and V in the same interactant. Since the knots of G_{RS} are determined only by the reversible reactions not involving $\square$, we know there is a reversible reaction having U and V on one side and some other species on the other side (e.g., $U + V \rightleftarrows T$). Hence the species graph contains an undirected path from between the species U and V (in the example, via the arrows $U \to T \to V$). Using such paths between species in the same interactant, we may trace a path on the species graph between any two species of the same knot (i.e., from X to Y).

Now consider the more general case of the species (X and Y) belonging to different knots of G_{RS}. Since they correspond to the same point of G_D, they must belong to knots in the same tree component of G_{RS}. Every pair of knots in this tree component is connected by lines of G_{RS}. These lines correspond to undirected lines of G_S. Hence we can pass between the species (X and Y) along a species path in G_S by passing along an interactant path within knots and by passing along reversible reactions between knots. When these links are added to the irreversible reaction steps on G_S, we obtain a directed cycle on G_S as a consequence of the directed cycle on G_D.

It now follows that whenever G_S is a tree, G_D is a tree. The points of G_D can then be numbered as in Example IV.8, so that every arrow of G_D passes from a lower to a higher number. From this property it follows that N is stable. We have now proved

$$\mathfrak{N}_s^L \supset \mathfrak{N}_{SST} \tag{IV.67}$$

where $\mathfrak{N}_{SST}$ is the set of networks whose semidirected species graphs are trees.

F. Thermodynamics and Lyapunov Functions

Thermodynamics proves the stability of equilibrium states with respect to possible energy density and concentration fluctuations in a closed system, by showing that such fluctuations would decrease the entropy density in violation of the second law. An extension of this idea to nonequilibrium systems was initiated by Gibbs and worked out in detail by Glansdorff and Prigogine.[64] The two most important results are the minimum entropy production theorem and the Glansdorff-Prigogine stability condition.

One of the main themes of this chapter is the great mathematical simplification that occurs when the state parameters are $(\mathbf{h},\mathbf{j})$ instead of the rate constants or the equivalent thermodynamic parameters. Note that $\mathbf{j}$ represents the point $\mathbf{E}\mathbf{j}$ in reaction velocity space. This section shows how the minimum entropy production theorem and the Glansdorff-Prigogine stability criterion are simplified by viewing them as theorems in reaction velocity space.

The dynamical state of the system for fixed $(\mathbf{h},\mathbf{j})$ is specified by $\mathbf{X} \in \Pi_X(\mathbf{C})$, where $\mathbf{C}=\underline{\gamma}\mathbf{h}^{-1}$ by (II.5) $(\mathbf{h}^{-1}\equiv(h_1^{-1},h_2^{-1},\ldots))$. As in Section II.E, let the first $r/2$ reactions be forward reactions, and the last $r/2$ reactions be the corresponding reverse reactions. Then $\boldsymbol{\nu}_i+\boldsymbol{\nu}_{i+r/2}=0$ for $1\leqslant i\leqslant r/2$. The subscript "rev" on a vector or array means that forward and reverse reactions have been interchanged. Thus

$$\underline{\boldsymbol{\nu}}+\underline{\boldsymbol{\nu}}_{\text{rev}}=0 \tag{IV.68}$$

$$\mathbf{v}_{\text{rev}}=\begin{pmatrix} 0 & \mathrm{I} \\ \mathrm{I} & 0 \end{pmatrix}\mathbf{v} \tag{IV.69}$$

hence

$$\underline{\boldsymbol{\nu}}\mathbf{v}=\frac{(\underline{\boldsymbol{\nu}}\mathbf{v}+\boldsymbol{\nu}_{\text{rev}}\mathbf{v}_{\text{rev}})}{2}=\frac{\underline{\boldsymbol{\nu}}(\mathbf{v}-\mathbf{v}_{\text{rev}})}{2} \tag{IV.70}$$

The condition for mass action kinetics is

$$\underline{\boldsymbol{\kappa}}-\underline{\boldsymbol{\kappa}}_{\text{rev}}=-\underline{\boldsymbol{\nu}} \tag{IV.71}$$

All thermodynamic properties of the system are determined once a chemical potential has been assigned to all internal and external species. Choosing

$$\beta\boldsymbol{\mu}^*=\beta\boldsymbol{\mu}_0^*+\ln\mathbf{X}^* \tag{IV.72}$$

gives the system the thermodynamic properties of an *ideal solution*. Vectors and arrays with an asterisk have been extended to include the external species; that is, $\boldsymbol{\mu}^*\in R^N$, where N is the total number of species (internal plus external). From the *Gibbs relation* between the *entropy density* s and the *internal energy density* e,

$$ds=\beta de-\beta\boldsymbol{\mu}^{*t}d\mathbf{X}^* \tag{IV.73}$$

we obtain an expression for the *irreversible entropy production density from chemical reactions*

$$\sigma \equiv \left(\frac{ds}{dt}\right)_e = -\beta\boldsymbol{\mu}^{*t}\dot{\mathbf{X}}^* = -\beta\boldsymbol{\mu}^{*t}\underline{\boldsymbol{\nu}}^*\mathbf{v} = -\frac{\beta}{2}\boldsymbol{\mu}^{*t}\underline{\boldsymbol{\nu}}^*(\mathbf{v}-\mathbf{v}_{\text{rev}}) \quad \text{(IV.74)}$$

The last two expressions in (IV.74) come from (II.2) and (IV.70). Let $\mathbf{k}^* \in R_+^r$ be the rate constants when the external species are included. The logarithm of the rate law (I.1) is then

$$\ln\mathbf{v} = \ln\mathbf{k}^* + \boldsymbol{\kappa}^{*t}\ln\mathbf{X}^* \quad \text{(IV.75)}$$

Interchanging forward and reverse reactions and subtracting gives

$$\ln\mathbf{v}/\mathbf{v}_{\text{rev}} = \ln\mathbf{k}^*/\mathbf{k}^*_{\text{rev}} - \underline{\boldsymbol{\nu}}^{*t}\ln\mathbf{X}$$

by (IV.71). Multiplying (IV.72) by $\underline{\boldsymbol{\nu}}^{*t}$ and adding to the equation above gives

$$\beta\,\underline{\boldsymbol{\nu}}^{*t}(\boldsymbol{\mu}^* - \boldsymbol{\mu}_0^*) = \ln\left(\frac{\mathbf{k}^*}{\mathbf{k}^*_{\text{rev}}}\right) - \ln\left(\frac{\mathbf{v}}{\mathbf{v}_{\text{rev}}}\right)$$

At equilibrium $\mathbf{v} = \mathbf{v}_{\text{rev}}$ and $\underline{\boldsymbol{\nu}}^*\boldsymbol{\mu}^* = 0$ (this is the condition $\Delta G = 0$). Hence the constants in this equation satisfy $-\beta\underline{\boldsymbol{\nu}}^{*t}\boldsymbol{\mu}_0^* = \ln(\mathbf{k}^*/\mathbf{k}^*_{\text{rev}})$, and we are left with

$$\beta\,\underline{\boldsymbol{\nu}}^{*t}\boldsymbol{\mu}^* = -\ln\frac{\mathbf{v}}{\mathbf{v}_{\text{rev}}} \quad \text{(IV.76)}$$

The entropy production density (IV.74) now becomes

$$\sigma = \frac{(\ln\mathbf{v}/\mathbf{v}_{\text{rev}})^t(\mathbf{v}-\mathbf{v}_{\text{rev}})}{2} \quad \text{(IV.77)}$$

Note that σ is completely determined by $\mathbf{v}$. This means that at the steady state $(\mathbf{h},\mathbf{j})$, σ is determined by $\mathbf{j}$ alone. The fact that $\mathbf{h}$ does not affect σ implies that σ cannot be related to the steady-state stability in general, because in general the stability problem depends on $\mathbf{h}$.

It is useful to have a feel for (IV.77). The entropy production σ vanishes only on $\mathcal{C}_E$, where $\mathbf{v} = \mathbf{v}_{\text{rev}}$. It diverges as $\mathbf{v}$ approaches the boundary of R_+^r, except where $\mathcal{C}_E$ meets the boundary (Fig. 3). Expanding the logarithm in (IV.77) in a Taylor series about any $\mathbf{v}_E \in \mathcal{C}_E$ gives

$$\sigma = \frac{(\mathbf{v}-\mathbf{v}_{\text{rev}})^t(\operatorname{diag}\mathbf{v}_E)^{-1}(\mathbf{v}-\mathbf{v}_{\text{rev}})}{2} \quad \text{(IV.78)}$$

For fixed $(\mathbf{h},\mathbf{j})$ and any $\mathbf{X} \in \Pi_X(\mathbf{C})$, the kth reaction velocity is given by the function

$$w_k(\mathbf{h},\mathbf{j},\mathbf{X}) = (\mathsf{E}\mathbf{j})_k \prod_{i=1}^{n}(h_i X_i)^{\kappa_{ik}} \quad \text{(IV.79)}$$

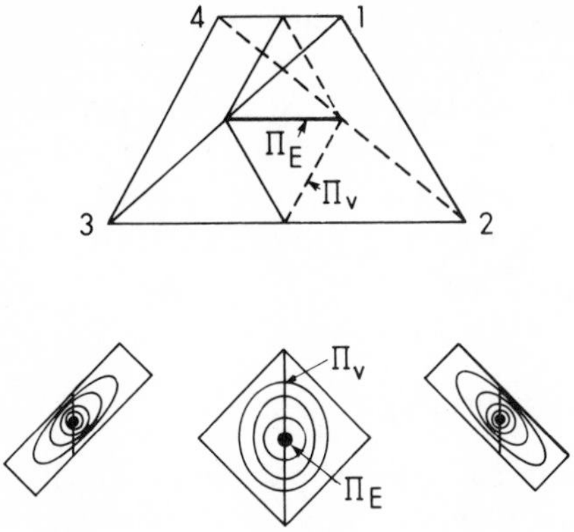

Fig. 3. The contours of constant σ for the reaction system of Example II.4, where $\underline{\nu}=(1 \quad 1 \quad -1 \quad -1)$, may be visualized in the tetrahedron $\mathbf{e}'\mathbf{v}=1$ shown; Π_E is a diagonal of the square Π_ν. The contours are sketched in three planes, all orthogonal to Π_E.

This expression is obtained from (II.31) using (II.36), (II.35), and (II.18). The set of restricted accessible reaction velocities is then

$$A(\mathbf{h},\mathbf{j}) \equiv \{\mathbf{w}(\mathbf{h},\mathbf{j},\mathbf{X}) \mid \mathbf{X} \in \Pi_X(\mathbf{C})\} \subset R_+^r$$

The dimension of this curved manifold is the same as the dimension of $\Pi_X(\mathbf{C})$, namely d. Let $\mathbf{w}^\sigma(\mathbf{h},\mathbf{j})$ be the point in $A(\mathbf{h},\mathbf{j})$ where σ attains its restricted absolute minimum; then the set of restricted minima is

$$M^\sigma = \{(\mathbf{h},\mathbf{w}^\sigma(\mathbf{h},\mathbf{j})) \mid \mathbf{h} \in R_+^n, \mathbf{j} \in R_+^f\} \subset R_+^n \times R_+^r$$

If $A(\mathbf{h},\mathbf{j})$ passes through any point $\mathbf{v}_E \in \mathcal{C}_E$, since $\sigma=0$ on $\mathcal{C}_E$ and $\sigma>0$ elsewhere, $\mathbf{w}^\sigma(\mathbf{h},\mathbf{j})=\mathbf{v}_E$. Since every point $v_E \in \mathcal{C}_E$ lies on some $A(\mathbf{h},\mathbf{j})$ having $w^\sigma(\mathbf{h},\mathbf{j})=v_E \in M^\sigma$, then $\mathcal{C}_E \subset M^\sigma$. These sets are illustrated in Fig. 4.

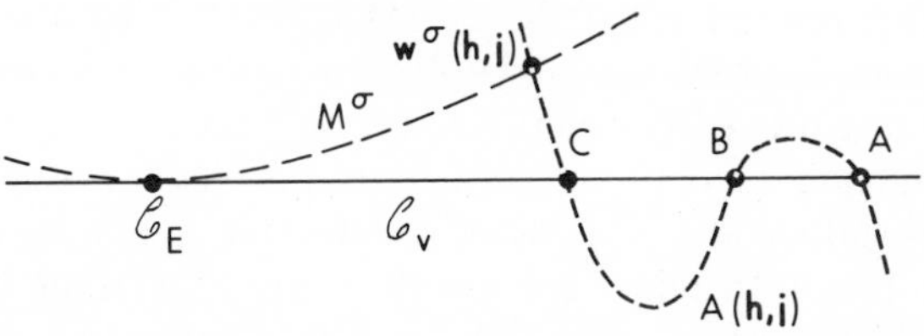

Fig. 4. For fixed $\mathbf{h} \in R_+^n$, the steady state $\mathbf{j}$ is the point A in $\mathcal{C}_v$. The line consisting of short dashes is the set of accessible states $A(\mathbf{h},\mathbf{j})$. There are two other accessible steady states at points B and C. The absolute minimum of σ on $A(\mathbf{h},\mathbf{j})$ determines the point $\mathbf{w}^\sigma(\mathbf{h},\mathbf{j})$ on M^σ, which is tangent to $\mathcal{C}_v$ at $\mathcal{C}_E$.

The key to the minimum entropy production theorem is the *tangentiality* of M^σ and $\mathcal{C}_v$ when they intersect in $\mathcal{C}_E$. To prove this, consider (IV.79) as a mapping $w: R_+^n \to R_+^r$, $\mathbf{X} \mapsto \mathbf{w}(\mathbf{h},\mathbf{j},\mathbf{X})$. (We keep $\mathbf{h},\mathbf{j}$ fixed.) The *derivative* of the mapping is the linear mapping Dw that approximates (is tangent to) w. This definition of a derivative is used in differential topology (see Chillingworth[65]). We represent Dw by the matrix $(\mathbf{Dw})$, which maps a small change in $\mathbf{X}$ into the resulting small change in $\mathbf{w}$. We calculate it by differentiating (IV.79) at the point $\mathbf{v} = \mathbf{w}(\mathbf{h},\mathbf{j},\mathbf{X})$ to get

$$(\mathbf{Dw}) = (\mathrm{diag}\,\mathbf{v})\,\underline{\boldsymbol{\kappa}}^t(\mathrm{diag}\,\mathbf{X})^{-1} \tag{IV.80}$$

The columns of $(\mathbf{Dw})$ span a linear subspace that is parallel to $A(\mathbf{h},\mathbf{j})$ at $\mathbf{w}(\mathbf{h},\mathbf{j},\mathbf{X})$. Similarly we regard σ as the mapping $\sigma: R_+^r \to R$, $\mathbf{v} \mapsto \sigma(\mathbf{v})$. Its derivative near equilibrium is obtained from (IV.78)

$$(\mathbf{D}\boldsymbol{\sigma}) = (\mathbf{v} - \mathbf{v}_{\mathrm{rev}})^t(\mathrm{diag}\,\mathbf{v}_E)^{-1} \tag{IV.81}$$

(This row vector is the transpose of $\boldsymbol{\nabla}\sigma$, the gradient of σ.) The condition that σ has a local extremum on $A(\mathbf{h},\mathbf{j})$ is that $(\mathbf{D}\boldsymbol{\sigma})$ is orthogonal to all columns of $(\mathbf{Dw})$, that is, $(\mathbf{D}\boldsymbol{\sigma})(\mathbf{Dw}) = 0$. Hence

$$(\mathbf{v} - \mathbf{v}_{\mathrm{rev}})^t(\mathrm{diag}\,\mathbf{v}_E)^{-1}(\mathrm{diag}\,\mathbf{v})\,\underline{\boldsymbol{\kappa}}^t(\mathrm{diag}\,\mathbf{X})^{-1} = 0$$

Multiply this on the right by $(\mathrm{diag}\,\mathbf{X})$ and let $\mathbf{q}^t(\mathbf{v},\mathbf{v}_E)$ be the row vector that remains on the left. The set

$$\{\mathbf{v} \in R_+^r \mid \mathbf{q}(\mathbf{v},\mathbf{v}_E) = 0\} \tag{IV.82}$$

is an approximation to M^σ that is valid near $\mathbf{v}_E$. Now consider the mapping $q: R_+^r \to R_+^n$, $\mathbf{v} \mapsto \mathbf{q}(\mathbf{v},\mathbf{v}_E)$. Using (IV.71), we evaluated Dq at $\mathbf{v}_E$ to obtain

$$(\mathbf{Dq}) = -\underline{\boldsymbol{\nu}}^t \tag{IV.83}$$

When the set (IV.82) meets $\mathcal{C}_E$ it must be orthogonal to the columns of $(\mathbf{Dq})$, that is, the rows of $\underline{\boldsymbol{\nu}}$, that is, S_ξ. Thus (IV.82) is parallel to S_v at $\mathbf{v}_E$, and so M^σ is tangent to $\mathcal{C}_v$.

This result just proved is exact. I prefer to think of it as the minimum entropy production theorem because it is the key to a more physical (approximate) statement that is usually called the minimum entropy production "theorem." Imagine a network where, for some $\mathbf{Ej}$ near $\mathcal{C}_E$, $\mathbf{w}^\sigma(\mathbf{h},\mathbf{j}) = \mathbf{Ej}$. Steady states sufficiently near equilibrium are asymptotically stable; hence as $\mathbf{v}$ approaches $\mathbf{Ej}$, σ approaches its minimum $\sigma_{\min}$. Thus $\sigma - \sigma_{\min}$ is a Lyapunov function. In the general case, when $\mathbf{Ej}$ is near $\mathcal{C}_E$, $\mathbf{w}^\sigma(\mathbf{h},\mathbf{j})$ is near $\mathbf{Ej}$, and the minimum of σ is close to steady state. If

$\mathbf{w}^{\sigma}(\mathbf{h},\mathbf{j}) \neq \mathsf{Ej}$, $\sigma - \sigma_{\min}$ cannot be a Lyapunov function, because if $\mathbf{v} = \mathbf{w}^{\sigma}(\mathbf{h}, \mathbf{j})$ initially, it must evolve to Ej and σ will increase.

However we can consider $\sigma - \sigma_{\min}$ as a Lyapunov function in the following more practical sense. Since M^{σ} is tangent to $\mathcal{C}_v$, the region where $\dot{\sigma} > 0$ becomes smaller as Ej approaches $\mathcal{C}_E$. Sufficently close to equilibrium, this region will be indistinguishable from steady state to experimental accuracy. If $\dot{\sigma} < 0$ in a larger region containing this small region, we can infer that the system must evolve to what is, practically speaking, the steady state. Then $\sigma - \sigma_{\min}$ is a Lyapunov function in this practical sense.

This practical form of the theorem can be proved by making approximations in the right places. For a short proof using the thermodynamic parameters see Nicolis and Prigogine.[2] We now prove it in reaction velocity space. From (IV.80) evaluated at equilibrium (this is a key approximation) instead of at the system's actual state, (II.2) and (IV.70) give

$$\dot{\mathbf{v}} = (\mathbf{Dw})\dot{\mathbf{X}} = \frac{(\operatorname{diag}\mathbf{v}_E)\,\underline{\boldsymbol{\kappa}}^t(\operatorname{diag}\mathbf{h})\,\underline{\boldsymbol{\nu}}(\mathbf{v} - \mathbf{v}_{\text{rev}})}{2} \tag{IV.84}$$

Interchanging forward and reverse reactions and using (IV.71) yields

$$\dot{\mathbf{v}} - \dot{\mathbf{v}}_{\text{rev}} = -\frac{(\operatorname{diag}\mathbf{v}_E)\,\underline{\boldsymbol{\nu}}^t(\operatorname{diag}\mathbf{h})\,\underline{\boldsymbol{\nu}}(\mathbf{v} - \mathbf{v}_{\text{rev}})}{2}$$

Differentiating (IV.78) and using (IV.81) then gives

$$\dot{\sigma} = (\mathbf{D}\boldsymbol{\sigma})(\dot{\mathbf{v}} - \dot{\mathbf{v}}_{\text{rev}}) = -\frac{(\mathbf{v} - \mathbf{v}_{\text{rev}})^t\,\underline{\boldsymbol{\nu}}^t(\operatorname{diag}\mathbf{h})\,\underline{\boldsymbol{\nu}}(\mathbf{v} - \mathbf{v}_{\text{rev}})}{2}$$

However $\dot{\mathbf{X}} = \frac{1}{2}\underline{\boldsymbol{\nu}}(\mathbf{v} - \mathbf{v}_{\text{rev}})$, so this becomes

$$\dot{\sigma} = -2\dot{\mathbf{X}}^t(\operatorname{diag}\mathbf{h})\dot{\mathbf{X}} \tag{IV.85}$$

The right-hand side is negative definite; thus where this equation is correct, $\dot{\sigma} < 0$ if $\dot{\mathbf{X}} \neq 0$.

If (IV.85) contained all the low-order terms in $\dot{\sigma}$ as a Taylor expansion in $\dot{\mathbf{X}}$, the minimum of σ would have to coincide exactly with the steady state. Since M^{σ} is very smooth and does not concide with $\mathcal{C}_v$ far from equilibrium, it is unlikely that $(M^{\sigma} \cap \mathcal{C}_v)\backslash\mathcal{C}_E$ contains any points near $\mathcal{C}_E$. That is, $\mathbf{w}^{\sigma}(\mathbf{h},\mathbf{j}) \neq \mathsf{Ej}$ except at equilibrium. Hence (IV.85) must be missing terms of the form $\sigma_0 + \mathbf{a}^t\dot{\mathbf{X}}$ that become dominant as $\dot{\mathbf{X}} \to 0$. These terms will cause σ to increase as the system evolves from $\mathbf{w}^{\sigma}(\mathbf{h},\mathbf{j})$ to Ej. Thus (IV.85) can be a valid approximation for $\dot{\sigma}$ only when $\dot{\mathbf{X}}$ is not too small (and also not too large). Near equilibrium this equation is valid over a considerable range of $\dot{\mathbf{X}}$. By considering it in the region where it is valid, we conclude that $\dot{\mathbf{X}}$ decreases until the missing constant and linear terms are significant. Sufficently close to equilibrium, this region is too small to be measured, and $\sigma - \sigma_0$ can be considered to be a Lyapunov function.

The general expression for entropy production is

$$\sigma = \chi' \mathbf{J}_N \tag{IV.86}$$

where χ is called a *thermodynamic force* and $\mathbf{J}_N$ is the nonequilibrium *thermodynamic flux*, as defined by

$$\chi \equiv \frac{\mathsf{F}_N^t \ln \mathbf{v}/\mathbf{v}_{\text{rev}}}{2} \tag{IV.87}$$

$$\mathbf{J}_N \equiv \frac{\mathsf{F}_N^t(\mathbf{v} - \mathbf{v}_{\text{rev}})}{2} \tag{IV.88}$$

This expression for $\mathbf{J}_N$ is equivalent to that of Section II.E. By using (IV.69) to verify that $\mathsf{F}_N\mathsf{F}_N^t(\mathbf{v} - \mathbf{v}_{\text{rev}}) = 2(\mathbf{v} - \mathbf{v}_{\text{rev}})$, one may easily verify that (IV.86) is equivalent to (IV.77).

Let χ° and $\mathbf{J}_N^\circ$ be steady-state quantities and let $\delta\chi$ and $\delta\mathbf{J}_N$ be deviations from steady state. Then σ in the general state may be expanded as

$$\sigma = \chi^{\circ t}\mathbf{J}_N^\circ + (\chi^{\circ t}\delta\mathbf{J}_N + \delta\chi^t J_N^\circ) + \delta\chi^t\delta\mathbf{J}_N \tag{IV.89}$$

The last term on the right is called the (density of) *excess entropy production* and is usually written $\delta^2\sigma$. We now express this quantity in terms of familiar matrices. Linearizing (differentiating) the mappings (IV.87) and (IV.88) gives the responses $\delta\chi$ and $\delta\mathbf{J}_N$ to a small change $\delta\mathbf{v}$:

$$\delta\chi = \mathsf{F}_N^t(\operatorname{diag}\mathbf{v})^{-1}\delta\mathbf{v}$$

$$\delta\mathbf{J}_N = \mathsf{F}_N^t\delta\mathbf{v}$$

Then

$$\delta^2\sigma = \delta\mathbf{v}(\operatorname{diag}\mathbf{v})^{-1}\mathsf{F}_N\mathsf{F}_N^t\delta\mathbf{v}$$

Since $\delta\mathbf{v} = (\mathbf{Dw})\delta\chi$, we may use (IV.80) at steady state to obtain

$$\delta^2\sigma = \delta\chi^t(\operatorname{diag}\mathbf{h})\,\underline{\boldsymbol{\kappa}}\mathsf{F}_N\mathsf{F}_N^t(\operatorname{diag}\mathsf{E}\mathbf{j})\,\underline{\boldsymbol{\kappa}}^t(\operatorname{diag}\mathbf{h})\delta\chi$$

Note that (IV.71) may be written $\underline{\boldsymbol{\kappa}}\mathsf{F}_N\mathsf{F}_N^t = -\underline{\boldsymbol{\nu}}$. Then taking the summation over the components of $\operatorname{diag}\mathsf{E}\mathbf{j}$ outside gives

$$\delta^2\sigma = \sum_{i=1}^{f} j_i\delta\chi^t(\operatorname{diag}\mathbf{h})\big(-\underline{\boldsymbol{\nu}}(\operatorname{diag}\mathbf{E}_i)\,\underline{\boldsymbol{\kappa}}^t\big)(\operatorname{diag}\mathbf{h})\delta\chi \tag{IV.90}$$

which, by (III.9) and (IV.2) is $\frac{1}{2}\delta\chi'\mathsf{R}(\mathbf{h},\mathbf{j})\delta\chi$. The matrix $\mathsf{R}(\mathbf{h},\mathbf{j})$ is defined by (III.68) with $\mathsf{Q} = (\text{diag}\,\mathbf{h})$. Hence $\delta^2\sigma > 0$ if and only if $\delta\chi'\mathsf{R}(\mathbf{h},\mathbf{j})\delta\chi > 0$.

We proved in Section IV.A that $L_M(\zeta,\mathbf{h})$, defined by (IV.1), is a Lyapunov function if and only if $\zeta' R(\mathbf{h},\mathbf{j})\zeta > 0$ for all $\zeta \in S_X \zeta \neq 0$. Hence $L_M(\delta\mathbf{X},\mathbf{h})$ is a Lyapunov function if and only if

$$\delta^2\sigma > 0 \qquad \text{(IV.91)}$$

for all sufficiently small deviations from steady state $\delta\mathbf{X} \in S_X$, $\delta\mathbf{X} \neq 0$. The steady state is therefore stable whenever (IV.91) is satisfied for all small deviations from steady state. This is the *Glansdorff-Prigogine stability criterion*.

Furthermore, since L_M has the mixing property discussed in Section IV.A, the network is mixing stable if and only if (IV.91) holds for all $(\mathbf{h},\mathbf{j}) \in D_R$. With this connection between excess entropy production and mixing stability established, some networks satisfying $\delta^2\sigma \geqslant 0$ can be found from the theory developed in Sections IV.A to IV.E. The network in Example IV.6 is asymptotically stable but cannot satisfy $\delta^2\sigma \geqslant 0$ everywhere because it is not mixing stable. Hence the Glansdorff-Prigogine criterion is a sufficient but not a necessary condition for stability. Complex balanced networks satisfy $\delta^2\sigma \geqslant 0$ everywhere.

G. Summary of the Lyapunov Functions and Some Speculations on the Road Ahead

The most important Lyapunov function for chemical networks that have been linearized about steady states is

$$L_M(\zeta,\mathbf{h}) \equiv \zeta'(\text{diag}\,\mathbf{h})\zeta$$

Complex balanced networks have this Lyapunov function. The Glansdorff-Prigogine criterion is a thermodynamic way of expressing necessary and sufficient conditions for the linearized system to have this Lyapunov function near a particular steady state. Since L_M does not depend on $\mathbf{j}$, L_M is potentially a Lyapunov function for all mixtures of networks having this Lyapunov function. When making mixtures of networks it is important to note that the conservation conditions are sometimes weakened. Networks that can tolerate complete removal of the conservation conditions without changing the properties of L_M are called mixing (asymptotically) stable. These networks may be mixed freely without considering the conservation condition constraints. The mixtures are stable, and, if certain other conditions are met, they are asymptotically stable. Other networks can tolerate the removal of some conservation constraints and not others. Hence the concept of "constrained mixing stability with respect to $\underline{\gamma}$" was developed in Section IV.A. If a network with conservation matrix $\underline{\gamma}$ is a mixture of extreme subnetworks that are constrained mixing stable with respect to $\underline{\gamma}$, the network is stable. This idea was extended to asymptotic stability in Section IV.A. The tests for mixing stability are simple and involve only the calculation of the eigenvalues of $\mathbf{S}_{\text{sym}}^{(1)}$. We do not yet understand what the mixing stability conditions say about network topology.

Another useful Lyapunov function is

$$L_D(\boldsymbol{\zeta}, \mathbf{h}, \mathbf{j}) \equiv \boldsymbol{\zeta}^t(\operatorname{diag}\boldsymbol{\rho}(\mathbf{j}))(\operatorname{diag}\mathbf{h})\boldsymbol{\zeta}$$

where $\boldsymbol{\rho}(\mathbf{j})$ is a vector function of $\mathbf{j}$; L_D was shown to be a Lyapunov function for all reversible mass action networks whose skeleton networks have knot graphs that are trees. The vector function $\boldsymbol{\rho}(\mathbf{j})$ was determined by the tree structure of the knot graph. Example IV.6 proved that this function is sometimes a Lyapunov function when L_M is not. Then we proved stability using L_D for all networks whose semidirected species graphs are trees. This stability proof extends to an asymptotic stability proof if (IV.66) holds. L_D should also work for other networks, provided $\boldsymbol{\rho}(\mathbf{j})$ is suitably defined.

Are there many asymptotically stable networks that require some other types of Lyapunov functions than L_M or L_D to prove asymptotic stability? I think the answer may well be "no," and I will support this answer by showing that when L_M does not work, instability is likely. In Example IV.5 we saw that 7 out of the 11 extreme subnetworks of the reversible Oregonator (with $f = 1$) were mixing stable. The two nonmixing ones were associated with instability. Thus L_M was capable of proving stability everywhere possible in a highly autocatalytic oscillatory network. In Example IV.2, we saw that the irreversible Michaelis-Menten mechanism is constrained mixing stable (and asymptotically stable) with respect to its own conservation matrix, but it is not mixing stable. Thus the conservation condition for enzyme $\mathrm{E} + \mathrm{F} = C$ assures asymptotic stability. However if this condition is violated by adding in other reactions, stability can be no longer deduced by considering the Michaelis-Menten subnetwork and using L_M. The question is whether there is a better Lyapunov function than L_M or whether instability occurs. To see that the latter is the case, note that when the reactions $\mathrm{E} + 2\mathrm{P} \rightleftarrows G$ are added, the network becomes unstable. This will be proved later. The instability occurs partly because the conservation condition $E + F = C$ is replaced with $E + F + G = C$. Note that if the Michaelis-Menten step had been mixing stable, we could have concluded that the new network is mixing stable because the new reactions form a complex balanced equilibrium extreme subnetwork. Stronger stability results than were obtained in Example IV.2 are not possible because of the existence of this closely related unstable network.

The instability in the previous example depends on more than just the breaking of the conservation condition $E + F = C$. For example, one can add the reactions E⇄G to the Michaelis-Menten step and instability is not produced, even though the conservation condition is modified in the same way as in the previous example. Asymptotic stability could be established in this case using (IV.32). The essential difference between these two examples is that the reaction E + 2P⇄G interacts with the Michaelis-Menten step at two places, E and P, thereby allowing a feedback cycle to form. The essential role played by feedback cycles is discussed in the next section. The reaction E⇄G does not form a new feedback cycle of interest.

This example suggests an interesting speculation about enzyme systems. We have seen that when the conservation condition $E + F = C$ of the Michaelis-Menten step is violated and the violation introduces a new feedback cycle, instability can occur. Let us speculate that instability frequently occurs in such situations. One can then argue that biochemical networks with this property have a lower evolutionary fitness because of the instability. Hence biochemical evolution should favor networks where each enzyme has its own independent conservation condition $E + F = C$. More complicated conservation conditions could also occur because of complexation between the enzyme and various substrates, acids, bases, or water. These modifications to the conservation condition would not produce instability if no feedback cycle were created. Thus we expect biochemical evolution to select networks where almost all enzymes catalyze a single reaction step and very few enzymes catalyze more than one reaction. This conclusion fits the widely accepted hypothesis of Beadle and Tatum[66] that each normal gene produces a single enzyme that regulates one step in the biosynthesis of a particular chemical. A few enzymes should catalyze more than one reaction because some instability is unavoidable in biochemical control mechanisms that are capable of switching. The facts seem to support the speculation that enzyme systems are prone to instability. Such instabilities would probably make evolution highly likely in almost any sufficiently complicated enzyme system.

These examples suggest that perhaps the broad outlines of the set of stable networks can be discerned with the Lyapunov functions that have been discussed. It is therefore time to see how many of the remaining networks can be proved to be unstable.

V. A DIAGRAMMATIC APPROACH TO INSTABILITY IN STOICHIOMETRIC NETWORKS

A. Motivation

A diagrammatic approach is useful if it saves computation, if it gives theoretical insight that an algebraic approach does not, or if the necessary and sufficient conditions for stability can be better expressed diagrammatically than algebraically. Since the stoichiometric network stability problem is still unsolved, we estimate the possible advantages of using diagrams by considering the next closest problem—sign stability.

The Quirk-Ruppert-Maybee (QRM) theorem (see Section III.5) gave necessary and sufficient conditions for sign semistability in algebraic form. We now show the conditions have a simple diagrammatic interpretation.

To determine whether any $n \times n$ matrix U is sign semistable, construct a *matrix diagram* $\mathfrak{D}_M$ as follows. Draw n dots on a piece of paper and label them 1 to n. If $U_{ij} > 0$, draw an arrow from the jth dot to the ith dot; if $U_{ij} < 0$, draw a *dashed* arrow from the jth dot to the ith dot. In graph theory terminology, this matrix diagram is a signed directed graph with loops.

Example V.1. The matrix diagram of

$$\mathsf{U} = \begin{pmatrix} -5 & -1 & 0 & 0 & 0 & 0 \\ 3 & 0 & 7 & 0 & 0 & 0 \\ 0 & -6 & 0 & 0 & 0 & 0 \\ 0 & 0 & 2 & 0 & 3 & 10 \\ 0 & 0 & 0 & 0 & 0 & 4 \\ 0 & 0 & 0 & -2 & 0 & 0 \end{pmatrix}$$

is shown in Fig. 5.

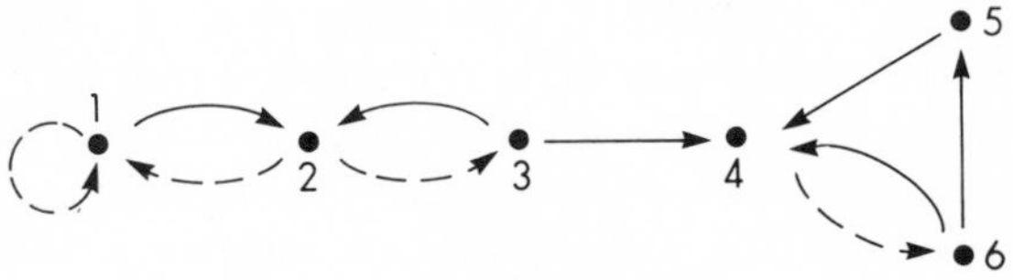

Fig. 5. A matrix diagram $\mathfrak{D}_M$.

For each nonzero product of k matrix elements $U_{i(1)i(2)}U_{i(2)i(3)} \cdots U_{i(k)i(1)}$, where $i(1), \ldots, i(k)$ is any sequence of distinct indices, there will be a sequence of arrows on the diagram from $i(1)$ to $i(k)$, from $i(k)$ to $i(k-1), \ldots,$ from $i(2)$ to $i(1)$. The sequence of arrows is called a *matrix diagram k-cycle*. In Section IV.D we called the product of matrix elements a matrix k-cycle and defined such a cycle to be positive (negative) if the product was positive (negative). Corresponding to a positive (negative) matrix k-cycle is a *positive* (*negative*) *matrix diagram k-cycle* that has an even (odd) number of dashed arrows.

Let us say a matrix diagram cycle is *destabilizing with respect to semistability* if the existence of such a cycle implies that U is not semistable. If a cycle is not destabilizing with respect to sign stability, it may be *safe with respect to sign stability* in the sense that every matrix that has the cycle is sign semistable, provided the matrix does not have any cycles that are destabilizing with respect to sign stability. The theorem may now be restated as follows.

The Quirk-Ruppert-Maybee (QRM) Theorem (Version 2). Negative matrix diagram 1-cycles and negative matrix diagram 2-cycles are safe with respect to sign semistability. All other matrix diagram cycles are destabilizing with respect to sign semistability (see Fig. 6).

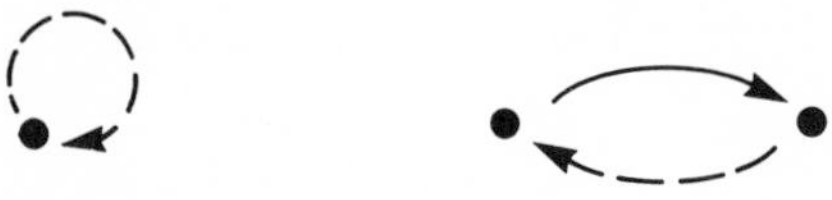

Fig. 6. The only matrix diagram cycles that are safe with respect to sign semistability.

As an example, note that all cycles in Fig. 5 are safe except for the negative cycle from 4 to 6 to 5 to 4. This cycle is destabilizing. Hence the matrix of Example V.1 is not sign semistable.

A matrix cycle or matrix diagram cycle may be interpreted as a *feedback cycle.* For example, when U has the matrix cycle $U_{12}U_{23}U_{31}$, a positive deviation of ζ_1 from steady state will contribute a term $U_{31}\zeta_1$ to $d\zeta_3/dt$. After a brief period ζ_3 will be larger or smaller ($U_{31} > 0$ or $U_{31} < 0$) than it otherwise would be, and this effect will be passed on to ζ_2 via the term $U_{23}\zeta_3$ in the expression for $d\zeta_2/dt$. The resulting change in ζ_2 will then be passed on to ζ_1 via the term $U_{12}\zeta_2$ in the expression for $d\zeta_1/dt$, thereby completing the feedback cycle. Since $U_{ij} > 0$ means that an increase in ζ_j increases ζ_i, we will say that X_j *promotes* X_i. A solid arrow is called a *promotion*. If $U_{ij} < 0$ we say X_j *inhibits* X_i. A dashed arrow is an *inhibition*. A positive (negative) matrix cycle corresponds to a *positive* (*negative*) *matrix diagram feedback cycle*. Thus a matrix diagram (feedback) cycle is positive (negative) if it has an even (odd) number of inhibitions. The QRM theorem says that the only safe feedback cycles are negative feedback cycles whose length is less than 3. Feedback cycles should also be involved in any explanation of chemical instabilities. Hence matrix diagram cycles are involved.

The necessary and sufficient conditions for sign stability as stated in Jefferies' theorem are also diagrammatic. The fact that these theorems on sign stability are most easily expressed diagrammatically strongly suggests that the same is probably true for stoichiometric network stability.

This speculation can be supported by examining the proof of instability given by Quirk and Ruppert. They show that the mere presence of certain cycles causes negative terms to appear in the polynomial that is obtained from a Hurwitz determinant, and that such terms can always make a Hurwitz determinant negative for a suitable choice of the parameters. The stoichiometric network stability problem can be approached in a similar manner. We should expect to be able to identify the negative terms in the polynomials obtained from the Hurwitz determinants (see Section IV.C) with matrix diagram cycles. We might then generalize the concepts of "safe" and "destabilizing" cycles to our problem.

If we use $\mathsf{M}_{\zeta R}$ in the network problem, the n points on the matrix diagram will correspond to the internal species X_i, $i = 1, \ldots, n$. Hence matrix diagram cycles are cycles among the species. These cycles are closely related to, but not exactly the same as, the *knot cycles* and *species cycles* discussed in Section IV.B and IV.E. An example will illustrate the difference.

Example V.2. The network of Example IV.6 is a knot-tree network. Consequently, by (IV.34), it is a species-tree network. The matrix diagram of $\mathsf{M}_{\zeta R} = -\mathsf{S}(\text{diag}\,\mathbf{h})$ is identical with the diagram of $-\mathsf{S}$ and is shown in Fig. 7. Note that the diagram contains two cycles among the species (X⇄Y and Y⇄Z) that are not equivalent to species cycles. Clearly, a species cycle will produce a matrix diagram cycle containing only promotions (solid arrows). A knot cycle will produce a matrix diagram cycle that must contain some promotions and may contain some inhibitions, but only between pairs of species within the same interactant. The two cycles of this example are not knot cycles because no promotions are present.

The stability of this irreversible mass action knot-tree network was proved in Example IV.6. Hence the positive 2-cycles of $\mathfrak{D}_M$ are not destabilizing with respect to stochiometric network linear semistability, even though the QRM theorem proves that these cycles are destabilizing with respect to sign semistability.

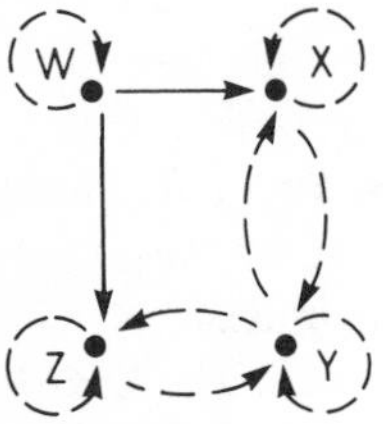

Fig. 7. The matrix diagram ($\mathfrak{D}_M$) of $\mathsf{M}_{\zeta R}$ for a knot-tree graph network.

Since all mass action species-tree graph networks are stable, we consider the matrix diagram cycles of such networks to be "safe" with respect to network stability. Hence instability requires a directed cycle on the semi-directed species graph. The theory of instabilities that will soon be developed is based on a refinement of the species cycle.

The stoichiometric network stability problem depends on the magnitudes of the elements of $-\mathsf{S}$, whereas the sign stability problem depends only on the signs of the elements. The introduction of these magnitudes into the problem will result in the species cycle being replaced by a more quantitative concept—the current cycle. All sources of instability that I have seen in chemical networks can be interpreted as coming from certain kinds of current cycles.

The next few subsections set up several related types of diagram for stoichiometric networks. Certain diagrams represent a quantitative generalization of $\mathfrak{D}_M$. Other diagrams make the current cycles readily apparent. The objective is to consolidate the theorems on instability around what I believe to be the key concept—the critical current cycle. Stability depends on the existence of critical (or close to critical) current cycles, and on the signs and magnitudes of other matrix diagram cycles that interact with critical current cycles in certain ways. I hope that the reader is now convinced that the further development of the subject will be simplified by the introduction of appropriate diagrams.

The original motivation for the diagrammatic approach of stoichiometric network analysis came from a curiosity about the significance of the Routh-Hurwitz conditions and a belief that an understanding of feedback cycles was essential to any theory of chemical instabilities. The work on sign stability and the theorems on Lyapunov functions in Section IV fit nicely into the philosophy of the approach; however the central ideas of what is now to be presented were developed before I was aware of either of these developments.

B. Diagrams

The basic concepts concerning diagrams are presented in a manner that stresses the relationship between diagrams and graph theory. Although graph theory is a major branch of mathematics, almost no work has been done on these diagrams. Most problems of graph theory can be generalized to the diagrams; thus a rich field of mathematics is potentially at hand.

A *graph* G is a set P of distinct *points* π_i, $i = 1, \ldots, n$, and a set L of unordered pairs of distinct points π_α—π_β, $\pi_\alpha, \pi_\beta \in P$, called *lines*. Let P^* be any set of n^* distinct subsets P_i of P. P^* may contain the empty set as an element. Formally, $P^* = \{P_i \mid i = 1, \ldots, n^*, P_i \subset P\}$. If we choose a set L^* of unordered pairs of "points" $P_\alpha, P_\beta \in P^*$, we obtain a graph G^* consisting of the "point" set P^* and the line set L^*. A line P_α—$P_\beta \in L^*$ is an unordered pair of *sets* of points of P, and may be written $\{\pi_{\alpha(1)}, \pi_{\alpha(2)}, \ldots, \pi_{\alpha(k)}\} - \{\pi_{\beta(1)}, \ldots, \pi_{\beta(l)}\}$. Although this is a relation between two "points" of P^*, we may also interpret it as a relation among the points of P. When all the lines of L^* of G^* are interpreted in terms of P rather than P^*, the result is called a set-graph diagram.

A *set-graph diagram* $\mathfrak{D}_{SG}$ is a set P of distinct points π_i, $i = 1, \ldots, n$, and a set L_{SG} of unordered pairs of sets of points of P, $\{\pi_{\alpha(1)}, \ldots\}$—$\{\pi_{\beta(1)}, \ldots\}$, called *set-graph lines* (SG-lines). The $\mathfrak{D}_{SG}$ may be represented on a piece of paper by drawing n distinct dots to represent the points π_i, and by drawing a branched line to represent each SG-line. One end of the line branches to $\pi_{\alpha(1)}, \pi_{\alpha(2)}, \ldots,$ and the other end branches to $\pi_{\beta(1)}, \pi_{\beta(2)}, \ldots$. For example, the SG-line $\{\pi_1, \pi_2\}$—$\{\pi_3, \pi_4, \pi_5\}$ is drawn as in Fig. 8. If a line of $\mathfrak{D}_{SG}$ joins a set P_1 to the empty set, one end of the line branches to all the points in P_1 and the other end branches to all the

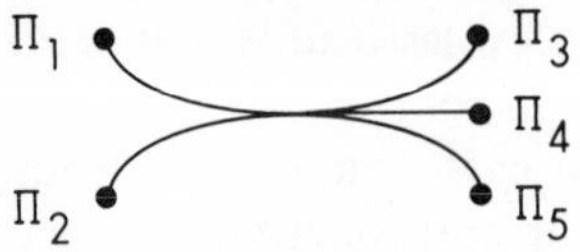

Fig. 8. The set-graph line $\{\pi_1, \pi_2\}$—$\{\pi_3, \pi_4, \pi_5\}$.

points in the empty set. Since the empty set has no points, the line terminates where there is no point. Logical consistency forces us to leave one end of these SG-lines dangling. Any attempt to avoid the unfamiliar by introducing a point representing Ø (Ø is a set—not a point) not only introduces an inconsistency, but produces unneccessary problems later on.

An example of G^* and its corresponding $\mathfrak{D}_{SG}$ is shown in Fig. 9. Mathematicians will recognize that Eulerian circuits and Hamiltonian cycles occur in SG-diagrams. The analogue of a directed graph is a directed SG-diagram.

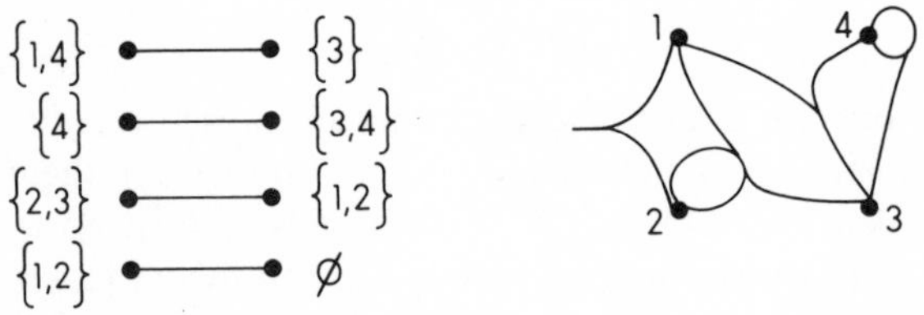

Fig. 9. A graph G^* (left) and its corresponding set-graph diagram $\mathfrak{D}_{SG}$ (right).

Set-graph diagrams can represent sets of chemical reactions. Recall that the interactant pseudograph G_I, defined in Section IV.B, is a graph whose lines are pairs of sets of species. Thus G_I is like G^*. We can construct an SG-diagram whose points are the chemical species and whose SG-lines branch at each end to all the species on each side of a reaction. The SG-diagram of the Oregonator is given in Fig. 10. It can be considered to have been derived from G_I in Section IV.B.

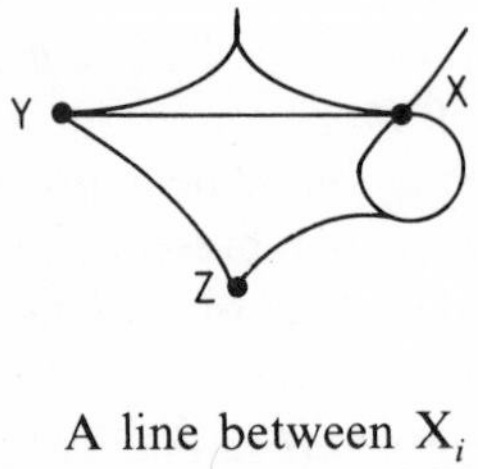

Fig. 10. The SG-diagram of the Oregonator.

A line between X_i and X_j on the semidirected species pseudograph G_{SS} means that X_i and X_j are on the opposite sides of a reaction. Hence a path can be traced between X_i and X_j along the corresponding SG-line of $\mathfrak{D}_{SG}$. Paths, trails, circuits, and cycles on G_{SS} have corresponding paths, trails, circuits, and cycles on $\mathfrak{D}_{SG}$. We made the convention that each reaction is represented on G_{SS} either by directed lines (the irreversible case) or by undirected lines (the reversible case). If we make the convention that irreversible and reversible reactions are to be distinguished by directed and undirected SG-lines on $\mathfrak{D}_{SSG}$, (called a *semidirected SG-diagram*) then there will be a one-to-one correspondence between directed cycles of G_{SS} and directed cycles of $\mathfrak{D}_{SSG}$. If $\mathfrak{D}_{SSG}$ has no directed cycles, it is a *tree*. From (IV.67) we conclude that *every network whose semidirected SG-diagram is a tree is stable.*

Conventionally, a *network* is defined to be a graph G, with a function $f: L \to S$, where L is the set of lines of G and S is any other set. (Usually S is R, the set of real numbers.) A network is usually a graph with numbers associated with the lines. This definition is often extended in an obvious way to directed graphs, multigraphs, and pseudographs.

Let us generalize the concept of a network to directed SG-diagrams (DSG-diagrams). Now all SG-lines are directed, and reversible reactions must be represented by two directed SG-lines. It will turn out that such a definition of a network is the sense in which a stoichiometric "network" is a network. Each DSG-line of a directed set-graph diagram $\mathfrak{D}_{DSG}$ has a *beginning* and an *end*. Let $b(l) \subset P$ be the set of points at the beginning of the DSG-line $l \in L$, and let $e(l) \subset P$ be the set of points at the end of l. The set of all *DSG-line beginnings* of $\mathfrak{D}_{DSG}$ is

$$B \equiv \{(l, \beta) \mid l \in L, \beta \in b(l)\}$$

and the set of *DSG-line ends* is

$$E \equiv \{(l, \varepsilon) \mid l \in L, \varepsilon \in e(l)\}$$

The set of *DSG-line termini* is

$$T \equiv \{(l, \tau, \sigma) \mid l \in L, (\tau, \sigma) \in (B \times \{-1\}) \cup (E \times \{1\})\} \qquad \text{(V.1)}$$

That is, to specify a DSG-line terminus $t = (l, \tau, \sigma) \in T$, one must specify a DSG-line $l \in L$, a point $\tau \in P$ to which the line connects, and whether the point occurs at the beginning ($\sigma = -1$) or end ($\sigma = 1$) of the line. A *directed set-graph network* (DSG-network) N_{DSG} is defined to be a DSG-diagram $\mathfrak{D}_{DSG}$, with a function $f: T \to S$, where T is the set of DSG-line termini of $\mathfrak{D}_{DSG}$ and S is any other set.

Consider a stoichiometric "network" N with mass action kinetics and no autocatalysis (no species occurs on both sides of any reaction). As before, we construct the SG-diagram and make it a directed SG-diagram by placing an arrow on each SG-line to indicate the direction of the reaction. Now forward and reverse reactions must be represented by *different* DSG-lines. Define $f: T \rightarrow \overline{R}_+$ by

$$f(l,\tau,\sigma) = \sigma\nu_{ij} \tag{V.2}$$

where i and j are such that R_j is the reaction associated with the DSG-line $l \in L$, and X_i is the species associated with the point $\tau \in P$. Since the direction of the line indicates the direction of the reaction, when X_i is on the left of the reaction ($\nu_{ij} < 0$), τ is at the line beginning ($\sigma < 0$), and $f(l,\tau,\sigma) \geqslant 0$. When X_i is on the right, $\nu_{ij} \geqslant 0$, $\sigma > 0$, and thus $f(l,\tau,\sigma) \geqslant 0$. Hence every mass action stoichiometric "network" with no autocatalysis is a DSG-network.

Example V.3. A proposed mechanism for the overall reaction $H_2 + Br_2 \rightarrow 2HBr$ is as follows (Ref. 67, p. 164):

$$Br_2 \rightarrow 2\,Br\cdot$$
$$Br\cdot + H_2 \rightarrow HBr + H\cdot$$
$$H\cdot + Br_2 \rightarrow HBr + Br\cdot$$
$$H\cdot + HBr \rightarrow H_2 + Br\cdot$$
$$Br\cdot + Br\cdot \rightarrow Br_2$$

Usually the concentrations of Br_2 and H_2 would be large compared to those of the other species and could be considered fixed on a short-time scale. Then we must treat them as external species. The DSG-network that results appears in Fig. 11.

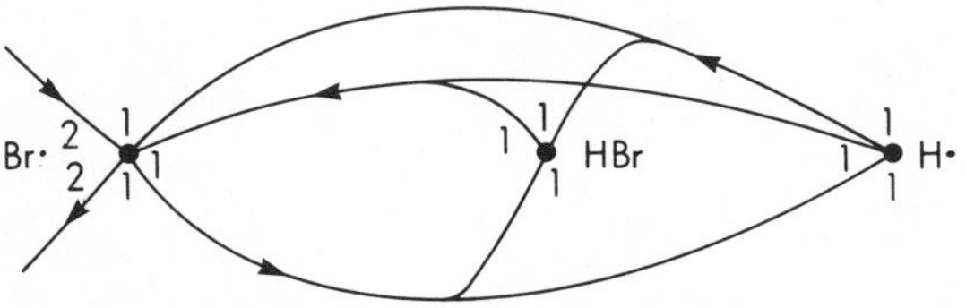

Fig. 11. The DSG-network (diagram) of a mass action stoichiometric "network."

To generalize (V.2) to autocatalytic reactions, note that the DSG-line corresponding to an autocatalytic reaction, such as

$$R_1 : X_1 + X_2 \rightarrow 2X_1 + X_3$$

has two line termini connecting to one point. In R_1 the point is π_1, which corresponds to X_1. Note that $\underline{\nu}$ contains only the net stoichiometry, and would be no different if the reaction were $X_2 \rightarrow X_1 + X_3$. Since the net reaction has X_1 on the right (making $\nu_{11} > 0$) we could associate ν_{11} with the end of the line and zero with the beginning; or, for arbitrary $\alpha \geqslant 0$, we could associate $\nu_{11} + \alpha$ with X_1 on the right and α with X_1 on the left. A generalization of (V.2) that is suitable for autocatalytic reactions is therefore

$$f_{ma}(l,\tau,\sigma) = \alpha_{ij} + H(\sigma\nu_{ij})\sigma\nu_{ij} \tag{V.3}$$

where $\alpha_{ij} \geqslant 0$ can be chosen arbitrarily and $H(x)$ is the Heaviside function, defined by $H(x) = 1$ if $x \geqslant 0$, $H(x) = 0$ otherwise. We will always choose α_{ij} to be the stoichiometry of X_i on the side of the reaction where $\sigma\nu_{ij} < 0$; then $\alpha_{ij} + \sigma\nu_{ij}$ must be the stoichiometry of X_i on the opposite side.

A stoichiometric "network" having power law kinetics, and possibly autocatalysis, corresponds to a DSG-network where $f_{pl}: T \to R^2$ is defined by

$$f_{pl}(l,\tau,\sigma) = (\alpha_{ij} + H(\sigma\nu_{ij})\sigma\nu_{ij}, H(-\sigma)\kappa_{ij}) \tag{V.4}$$

where i and j are determined as in (V.2). Note that when $\nu_{ij} > 0$ and $\kappa_{ij} > 0$, R_j will produce X_i at a rate that depends on X_i. Such reactions must be written with X_i on the left (making them autocatalytic), so that the DSG-line has a beginning at X_i. Otherwise there would be no line beginning to map into κ_{ij} on the DSG-diagram. See Fig. 12 for an illustration.

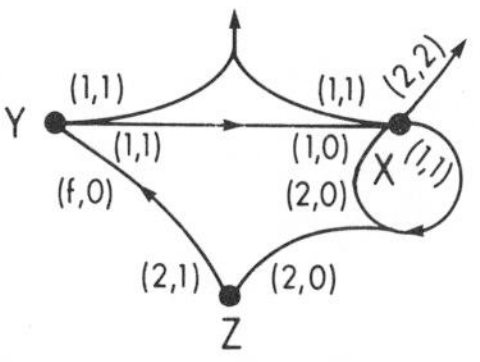

Fig. 12. The DSG-network of the Oregonator (defined in Example II.1). This power law network does not have mass action kinetics.

The general stoichiometric "network," defined in Section II.A, becomes a DSG-network if we define

$$f_g(l,\tau,\sigma) = (f_{ma}(l,\tau,\sigma), H(-\sigma)v_j(\mathbf{X},\mathbf{k})) \tag{V.5}$$

where v_j is the rate function associated with R_j. The diagram $\mathfrak{D}_{DSG}$ has the set of points $\{X_1, \ldots, X_n\}$, the set of lines $\{R_1, \ldots, R_n\}$ where R_i is interpreted as a directed relation between the species on the left and right of the reaction. Associated with each line terminus is a point of $R \times S^*$, where S^* is a function space containing the rate function $v_j(\mathbf{X},\mathbf{k})$.

The linearized steady-state stability problem for the general stoichiometric network depends on $\mathbf{v}(\mathbf{X},\mathbf{k})$ only through several first derivatives of $\mathbf{v}$ evaluated at steady state. These derivatives are given by the effective power (matrix) function $\underline{\kappa}(\mathbf{p})$, defined by (III.3). Hence the general linear stability problem leads to the DSG-network with

$$f_{gl}(l,\tau,\sigma) = (f_{ma}(l,\tau,\sigma), H(-\sigma)\kappa_{ij}(\mathbf{p})) \tag{V.6}$$

where the image set of f_{gl} is again R $\times$ (a function space), and i and j are determined by the condition that l is R_j and τ is X_i. Since the general case can be obtained from (V.4) by allowing $\underline{\kappa}$ to depend on some additional parameters, we base the theory of instability on the DSG-networks defined by (V.4).

We now define the *network diagram* $\mathfrak{D}_N$. It is equivalent to the DSG-network defined by (V.4) but is a more practical notation for chemical networks. The DSG-lines are replaced with *branched arrows* A_j (corresponding to R_j); the pairs of numbers appearing on the DSG-network are represented by drawing *barbs* and *feathers* on the *heads* and *tails* of the arrows, respectively. This notation is practical only when these numbers are small positive integers; however we use network diagrams to illustrate the general case, even though we cannot draw them in general.

The three important numbers appearing on the termini of DSG-lines are given by (V.4). A new notation for them is:

$$B(X_i, A_j) = \alpha_{ij} + H(\nu_{ij})\nu_{ij}$$

$$T(X_i, A_j) = \alpha_{ij} - H(-\nu_{ij})\nu_{ij}$$

$$L(X_i, A_j) = \kappa_{ij} \tag{V.7}$$

The end that meets X_i of the DSG-line representing R_j, has the associated pair of numbers $(B(X_i, A_j), 0)$. In the new notation, the head of A_j at X_i is drawn with $B(X_i, A_j)$ barbs. Thus the diagram below shows the DSG-line ends and their respective representations on $\mathfrak{D}_N$:

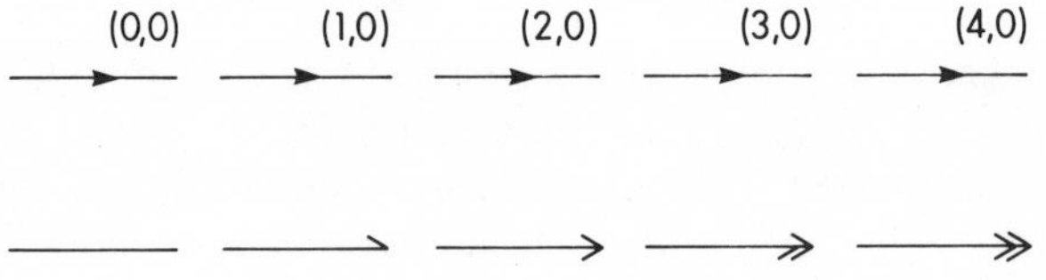

These arrow heads have 0, 1, 2, 3, and 4 barbs. These numbers are the stoichiometries of X_i on the right of the reactions.

Consider now the beginning (at X_i) of the DSG-line representing R_j. The associated pair of numbers is $(T(X_i, A_j), L(X_i, A_j))$. If $T \geqslant L$, these may be represented on the tail of A_j by drawing $L(X_i, A_j)$ feathers on the left, and $T(X_i, A_j) - L(X_i, A_j)$ feathers on the right. Hence the correspondence between notations is as follows:

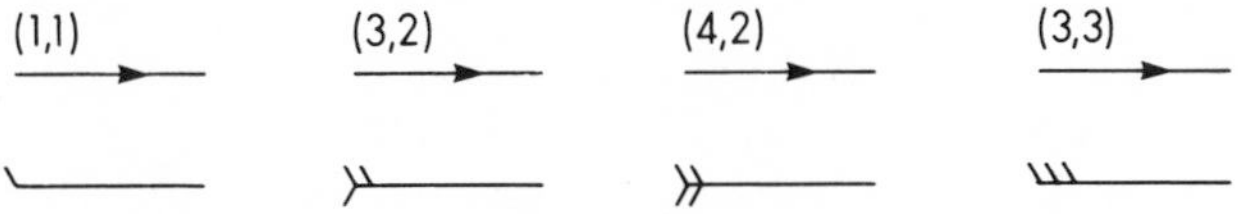

Note that the stoichiometric coefficient of X_i is then the total number of feathers T; the order of kinetics of X_i is the number of left feathers L. These conclusions follow from (V.7), provided α_{ij} is chosen properly in the autocatalytic case and that $\alpha_{ij} = 0$ when R_j is not autocatalytic. When $T < L$ this convention cannot be used; however the reaction stoichiometries may always be multiplied by a positive constant so that $T \geqslant L$ and the scaling will only affect the numerical values of the rate constants.

Note that no DSG-line beginning can have $(0, 0)$, because then X_i does not play a role in the reaction, and the DSG-line beginning could not touch X_i. Hence the arrow tail with no feathers (left or right) cannot occur. Thus no ambiguity will result if we make the convention that the single feather in the $(1, 1)$ case may be drawn straight back, rather than to the left. This convention means that tails that appear to have no feathers should be considered to have $(T, L) = (1, 1)$. Figure 13 summarizes these conventions, and Fig. 14 gives $\mathfrak{D}_N$ for the Oregonator—note how much tidier it is than the DSG-network of Fig. 12.

Henceforth the network diagram $\mathfrak{D}_N$ is used to represent DSG-networks of the type specified by (V.4). We will always assume that the stoichiometry has been scaled so that $|\nu_{ij}| \geqslant \kappa_{ij}$ for all i, j, and with proper

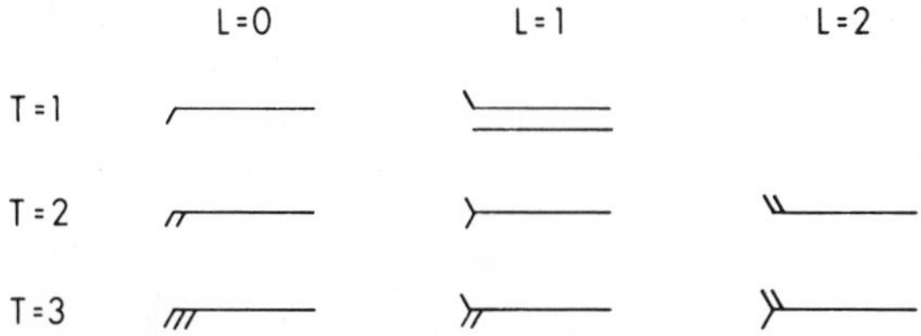

Fig. 13. Arrow tails which correspond to (T, L) on the DSG-line: T gives the stoichiometry and L gives the order of kinetics of the reaction in X.

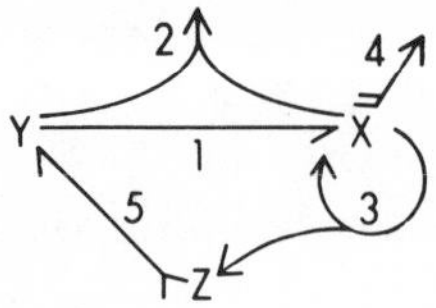

Fig. 14. The network diagram $\mathfrak{D}_N$ of the Oregonator for $f = 1$. The numbers label the reactions $R_1, \ldots, R_5$.

treatment of autocatalysis this implies $T \geqslant L$ at every arrow tail. The theorems that follow are valid for all nonnegative real values of T, L, and B, except where stated explicitly. The use of network diagrams to illustrate the theorems must not be interpreted by the reader to mean that the theorems are valid only when the diagrams can be drawn.

C. Current Diagrams and Current Matrix Diagrams

A *current diagram* $\mathfrak{D}_C$ is a network diagram $\mathfrak{D}_N$ with the property that for $i = 1, \ldots, n$,

$$\sum_{j=1}^{r} B(X_i, A_j) = \sum_{j=1}^{r} T(X_i, A_j) \qquad \text{(V.8)}$$

That is, the sum of the barbs on all the incoming arrow heads at X_i equals the sum of the total feathers on all of the leaving arrow tails at X_i. Substituting (V.7) into this equation gives

$$\sum_{j=1}^{r} \nu_{ij} = 0 \qquad \text{(V.9)}$$

which implies $\underline{\nu}\mathbf{e} = 0$. Hence $\mathbf{e}$ is a steady-state velocity vector. Since this argument may be reversed, we conclude that $\mathbf{e} \in \mathcal{C}_v$ if and only if $\mathfrak{D}_N$ is a current diagram. When $\mathbf{e} \in \mathcal{C}_v$, all reactions have the same rate at steady state. Then the relative numbers of species produced and consumed over any time interval are proportional to the reaction stoichiometries. Hence the current diagram represents a steady state in which the production and consumption of every species by every reaction is proportional to the numbers of barbs B and total feathers T at the species for the reaction. The diagram helps visualize the flow at steady state.

One can construct a current diagram corresponding to any steady state $\mathbf{v}^0 \in \mathcal{C}_v$ as follows. Define the *current stoichiometric matrix* $\underline{\nu}^*$ by

$$\underline{\nu}^* = \underline{\nu}\,\mathrm{diag}\,\mathbf{v}^0 \qquad \text{(V.10)}$$

Since $\underline{\nu}\mathbf{v}^0 = 0$, $\underline{\nu}^*\mathbf{e} = 0$. Hence any network N^* with the stoichiometric matrix $\underline{\nu}^*$ has the steady state $\mathbf{e}$. The network diagram of N^* must be a current diagram, and N^* will be equivalent to the original network if $\underline{\boldsymbol{\kappa}}^* = \underline{\boldsymbol{\kappa}}$. A particular network with $\underline{\boldsymbol{\nu}}$ can have many current stoichiometric matrices $\underline{\boldsymbol{\nu}}^*$, and in each case $\mathbf{e}$ corresponds to a different current of $\underline{\boldsymbol{\nu}}$. Since $\underline{\boldsymbol{\nu}}^*$ is a stoichiometric matrix, the asterisk is not really needed. When we wish to study a particular current of a network we may assume, without loss of generality, that $\mathbf{e} = \mathsf{E}\mathbf{j}$ and that $\underline{\boldsymbol{\nu}}\mathbf{e} = 0$. This convention can be stated by saying that $\underline{\boldsymbol{\nu}}$ is the current stoichiometric matrix.

A current diagram of an extreme network is called an *extreme current diagram* $\mathfrak{D}_{EC}$. If E^i is an extreme subnetwork of N, the corresponding extreme current diagram is obtained by letting $\underline{\boldsymbol{\nu}}^* = \underline{\boldsymbol{\nu}}(\operatorname{diag}\mathbf{E}_i)$. In chemistry, the elements of $\underline{\boldsymbol{\nu}}$ and $\mathbf{E}_i$ are always small integers (except in certain models like the Oregonator, where $f \in R$); hence $\mathfrak{D}_{EC}$ can always be drawn. No problems will arise if the reaction stoichiometries are scaled differently in different subnetworks of N. Figure 15 illustrates some current diagrams.

When a network has a small number of extreme subnetworks, the best way to find them is usually by examining $\mathfrak{D}_N$. The network diagram can be modified by deleting reactions and multiplying reaction stoichiometries until it evolves into the $\mathfrak{D}_{EC}$ for an extreme network.

Example V.9. Let us construct the extreme current diagrams of the Oregonator from $\mathfrak{D}_N$ in Fig. 14. If R_5 is deleted, the steady-state condition (V.8) at Z can only be met if R_3 is deleted, and the steady-state condition at Y can only be met if both R_1 and R_2 are deleted. Then only R_4 remains and it must be deleted to satisfy the steady-state conditions at X. Hence we cannot delete R_5 or R_3 or both of R_1 and R_2. If we delete R_2, the steady-state condition is satisfied at all species. This yields $\mathfrak{D}_{EC}$ for E^2. If we delete R_1, we can satisfy the steady-state condition at X only by deleting R_4. This yields $\mathfrak{D}_{EC}$ for E^1. There are no other possibilities, so N has only these two extreme subnetworks, which appear in Fig. 15.

Suppose we fix $\mathbf{j} \in R^f_+$ and ask whether all the steady states corresponding to $\mathbf{h} \in R^n_+$ are stable when the dynamics are linearized about steady state. From (III.11), the matrix of the linearized system is $-\mathsf{V}(\mathbf{j})\operatorname{diag}\mathbf{h}$, where $\mathsf{V}(\mathbf{j})$ is defined by (III.10). The solution to this problem is therefore completely determined by $\mathsf{V}(\mathbf{j})$. We should be able to determine $\mathsf{V}(\mathbf{j})$ from

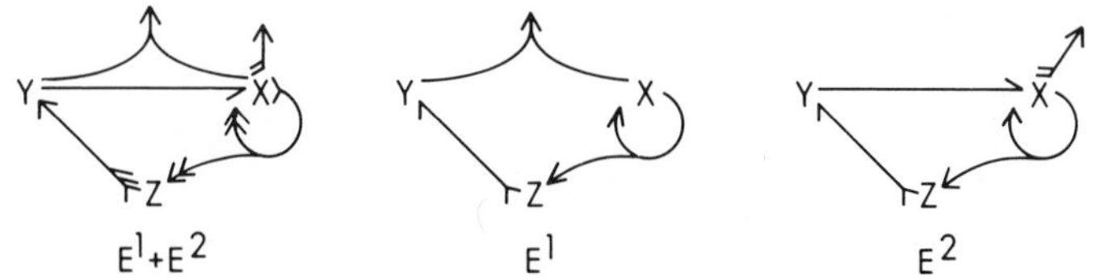

Fig. 15. Three current diagrams $\mathfrak{D}_C$ of the Oregonator. The diagrams correspond to $\mathbf{j} = (1,1)^t$, $(1,0)^t$, and $(0,1)^t$, respectively. Thus the left-hand diagram is the "sum" of the other two diagrams.

the corresponding current diagram because if we draw the current diagram using $\underline{\boldsymbol{\nu}}^*$ given by (V.10), where $\mathbf{v}^0 = \mathbf{Ej}$, then from (III.10), $-V(\mathbf{j}) = \underline{\boldsymbol{\nu}}^* \underline{\boldsymbol{\kappa}}^t$. In other words, if $\underline{\boldsymbol{\nu}}$ is the stoichiometric matrix of the current $\mathbf{e} = \mathbf{Ej}$, then

$$-\mathsf{V}(\mathbf{j}) = \underline{\boldsymbol{\nu}}\, \underline{\boldsymbol{\kappa}}^t \tag{V.11}$$

Each element of $-\mathsf{V}(\mathbf{j})$ is the product of two numbers associated with the branched arrows on $\mathfrak{D}_C$. Let us adopt conventions that will enable us to represent $-\mathsf{V}(\mathbf{j})$ in the most efficient form diagrammatically and then look for rules to obtain the diagrammatic form of $-\mathsf{V}(\mathbf{j})$ from $\mathfrak{D}_C$.

The diagrammatic form of $-\mathsf{V}(\mathbf{j})$, called the *current matrix diagram* $\mathfrak{D}_{CM}$, is defined as the diagram that results when one makes the following modifications to the matrix diagram $\mathfrak{D}_M$ of $-\mathsf{V}(\mathbf{j})$, defined in Section V.A.

1. The arrow (directed line) $X_j \to X_i$ of $\mathfrak{D}_M$ represents the sign of $-V_{ij}(\mathbf{j})$, when $V_{ij}(\mathbf{j}) \neq 0$. Now make the number of barbs on the arrow correspond to $|V_{ij}(\mathbf{j})|$ (as for $\mathfrak{D}_N$).
2. In place of the point π_i representing the species X_i, write the symbol that represents the reciprocal steady-state concentration of X_i in the vector $\mathbf{h}$. I prefer to use uppercase Roman letters X, Y, . . . for species, uppercase italic letters $X, Y, \ldots$ for their concentrations as dynamical variables, and lowercase italic letters $x, y, \ldots$ for the parameters representing the corresponding reciprocal steady-state concentrations.
3. Matrix diagrams of chemical networks have dashed loops at almost all species (see, e.g., Fig. 7). Delete all such loops (dashed or solid) and write the value of the corresponding diagonal element of $\mathsf{V}(\mathbf{j})$ in front of the parameter chosen in (2).

Example V.5. The Oregonator with $f = 1$ has two currents so, by (IV.11), $\mathsf{V}(\mathbf{j}) = j_1\mathsf{S}^{(1)} + j_2\mathsf{S}^{(2)}$ and the matrices $\mathsf{S}^{(1)}$ and $\mathsf{S}^{(2)}$ are given in Example III.1. Take $\mathbf{h} = (x, y, z)$. Then when $\mathbf{j} = (1, 1)$,

$$-\mathsf{V}(1,1) = \begin{pmatrix} -3 & 0 & 0 \\ -1 & -2 & 2 \\ 4 & 0 & -4 \end{pmatrix}$$

and $\mathfrak{D}_{CM}$ is shown in Fig. 16*a*. When $\mathbf{j} = (1, 0)$, $\mathbf{V}(\mathbf{j}) = \mathsf{S}^{(1)}$ and $\mathfrak{D}_{CM}$ appears in Fig. 16*b*. The case $\mathbf{j} = (0, 1)$ appears in Fig. 16*c*.

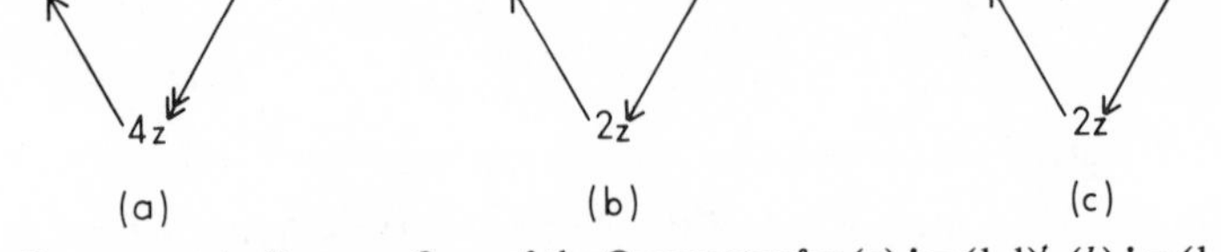

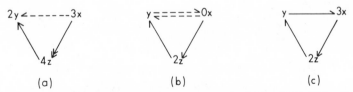

Fig. 16. Current matrix diagrams $\mathfrak{D}_{CM}$ of the Oregonator for (*a*) $\mathbf{j} = (1, 1)^t$, (*b*) $\mathbf{j} = (1, 0)^t$, and (*c*) $\mathbf{j} = (0, 1)^t$.

The sign semistability of $-\mathsf{V}(\mathbf{j})$ may be determined from the diagrammatic rules of Section V.A. Diagram $\mathfrak{D}_{CM}$ is treated like $\mathfrak{D}_M$, except that there are no 1-cycles. Instead, a positive number at a point is safe and a negative number is destabilizing. From Fig. 16, $-\mathsf{V}(1,1)$ is sign semistable, whereas $-\mathsf{V}(1,0)$ and $-\mathsf{V}(0,1)$ are not. The conditions of Jefferies' theorem are also satisfied trivially, so $-\mathsf{V}(1,1)$ is sign asymptotically stable. However sign asymptotic stability implies D-stability (Ref. 59, Theorem 7-4-1), so $-\mathsf{V}(1,1)$ is D-stable. We showed just prior to (III.74) that the current $\mathbf{j}$ is linearly asymptotically stable if and only if $-\mathsf{V}(\mathbf{j})$ is D-stable. Hence $\mathbf{j}=(1,1)$ is a linearly asymptotically stable current of the Oregonator.

Consider now the "addition" of current matrix diagrams. From (III.11),

$$-\mathsf{V}(1,1)=-\mathsf{V}(1,0)-\mathsf{V}(0,1)$$

Thus the "sum" of Fig. 16*b* and 16*c* should "equal" 16*a*. The rules for adding current matrix diagrams are as follows:

1. Only arrows in the same direction between the same points may be added.
2. Arrows are added algebraically with inhibitions (dashed arrows) having the opposite signs from promotions (solid arrows). The absolute value of each arrow is the number of barbs.
3. The numbers at the points are added algebraically.

The rules for constructing $\mathfrak{D}_{CM}$ from $\mathfrak{D}_C$ may be obtained by expressing (V.11) in terms of B, L, and T using (V.7). The number appearing in front of h_i is

$$V(\mathbf{j})_{ii}=\sum_{k=1}^{r}\left(T(\mathrm{X}_i A_k)-B(\mathrm{X}_i,A_k)\right)L(\mathrm{X}_i,A_k) \tag{V.12}$$

The TL terms are positive and correspond to a safe negative 1-cycle in the sign stability problem of $-\mathsf{V}(\mathbf{j})$. Later we will see that these terms are the principal source of stability in stoichiometric networks. The BL terms are negative and correspond to a destabilizing positive 1-cycle in the sign stability problem. Such terms are nonzero only when there is autocatalysis ($B(\mathrm{X}_i A_k)\neq 0$).

Let $\underline{\boldsymbol{\nu}}$ be the current stoichiometric matrix. When $\nu_{ik}>0$, R_k makes the contribution

$$b_{ijk}\equiv\nu_{ik}\kappa_{jk}=L(\mathrm{X}_j,A_k)B(\mathrm{X}_i,A_k) \tag{V.13a}$$

to $(\underline{\boldsymbol{\nu\kappa}}^t)_{ij}$. This contribution appears as a solid arrow of $\mathfrak{D}_{CM}$ having b_{ijk} barbs and going from the reactant X_j of R_k to the product X_i of R_k. When

$\nu_{ik} < 0$, R_k makes the contribution

$$b_{ijk}^{-} \equiv \nu_{ik}\kappa_{jk} = -L(X_j, A_k)T(X_i, A_k) \qquad \text{(V.13b)}$$

which appears as a dashed arrow of $\mathcal{D}_{CM}$ having b_{ijk}^{-} barbs and going from the reactant X_j to the reactant X_i of R_k.

An efficient set of rules for constructing $\mathcal{D}_{CM}$ from $\mathcal{D}_C$ is the following. The rules may be applied to any $\mathcal{D}_N$, although the resulting diagram is only a $\mathcal{D}_{CM}$ if $\mathcal{D}_N$ is also a current diagram.

1. For each species X_i, sum $L(X_i, A_k)T(X_i, A_k)$ over all outgoing reactions R_k and write this number at X_i. After the number write the symbol for h_i.
2. For every pair of species that (both) appear on the left of a reaction R_k, draw an inhibition (dashed arrow) from h_i to h_j with $L(X_i, A_k)$ $T(X_j, A_k)$ barbs and a similar inhibition with i and j interchanged.
3. For every pair of species that appear on opposite sides of R_k, draw a promotion (solid arrow) from the species on the left (X_i, say) to the species on the right (X_j, say) with $L(X_i, A_k)T(X_j, A_k)$ barbs.
4. Redraw the diagram with the arrows added algebraically and the 1-cycles combined with the numbers at each point.

These rules are simple to remember because in every case L is associated with the species that will appear at the tail of the arrow of $\mathcal{D}_{CM}$, and B or T with the species that will appear at the head of the arrow of $\mathcal{D}_{CM}$. The pattern of arrows produced by a typical reaction is shown in Fig. 17; Fig. 18 shows a very complicated case with autocatalysis. These examples serve to illustrate the rules; however the resulting diagram is not a $\mathcal{D}_{CM}$ because $\mathcal{D}_N$ is not a current diagram. Figure 19 shows how the rules yield the $\mathcal{D}_{CM}$'s of Fig. 16 from the $\mathcal{D}_C$'s of Fig. 14.

We have seen in this section that every current can be drawn as a current diagram $\mathcal{D}_C$. The stoichiometries of this diagram indicate the production and consumption of every species by every reaction at steady state. The current diagrams of the extreme currents, $\mathcal{D}_{EC}$, can often be constructed directly from the network diagram more easily than by using the algebraic procedure of Section II.C. $\mathcal{D}_C$ represents the set of steady states with a particular $\mathbf{j}$. The stability of these states is determined by the matrix $-\mathbf{V}(\mathbf{j})$, which can be represented as a current matrix diagram $\mathcal{D}_{CM}$. Some very simple rules enable one to draw $\mathcal{D}_{CM}$ using $\mathcal{D}_C$. The sign

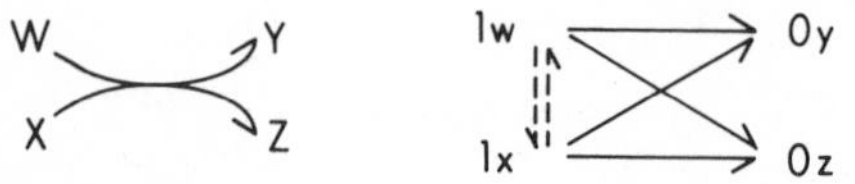

Fig. 17. $\mathcal{D}_N$ and "$\mathcal{D}_{CM}$" of the reaction $W + X \rightarrow Y + Z$ with mass action kinetics.

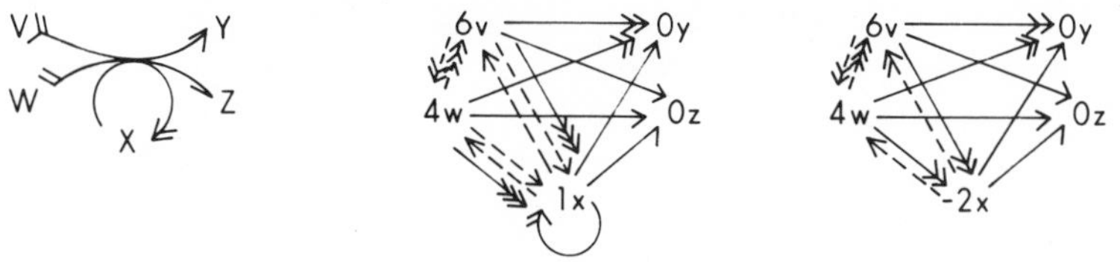

Fig. 18. $\mathfrak{D}_N$ and "$\mathfrak{D}_{CM}$" of the reaction 3V + 2W + X → 3X + 2Y + Z with the rate law V^2W^2X. The middle diagram shows the arrows before addition.

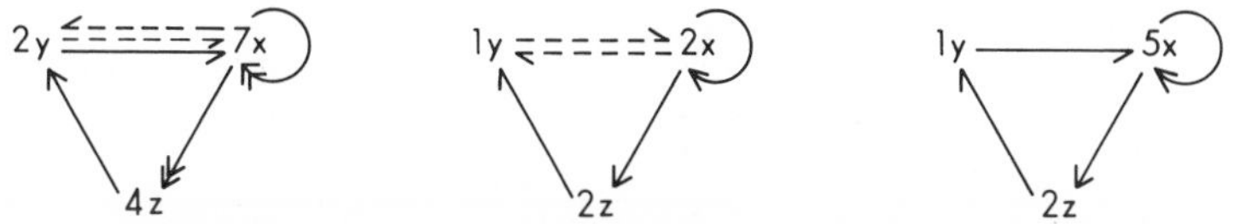

Fig. 19. Intermediate steps in obtaining Fig. 16 from Fig. 15.

stability of $-V(\mathbf{j})$ may be read easily from $\mathfrak{D}_{CM}$. If sign asymptotic stability holds, the *current* is also linearly asymptotically stable. The problem is rarely this simple. When sign asymptotic stability does not hold, we must examine $\mathfrak{D}_{CM}$ in more detail to determine network stability.

D. Diagrammatic Stability Polynomials

The expressions for α_i or Δ_i as polynomials in $\mathbf{h}$ and $\mathbf{j}$, or in $\mathbf{h}$ for fixed $\mathbf{j}$, are called *stability polynomials.* The coefficients in the first case are determined by $-\mathsf{S}^{(1)}, \ldots, -\mathsf{S}^{(f)}$, and in the latter case by $-\mathsf{V}(\mathbf{j})$. Since these matrices are represented by their current matrix diagrams, it is possible to represent any coefficient in these polynomials as a sum of diagrams. Computer studies have shown that networks seem to be linearly exponentially unstable whenever any efficient is negative. Hence we work toward obtaining necessary and sufficient conditions for the existence of a negative coefficient. If such a coefficient exists, it can always be represented as a sum of diagrams; hence stability depends on whether certain types of diagram can be constructed from $\mathfrak{D}_{CM}$. What do the diagrams that sum to a negative coefficient in a stability polynomial look like? What features of $\mathfrak{D}_{CM}$, and ultimately $\mathfrak{D}_C$, make such diagrams possible? Answers to these questions are developed in this and succeeding sections.

Consider α_i as a polynomial in $\mathbf{h}$ with $\mathbf{j}$ fixed. Write the characteristic equation as

$$\det|\lambda \mathsf{I} + \mathsf{V}(\mathbf{j})\operatorname{diag}\mathbf{h}| = 0 \tag{V.14}$$

and recall that α_i is the coefficient of λ^{n-i} in this expansion. The $n-i$ factors of λ come from the diagonal elements of $n-i$ distinct columns.

From the remaining i columns come factors of h_k, say $h_{\gamma(1)} \cdots h_{\gamma(i)}$. Since the jth column of $\mathsf{V}\,\mathrm{diag}\,\mathbf{h}$ is associated with h_j, these i factors are all different. Hence the general form of α_i is

$$\alpha_i(\mathbf{h},\mathbf{j}) = \sum_{\gamma \in A(n,i)} h_{\gamma(1)} \cdots h_{\gamma(i)} \beta_i(\gamma,\mathbf{j}) \tag{V.15}$$

where

$$\beta_i(\gamma,\mathbf{j}) \equiv \begin{vmatrix} V_{\gamma(1)\gamma(1)}(\mathbf{j}) \cdots & V_{\gamma(1)\gamma(i)}(\mathbf{j}) \\ \vdots & \\ V_{\gamma(i)\gamma(1)}(\mathbf{j}) \cdots & V_{\gamma(i)\gamma(i)}(\mathbf{j}) \end{vmatrix} \tag{V.16}$$

The sum is taken over the $\binom{n}{i}$ ways to choose i distinct integers $\gamma(1) \cdots \gamma(i)$ from the n distinct integers $1, \ldots, n$. We treat such a choice as an i-vector and let $A(n,i)$ be the set of such vectors.

The network is exponentially unstable if there exists i, $\mathbf{j}$, and $\boldsymbol{\gamma} \in A(n,i)$ such that $\beta_i(\gamma,\mathbf{j}) < 0$. To prove this, chose any $\varepsilon > 0$, let $h_{\gamma(1)} = h_{\gamma(2)} = \cdots = h_{\gamma(i)} = \varepsilon^{-1}$, and let all other components of h equal 1. Then $\alpha_i(\mathbf{h},\mathbf{j})$ is a polynomial in ε^{-1} with $\beta_i(\gamma,\mathbf{j})$ as the coefficient of the algebraically lowest power term, ε^{-i}. As $\varepsilon \to 0$ this term dominates all other terms and makes α_i negative. The network is exponentially unstable by (III.54).

The algebraic condition $\beta_i(\gamma,\mathbf{j}) < 0$ gives little insight into the network features that could cause instability. We now develop the machinery to interpret this condition diagrammatically by constructing a diagrammatic representation of $\det \mathsf{V}$. The same technique also works for $\beta_i(\gamma,\mathbf{j})$. Recall that

$$\det \mathsf{V} \equiv \sum_{\gamma \in P} \mathrm{sgn}(\gamma) V_{1\gamma(1)} V_{2\gamma(2)} \cdots V_{n\gamma(n)} \tag{V.17}$$

where P is the set of $n!$ permutations (considered as n-vectors) of the integers $1, \ldots, n$, and $\mathrm{sgn}(\gamma)$ is the *sign* of the permutation γ, which is by definition $+1$ (-1) if γ is expressible as an even (odd) number of transpositions. Every permutation can be written as a product of cyclic permutations; also, a cyclic permutation of i objects can be written as $i-1$ transpositions. Hence the number of transpositions of $1, \ldots, n$ in γ is $n-c$, where c is the number of cyclic permutations in the cyclic decomposition of γ. Therefore the factor $\mathrm{sgn}(\gamma)$ in (V.17) may be replaced by $n-c$ factors of -1. However $(-1)^{n-c} = (-1)^{n+c}$, so we may associate a minus sign with each of the factors V_{ij} and a minus sign with each cycle. Then

$$\det \mathsf{V} = \sum_{\gamma \in P,\ \gamma = \mathbf{a};\,\mathbf{b};\,\mathbf{c}} C(\mathbf{a}) C(\mathbf{b}) C(\mathbf{c}) \cdots \tag{V.18}$$

where

$$C(\mathbf{a}) = -(-V_{a(1)a(2)})(-V_{a(2)a(3)}) \cdots (-V_{a(k)a(1)}) \qquad \text{(V.19)}$$

and "$\gamma = \mathbf{a}; \mathbf{b}; \mathbf{c}$" means that the cyclic decomposition of γ is given by the vectors $\mathbf{a}, \mathbf{b}, \mathbf{c}, \ldots$. Each integer $i \in [1, n]$ occurs as a component of exactly one of the vectors $\mathbf{a}, \mathbf{b}, \mathbf{c} \ldots$.

This result has a very simple diagrammatic interpretation. Each factor $-V_{a(i)a(i+1)}$ on the right of (V.19) is represented by an arrow of the current matrix diagram $\mathfrak{D}_{CM}$ corresponding to $\mathbf{j}$ (i.e., the diagram of $-\mathbf{V}(\mathbf{j})$). The arrows that correspond to (V.19) therefore form a matrix diagram cycle. Thus

$$C(\mathbf{a}) = -(\text{a matrix diagram feedback cycle}) \qquad \text{(V.20)}$$

Then (V.18) says that $\det \mathbf{V}$ is the sum of all products of matrix diagram cycles such that each product involves each of the n points exactly once, and a factor of -1 enters for each cycle in the product.

Equations (V.18) and (V.20) imply that the basic building blocks of $\det \mathbf{V}$ (hence of α_i and Δ_i) are not matrix diagram cycles but matrix diagram cycles with a minus sign. These *sign-switched cycles* are represented diagrammatically as *polygons*, with dashed or solid edges according to whether the corresponding arrows on $\mathfrak{D}_{CM}$ are dashed or solid. A barb may be added to distinguish between polygons representing current matrix cycles with arrows pointing in opposite directions, when necessary. The polygon notation is illustrated in Fig. 20. For given $\mathbf{j}$, the polygon represents the number $C(\mathbf{a}) = -(-V^{(\mathbf{j})}_{a(1)a(2)}) \cdots (-V^{(\mathbf{j})}_{a(k)a(1)})$, which is minus the number represented by the corresponding product of arrows of $\mathfrak{D}_{CM}$.

Since the polygon notation is inadequate for 2-cycles, we extend the notation as follows. A positive (feedback) 2-cycle consisting of solid arrows is drawn as a solid line segment (the 2-polygon); a negative 2-cycle is drawn as a dashed line segment; and a positive 2-cycle consisting of two dashed arrows is drawn as a line segment with a small cross (Fig. 21).

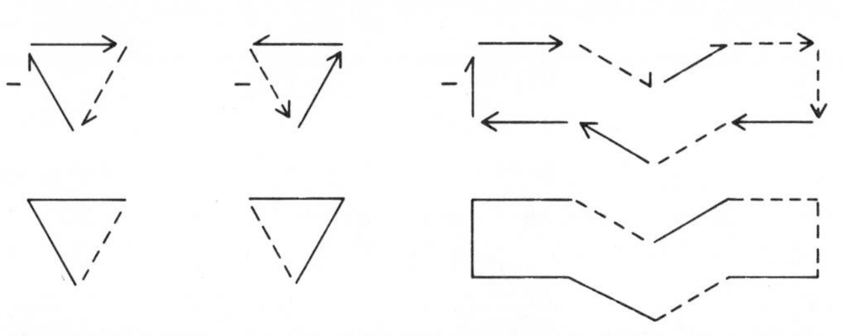

Fig. 20. Polygon notation for sign-switched cycles. The arrows of $\mathfrak{D}_{CM}$ with minus signs (top) and the corresponding polygons without minus signs (below) represent the same numbers.

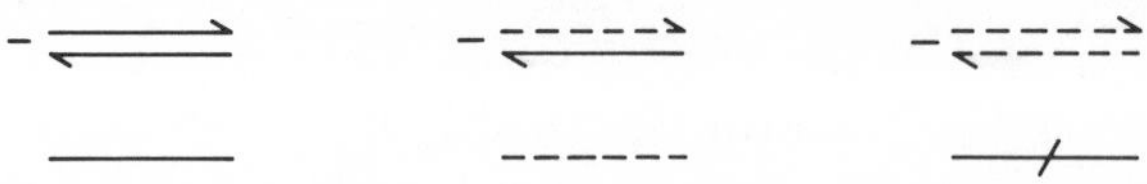

Fig. 21. Notation for sign-switched 2-cycles. The arrows of $\mathcal{D}_{CM}$ (top) are drawn (below) in polygon notation.

Since an SS-cycle or polygon represents $C(\mathbf{a})$, the polygon is *positive* (*negative*) if $C(\mathbf{a}) > 0$ [$C(\mathbf{a}) < 0$]. The corresponding feedback (matrix diagram) cycle has the opposite sign. These conventions make it easy to determine the sign of a cycle. A dashed edge of a polygon always represents an inhibition ($\mathsf{M}_{ij} < 0$) and a solid edge represents a promotion ($\mathsf{M}_{ij} > 0$) or, in the case of 2-cycles, possibly two inhibitions. Hence the feedback is positive (negative) if there are an even (odd) number of dashed edges. Since the polygon has the opposite sign from the feedback, *a polygon is positive* (*negative*) *if it has an odd* (*even*) *number of dashed edges.*

The numbers at each species (V_{ii} at X_i) on $\mathcal{D}_{CM}$ are represented as a large dot if $C(a) = V_{a(1)a(1)} > 0$, and as a small circle if $C(a) < 0$. These objects ars called *positive* and *negative stabilizers*, respectively. A negative stabilizer is called a *destabilizer* (Fig. 22).

Fig. 22. A stabilizer and a destabilizer (below) with the corresponding objects of $\mathcal{D}_{CM}$ (above).

Each term in the sum (V.18) for $\det \mathsf{V}$ is a product of stabilizers and polygons. Since each $i \in 1, \ldots, n$ appears in exactly one factor $C(\mathbf{a})$, the vertices of the polygons and the stabilizers can be placed at n points representing the n species. No two vertices can occupy the same point, and every point is occupied by a vertex (either a stabilizer or vertex of a polygon). Hence the product $C(\mathbf{a})C(\mathbf{b})C(\mathbf{c}) \cdots$ is now represented diagrammatically as the *nonoverlapping stabilizer-polygon diagram* $\mathcal{D}_{NOSP}$ $(\mathbf{a}, \mathbf{b}, \mathbf{c}, \cdots)$. Since $\det \mathsf{V}$ is the sum over all permutations γ, and since the permutations are in one-to-one correspondence with such diagrams, $\det \mathsf{V}$ is the sum of all such diagrams.

$$\det \mathsf{V}(\mathbf{j}) = \sum_{\gamma \in P} \mathcal{D}_{NOSP}(\gamma, \mathbf{j}) \qquad \text{(V.21)}$$

The sign of $\mathcal{D}_{NOSP}(\gamma, \mathbf{j})$ is the sign of $C(\mathbf{a})C(\mathbf{b}) \cdots$. A factor of -1 enters from each cycle having an even number of dashed edges, and from each destabilizer. Hence $\mathcal{D}_{NOSP}(\gamma, \mathbf{j}) < 0$ if and only if the sum of the number of destabilizers and the number of polygons having an even number of dashed edges is odd. The sign of any term can thus be read from its diagram.

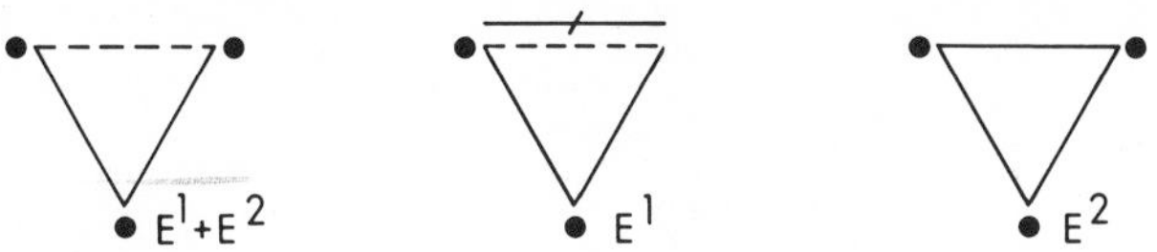

Fig. 23. $\mathcal{D}_{ASP}$ of the Oregonator currents $\mathbf{j} = (1, 1)^t$, $\mathbf{j} = (1, 0)^t$ and $\mathbf{j} = (0, 1)^t$, respectively. Compare with Fig. 16.

To construct the coefficients $\beta_i(\gamma, \mathbf{j})$ for any network, one first constructs the diagram $\mathcal{D}_{ASP}$ of *all stabilizers and polygons* (Fig. 23). Negative factors can enter into terms in β_i only from destabilizers (of which the Oregonator has none—nor has any other realistic chemical network) and from polygons with $0, 2, 4, \ldots$ dashed edges. Thus, for example, only the triangle in E^2 and the crossed line in E^1 can contribute negative factors to any term of β_i for the three Oregonator currents of Fig. 23.

When the determinant expression (V.16) for $\beta_i(\gamma, \mathbf{j})$ is expanded by the same method we used for det V, the result is a sum similar to (V.21); however only the i points $\gamma(1), \ldots, \gamma\,(i)$ are involved. Since $\mathcal{D}_{NOSP}(\gamma, \mathbf{j})$ is a nonoverlapping stabilizer polygon diagram on precisely these points,

$$\beta_i(\gamma, \mathbf{j}) = \sum \mathcal{D}_{NOSP}(\gamma, \mathbf{j}) \qquad \text{(V.22)}$$

where the sum is taken over all diagrams on these points.

The types of terms in this sum can be represented as in Fig. 24. The diagrams in these sums are called *supergraphs*, and implicit summation over all equivalent diagrams is assumed. The star represents a stabilizer of either type, while the heavy polygons represent polygons with all types of edge.

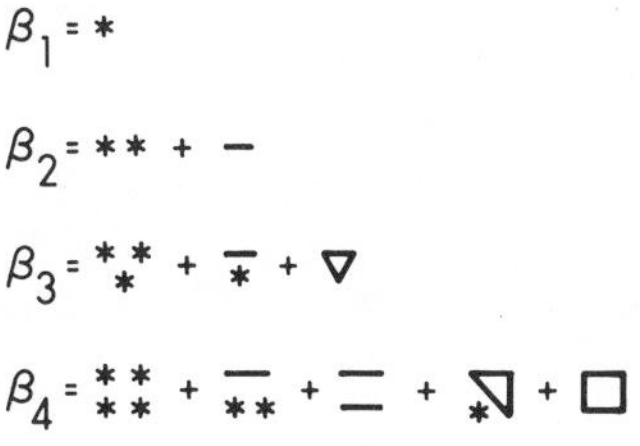

Fig. 24. Supergraph representation of $\beta_i(\gamma, \mathbf{j})$.

Example V.6. We attempt to prove unstable each of the Oregonator currents of Fig. 23 by searching for some i, γ such that $\beta_i(\gamma, \mathbf{j}) < 0$. To do this we look for polygons that can contribute negative factors to terms in $\beta_i(\gamma, \mathbf{j})$. The only such polygons are the line segment in E^1 and the triangle in E^2. The line segment connects the points x and y. Consider β_2 for these two points. The line segment is a negative term in β_2, and there are no other terms because no other set of objects from $\mathcal{D}_{ASP}$ can produce a $\mathcal{D}_{NOSP}$ on these two points. Hence

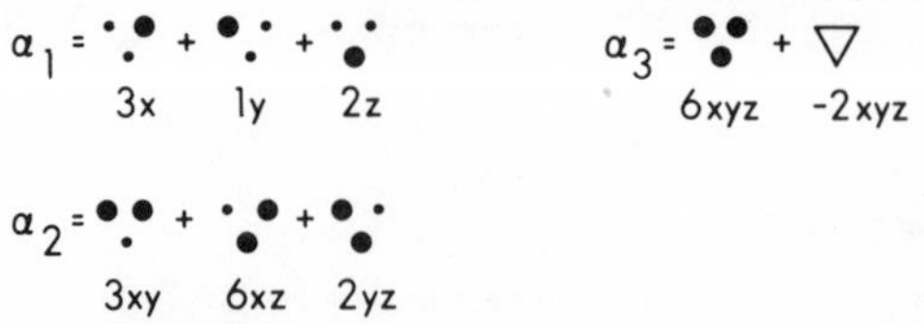

Fig. 25. Expansions of some stability polynomials of E^2 of the Oregonator as sums of nonoverlapping stabilizer-polygon diagrams. The available stabilizers and polygons are given in Fig. 23. To determine their algebraic values, see Fig. 16.

$\beta_2 < 0$; thus we have proved that E^1 is exponentially unstable. Now consider the triangle in E^2. It appears as a negative term in β_3; however there is also a positive term coming from the three stabilizers. We must compare the magnitudes of these terms by consulting $\mathfrak{D}_{CM}$ in Fig. 16. The stabilizers at x, y, z are, respectively, 3, 1, 2, so their product is 6. The arrows of the 3-cycle in the order $x \to z \to y \to x$ have 2, 1, 1 barbs, respectively, so the triangle term is -2. Hence $\beta_3 = 6 - 2 = 4 > 0$. We have failed to prove E^2 unstable. This result was expected because this current was proved to be mixing stable in Example IV.1.

We say that $\mathfrak{D}_{NOSP}(\gamma, \mathbf{j})$ is *interpreted including* **h** if the diagram is interpreted to mean $h_{\gamma(1)}h_{\gamma(2)} \cdots h_{\gamma(i)}C(\mathbf{a})C(b) \cdots$ instead of $C(\mathbf{a})$ $C(b) \cdots$. The rule for adding parameters is simple. Each vertex of a polygon and each stabilizer should be accompanied by its parameter h_i. Since $\alpha_i(\mathbf{h}, \mathbf{j})$ in (V.15) is the sum over all choices of i points, α_i may be represented by sum of all $\mathfrak{D}_{NOS_P}$'s on any i of the n points. The diagrams must be interpreted including **h**. The supergraph representation of α_i is identical with that for β_i in Fig. 24, except that besides including **h**, the implicit summations are also over all choices of i points. When drawing the diagrams, we use a light dot to indicate the points that are unoccupied. Figure 25 gives an example.

The Hurwitz determinants are sums of products of the α_i's. Hence a stabilizer-polygon diagram (SPD) expansion of Δ_i can be obtained. In making such expansions we treat each polygon and stabilizer as though it were a separate algebraic variable. Figure 26 shows that cancellation of terms in the SPD-expansion occurs (see Clarke[36]). From these cancellations we will later draw some general conclusions that are independent of the algebraic values of the polygons. Since the Hurwitz determinants are

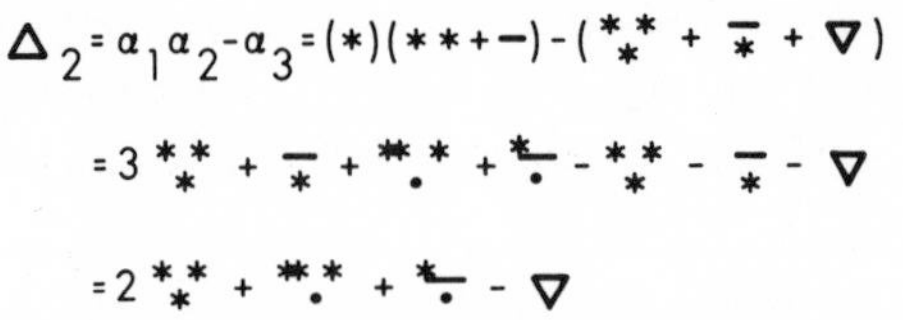

Fig. 26. Cancellation of terms in the SPD-expansion of Δ_2.

much more complicated than the α_i's, let us push our understanding of stability as far as possible using the α_i's before taking a more detailed look at the Δ_i's.

E. Theorems on Semistability

Our immediate objective is to find necessary and sufficient conditions for the linear semistability (or equivalently linear exponential instability) of stoichiometric networks. The analogous problem for sign stability was solved by the QRM theorem (see Section III.E). This problem is undoubtedly the easiest network stability problem. The theorems that will be proved do not reach this objective, although they make significant progress toward it.

First consider the stability problem for a particular current; that is, **j** is fixed and **h** takes all values in R^n_+. The The current stability problem is otherwise analogous to the network stability problem and hence we use analogous terminology. Theorem V.1 gives a sufficient condition for semistability (because stability implies semistability).

Theorem V.1. For a network N, the current $\mathbf{j} \in \overline{R}^f_+$ is linearly stable (or semistable) if every principal minor of $V_{sym} \equiv \frac{1}{2}(V(\mathbf{j}) + V(\mathbf{j})^t)$ is nonnegative.

Proof. The argument is a slight modification of that given in Section IV.A. By (III.9) and (III.10) $V(\mathbf{j}) = \sum j_i S^{(i)}$, so (IV.2) becomes

$$\zeta^t R \zeta = 2\zeta^t(\mathrm{diag}\,\mathbf{h})V_{sym}(\mathrm{diag}\,\mathbf{h})\zeta$$

where R is defined by (III.68) using $Q = \mathrm{diag}\,\mathbf{h}$. If the principal minors of V_{sym} are nonnegative, then V_{sym} is positive semidefinite, and $\zeta^t R \zeta \geqslant 0$ for ζ and all $\mathbf{h} \in R^n_+$. Hence dL_Q/dt (see III.69) cannot increase, and the steady state is stable. ∎

Theorem V.2. For a network N, the current $\mathbf{j} \in \overline{R}^f_+$ is not linearly semistable if any principal minor of $V(\mathbf{j})$ is negative.

Proof. Each principal minor is $\beta_i(\gamma, \mathbf{j})$, for some i and γ, by (V.16). Following this equation we proved that the network is exponentially unstable if any $\beta_i(\gamma, \mathbf{j}) < 0$ by showing that parameters could be chosen to make $\alpha_i < 0$. ∎

Currents that do not meet the conditions of either of these theorems are difficult to treat. The failure of the first theorem means that the mixing stability Lyapunov function will not work. The failure of the second theorem means that that α_i's are always nonnegative and that further testing for instability requires consideration of the Hurwitz determinants. Let us call these *Hurwitz nonmixing* (HNM) *currents*.

To see how both theorems can fail to apply to some $\mathsf{V}(\mathbf{j})$, consider the expansions of the minors using nonoverlapping stabilizer-polygon diagrams. The failure of the second theorem implies that every $\beta_i(\gamma, \mathbf{j})$ in the expansions of $\mathsf{V}(\mathbf{j})$ is positive. The transpose matrix has the same expansions. However the principal minors of V_{sym} are the sums of the corresponding expansions for V and V' plus cross terms. The cross terms consist of matrix feedback cycles built from the arrows of both V and V'. If new terms containing positive feedback cycles (which give a negative contribution to the minors) are sufficiently large, the first theorem can also fail. For example, the network of Example III.4 is not mixing stable and has all $\beta_i(\gamma) \geqslant 0$. It is a Hurwitz nonmixing network.

Finally, we can reexpress our earlier results on the stability of semi-directed species trees in terms of the network diagram. An *irreversible cycle* of $\mathfrak{D}_N$ is a non-self-intersecting closed path on $\mathfrak{D}_N$ in the direction of the reactions that has the following property. If any step in the cycle (say, $X \to Z$) involves the same reaction as any other step (say, $W \to Y$) or the reverse reaction of any other step, the cycle is an irreversible cycle only if the pairs of species in these two steps are not identical; that is, $\{X, Z\} \neq \{W, Y\}$. Note that $\{X, Z\} \neq \{X, Y\}$ if $Z \neq Y$.

Theorem V.3. A mass action network is linearly stable (hence semistable) if $\mathfrak{D}_N$ has no irreversible cycles.

Proof. The definition of irreversible cycle includes all cycles of $\mathfrak{D}_N$ except cycles between the species of a pair (e.g., $X \rightleftarrows Y$) using both directions of a reversible reaction. The method of constructing $\mathfrak{D}_N$ and G_{SS} results in a one-to-one matching between the irreversible cycles of $\mathfrak{D}_N$ and the directed cycles of G_{SS}. The theorem is a restatement of (IV.67). ■

F. Current Cycles

A *current cycle* is an irreversible cycle on a current diagram $\mathfrak{D}_C$. The current cycle can be traced on $\mathfrak{D}_C$ along the reactions in the proper direction from species to species until the path arrives back at the first species. Figure 27 shows two current cycles on current diagrams of the Oregonator.

If a network does not have a current cycle in any of its current diagrams, $\mathfrak{D}_N$ usually does not have an irreversible cycle. I say "usually" because there are pathological cases such as the following: $\mathfrak{D}_N$ may have a reaction that does not appear in any current because of unusual stoichiometry. If all the irreversible cycles of $\mathfrak{D}_N$ involve this reaction, there cannot be a current cycle in any current of the network because no current has this reaction. Yet $\mathfrak{D}_N$ can have irreversible cycles. These pathological cases can be eliminated by deleting from the network the reaction that can never

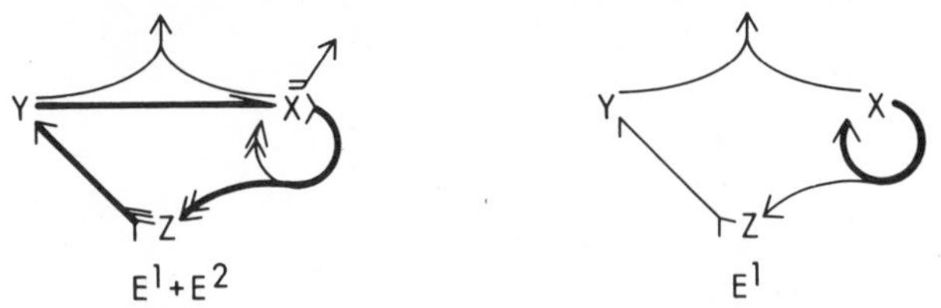

Fig. 27. The only current cycles that can be traced on the three diagrams of Fig. 15 are a 3-cycle in $E^1 + E^2$ shown on the left by a heavy line, and a 1-cycle at X in each diagram. This cycle is illustrated on the right for E^1.

occur. We then obtain a network diagram whose reactions all appear in some current. It is always possible to choose a current that has a nonzero velocity for every reaction. If $\mathfrak{D}_N$ has an irreversible cycle, this current must have a current cycle. Conversely, if no current diagram has a current cycle, $\mathfrak{D}_N$ cannot have an irreversible cycle (provided we have deleted the pathological reactions), and then the network is stable by Theorem V.3.

Current cycles are therefore necessary for exponential instability. We now construct a theory of instability in which current cycles play a key role.

Exponential instability can often be proved by showing that a principal minor of $\mathbf{V(j)}$ is negative (Theorem V.2); that is, by showing that $\beta_i(\gamma, \mathbf{j}) < 0$ for some i and γ. To make $\beta_i(\gamma, \mathbf{j})$ negative, we must have negative terms in the SPD-expansion of β_i. These terms must contain negative sign-switched cycles, that is, positive feedback cycles. One type of positive feedback cycle consists of solid arrows only. It corresponds to a current cycle. Thus current cycles can produce a negative term in $\beta_i(\gamma, \mathbf{j})$ and destabilize directly.

A destabilizer at $X_{\gamma(1)}$ makes $\beta_1(\gamma, \mathbf{j}) < 0$, and then the network is unstable. A destabilizer can come only from a solid arrow loop at $X_{\gamma(1)}$ on the matrix diagram. This loop comes from a current 1-cycle of $\mathfrak{D}_C$.

The remaining positive feedback cycles must all have a positive even number of dashed arrows on $\mathfrak{D}_{CM}$. Among these cycles are found the positive feedback $2k$-cycles consisting of dashed arrows *only*, such as those appearing in Fig. 7 (case $2k = 2$). These cycles cannot destabilize a network without a current cycle; however we will discover later on that they can destabilize when they receive "assistance" from a current cycle.

Finally there are positive feedback cycles that have both dashed arrows and solid arrows. When such cycles can be constructed from $\mathfrak{D}_{CM}$, a corresponding *knot cycle* must exist on the knot graph. The solid arrows correspond to lines between knots (or between a knot and itself) and the dashed arrows make the transitions between species within the knots. This

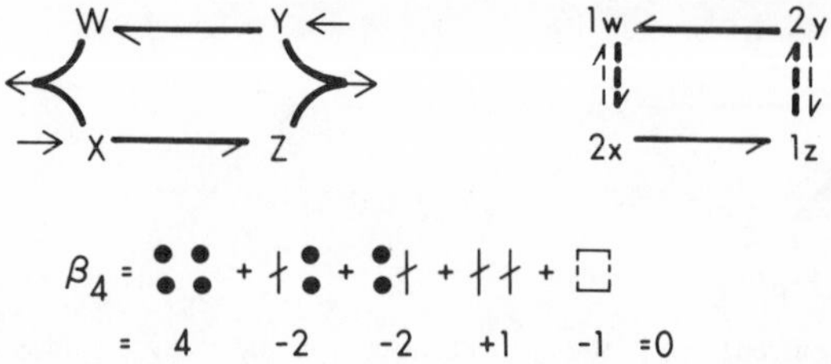

Fig. 28. A noncurrent knot cycle (heavy lines) on $\mathcal{D}_C$ (left) and the corresponding feedback cycle (heavy lines) on $\mathcal{D}_{CM}$ (right).

type of knot cycle is called a *noncurrent* (NC) *knot cycle*. A current cycle also produces a knot cycle with no dashed arrows called a *current* (C) *knot cycle*.

Figure 28 shows the current diagram of a network that has no current cycle. Hence this network is linearly stable. On the current diagram, the parts of the arrows that correspond to a noncurrent knot cycle are darkened. Note how the "path" travels between the species W and X. This is not a legitimate path for current cycles, but it is a legitimate path for knot cycles. Let us call it a *knot path* on the current diagram. From this point on, we no longer need to refer to knot graphs because we can easily visualize knot cycles on the current diagram. The right-hand side of Fig. 28 shows the feedback cycle of $\mathcal{D}_{CM}$ that results from this NC-knot cycle. The SPD-expansion shows that the NC-knot cycle comes very close to producing an instability. If β_4 were negative instead of zero, instability would result.

In the example of Fig. 28, $\alpha_4 = \beta_4 = 0$ and rank $\underline{\nu} = 4$. Hence this network has a zero eigenvalue and is not asymptotically stable. This example proves that asymptotic stability is not a general consequence of the absence of current cycles.

A current i-cycle $X_{\gamma(1)}, \ldots, X_{\gamma(i)}, X_{\gamma(1)}$ is *strong*, *critical*, or *weak* if $\beta_i(\gamma, \mathbf{j}) < 0$, $\beta_i(\gamma, \mathbf{j}) = 0$, or $\beta_i(\gamma, \mathbf{j}) > 0$, respectively. Since every species must be consumed by a reaction that makes a positive contribution to the species' stabilizer, a stabilizer can vanish only when the species reproduces itself autocatalytically. We treat autocatalysis as a 1-cycle, and say the 1-cycle is strong, critical, or weak if $\beta_1 < 0$, $\beta_1 = 0$, or $\beta_1 > 0$, respectively. Then Theorem V.2 implies the following result.

Theorem V.4. A network is exponentially unstable if it has a strong current cycle.

A destabilizer is a strong current 1-cycle; thus the presence of a destabilizer implies exponential instability.

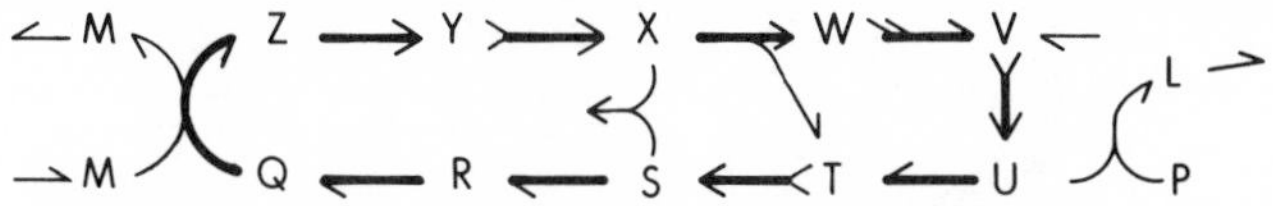

Fig. 29. A nonsimple current cycle (heavy line) that has a short circuit because the reaction $X + S \to \square$ consumes two species in the cycle, and a second short circuit because the reaction $X \to 2W + T$ produces the species T, which is farther on along the cycle.

Under what conditions is a current *i*-cycle strong, critical, or weak? This question has an easy answer when the current cycle is "simple." A current cycle is said to be *simple* if no single arrow of $\mathfrak{D}_{CM}$ short-circuits the cycle. Short circuits can be caused by a reaction that consumes two species in the cycle (Fig. 29), or by one reaction in the cycle having a product that is farther along the cycle. For example, the current cycle $X \to Z \to Y \to X$ at the left in Fig. 27 is not simple because the reaction $2X \to 4X + 4Z$ short-circuits X to X. This short circuit is itself the simple current cycle at the right in the same figure. Note that a cycle that has short circuits involving more than one reaction can be simple. Such short circuits are unimportant because they involve intermediate species that are not part of the cycle; thus diagrams involving such species do not occur in the expansion of $\beta_i(\gamma, \mathbf{j})$.

The following terminology helps us discuss how *noncyclic reactions* interact with the *cycle reactions* and *cycle species* of a particular current cycle. An *entrance reaction* is a reaction that produces a species of the cycle but does not consume any cycle species. An *exit reaction* consumes a cycle species but does not produce one. A *tangent reaction* is any cycle reaction whose form is not $X \to Y$. For example, in Fig. 29 $\square \to V$ is an entrance reaction, $P + U \to L$ is an exit reaction, and $Q + N \to M + Z$ and $X \to T + W$ are the only tangent reactions. The order of kinetics of an exit reaction is called the *exit order* and the order of kinetics of a cycle reaction is called the *cycle order*.

Theorem V.5: The Simple Current Cycle Theorem. If $X_{\gamma(1)} \cdots X_{\gamma(i)}$ is a simple current cycle, then

1. The cycle is critical if
   a. The cycle has no entrance reactions, and
   b. for every species $X_{\gamma(j)}$ in the cycle, the exit order of all exit reactions (if any) at $X_{\gamma(j)}$ equals the cycle order at $X_{\gamma(j)}$.
2. The cycle can only be weakened by
   c. Entrance reactions.
   d. Exit reactions whose exit order exceeds the cycle order.
3. The cycle can only be strengthened by
   e. Exit reactions whose exit order is less than the cycle order.

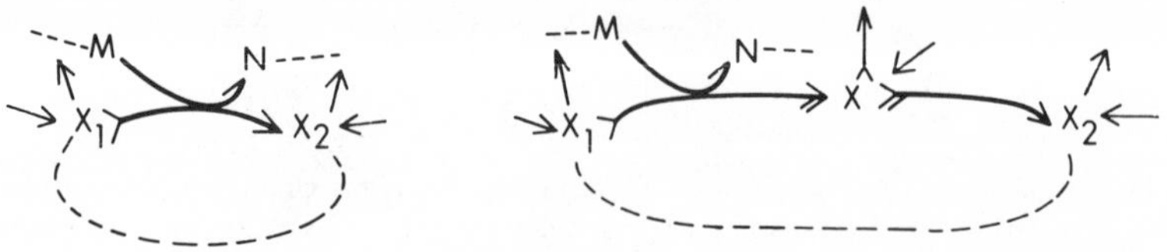

Fig. 30. Adding a new species X between X_1 and X_2 in a simple current cycle.

Proof. $\beta_i(\gamma, \mathbf{j})$ is the sum of all nonoverlapping stabilizer-polygon diagrams on the points $h_{\gamma(1)} \cdots h_{\gamma(i)}$. Since the cycle is simple, no polygon can be constructed using any proper subset of the species. Hence $\beta_i(\gamma, \mathbf{j})$ consists of only two terms. One is the product of stabilizers $S_i = s_{\gamma(1)} \cdots s_{\gamma(i)}$ for all species, and the other is the polygon P_i corresponding to the cycle.

Figure 30 illustrates how a current cycle with the species X_1 and X_2 may be extended by adding a new species X between them. Any number of entrance and exit reactions may involve X; the kinetics of all reactions at X are arbitrary. The only restriction is that the total numbers of barbs and feathers at X must be equal [condition (V.8)]. Conversely, one may delete a species to make the cycle smaller. The theorem is proved by deleting all species in the cycle except one, then building up the cycle again by adding one species at a time. It does not matter how entrance, exit, and tangent reactions are treated when species are deleted and added, because only the cycle arrows and the stabilizers of cycle species affect β_i. When a species is added to or deleted from a simple current cycle, the cycle remains a simple current cycle.

Consider a simple current cycle with j species, and let S_j and P_j be the stabilizer and polygon terms in β_j for the cycle. Let us calculate the new polygon term that results after the species X has been added as in Fig. 30. For the arrow in the cycle, let L_1 be the number of left feathers at X_1 and let B_2 be the number of barbs at X_2. Then the arrow from X_1 to X_2 had the value $L_1 B_2$ prior to the addition of X. Afterward this arrow is replaced by two arrows. If B is the number of barbs on the new cycle arrow producing X, and if L is the number of left feathers on the cycle arrow removing X, the two new arrows have values $L_1 B$ and LB_2. Hence P_j previously contained the factor $L_1 B_2$, which, on the addition of X, has been replaced by $(L_1 B)(LB_2)$. The new polygon term is thus $P_{j+1} = BLP_j$.

To calculate the new stabilizer term, first note that because the new cycle is simple, an autocatalytic 1-cycle cannot occur at X. Then expression (V.12) gives the following expression for the stabilizer at X:

$$s_X = \sum_{k=1}^{r} T(X, A_k) L(X, A_k) \tag{V.23}$$

Consider the case where conditions 1a and 1b of Theorem V.5 are satisfied. Since every exit order equals the cycle order, (V.23) factors to become $s_X = L\sum T(\mathrm{X}, A_k)$. Since there are no entrance reactions, (V.8) becomes $B = \sum T(\mathrm{X}, \mathrm{A}_k)$. Hence $s_X = BL$, and the new stabilizer term in β_{j+1} is $S_{j+1} = BLS_j$. Then

$$\beta_{j+1} = S_{j+1} + P_{j+1} = BL(S_j + P_j) = BL\beta_j \qquad \text{(V.24)}$$

Hence β_{j+1} vanishes if and only if β_j vanishes. Now consider a current 1-cycle at X satisfying 1a and 1b. The stabilizer at X has the same contribution from the outgoing reactions as was calculated above, namely BL. The current cycle is a single arrow that may be considered as a positive feedback 1-cycle. This arrow has the value BL (calculated above) and is subtracted from the contribution from the outgoing reactions to yield $\beta_1 = BL - BL = 0$. Then, from (V.24) applied many times, $\beta_j = 0$. Case 1 is proved.

To prove case 2, consider a cycle containing entrance reactions and/or exit reactions at X whose exit order exceeds the cycle order. Then $L(\mathrm{X}, A_k) \geqslant L$ for all A_k consuming X. Strict inequality holds when the exit order exceeds the cycle order. From (V.23) and (V.8),

$$s_X \geqslant L\sum_k T(\mathrm{X}, A_k) = L\Big(B + \sum{}' B(\mathrm{X}, A_k)\Big) \geqslant LB \qquad \text{(V.25)}$$

where the prime on the second summation means that the cycle reaction producing X is omitted. When 2c and/or 2d apply at X, one of the inequalities in (V.25) is strict. Thus $S_{j+1} > BLS_j$ and

$$\beta_{j+1} = S_{j+1} + P_{j+1} > BL(S_j + P_j) = BL\beta_j \qquad \text{(V.26)}$$

Hence situations 2.c or 2.d at X increase β; thus the cycle becomes weaker.

To prove case 3, consider that there are no entrance reactions at X, that $L(\mathrm{X}, A_k) \leqslant L$ for all A_k, and that strict inequality holds for some A_k. From (V.23) and (V.8)

$$s_X < L\sum_k T(\mathrm{X}, A_k) = LB$$

Then, as above,

$$\beta_{j+1} = S_{j+1} + P_{j+1} < BL(S_j + P_j) = BL\beta_j \qquad \text{(V.27)}$$

Hence situation 3.e decreases β, making the cycle stronger.

We have shown that situations 2.c and 2.d make the cycle weaker, that situation 3.e makes the cycle stronger, while situations 1.a and 1.b do not make the cycle stronger or weaker. Every current cycle satisfies 1.a or 2.c;

every exit reaction must have an exit order that equals, is less than, or is greater than the cycle order. Since every feature that can affect $\beta_i(\gamma, \mathbf{j})$ is covered by these cases, situations 1.b and 2.c are the *only* way a simple current cycle can be weakened and situation 3.e is the *only* way it can be strengthened. ■

The following example illustrates how the strengthening and weakening situations can cancel to produce a critical simple current cycle.

Example V.7. The middle current diagram in Fig. 31 has the simple current cycle W→X →Y→Z→W. Part 2 of Theorem V.5 applies at W and Y, thereby increasing β_4. Part 3 applies at X and Z, thereby decreasing β_4. The stabilizers are 4, 6, 4, 6, respectively, and the arrows are also 4, 6, 4, 6. Thus $\beta_4 = 0$. The effects of parts 2 and 3 of the theorem have canceled. This current is not extreme, and is one-tenth of the sum of the two extreme current diagrams at the left and right. Numbers indicate T and B for each arrow tail and head. The two extreme currents satisfy part 2 of the theorem at W and Y and part 1 at only one species. Thus we suspect (and a calculation verifies) that $\beta > 0$ for the simple current cycle WXYZW in both extreme currents. It turns out that β_4 as a function of $\mathbf{j}$ is always nonnegative and has a minimum of zero halfway between these two extreme currents. Incidentally, this is the simplest example I have been able to construct where $\beta_i(\gamma, \mathbf{j})$ goes through a minimum and vanishes at the minimum. Note that the example is very unrealistic.

We now obtain necessary and sufficient conditions for the vanishing of any $\beta_i(\gamma, \mathbf{j})$ where we let $\Sigma \equiv \{X_{\gamma(1)}, \ldots, X_{\gamma(i)}\}$ be *any* subset of the species of any network $N \in \mathfrak{N}$. We begin by defining a new network N^* whose species set is Σ, and whose reactions are obtained from the reaction of N by treating all species not in Σ as external species (see Section II.A). A simple way to picture the corresponding current of N^* is to begin with the current diagram $\mathfrak{D}_C$ of N, to erase all species not in Σ, and to erase all arrow branches to and from these species. The result is $\mathfrak{D}_C^*$. Since this procedure does not alter the arrow tails emanating from any species in Σ, the stabilizers of these species will be the same in $\mathfrak{D}_{CM}$ and $\mathfrak{D}_{CM}^*$. Neither does the procedure affect the arrows of $\mathfrak{D}_{CM}$ between pairs of species in Σ. Hence all the stabilizers and arrows of $\mathfrak{D}_{CM}$ that occur in $\beta_i(\gamma, \mathbf{j})$ appear on $\mathfrak{D}_{CM}^*$. Let $\mathbf{V}^*(\mathbf{j})$, $\underline{\boldsymbol{\nu}}^*$, $\underline{\boldsymbol{\kappa}}^*$ refer to N^*. Since N^* has the i species of Σ, by (V.16)

$$\beta_i(\gamma, \mathbf{j}) = \det \mathbf{V}^*(\mathbf{j}) \tag{V.28}$$

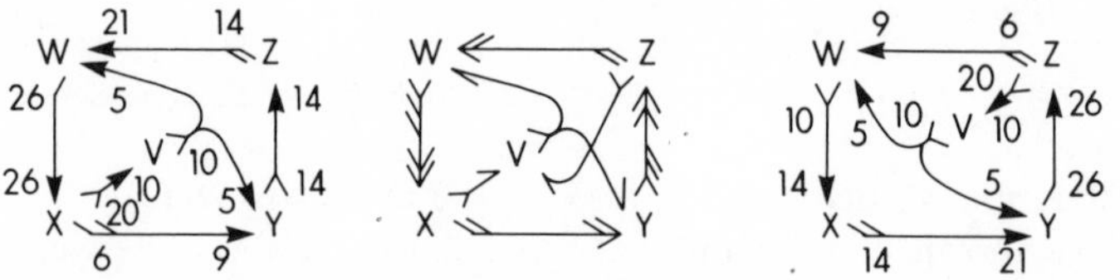

Fig. 31. The middle $\mathfrak{D}_C$ is a mixture of equal parts of the $\mathfrak{D}_{EC}$'s at the right and left. For the simple current cycle WXYZW, β_4 has a minimum as a function of $\mathbf{j}$ at the middle $\mathfrak{D}_C$.

If $\underline{\nu}^*$ is the corresponding current stoichiometric matrix (i.e., $\underline{\nu}^*\mathbf{e} = 0$ and $\mathbf{e} = \mathsf{E}\mathbf{j}$), then by (V.11) $\beta_i(\underline{\gamma}, \mathbf{j})$ vanishes if and only if $\det \underline{\nu}^*\underline{\kappa}^* = 0$.

Theorem V.6. $\operatorname{Det} \underline{\nu\kappa}' = 0$ if and only if at least one of the following conditions hold:

1. The rows of $\underline{\nu}$ or $\underline{\kappa}$ are linearly dependent.
2. There exists $\mu \in R^n$ such that $\underline{\kappa}'\mu$ is a nonzero vector in S_v (the null space of $\underline{\nu}$).

Proof. In case 1 there exists $\lambda \in R^n$ such that either $\lambda'\underline{\nu} = 0$ or $\lambda'\underline{\kappa} = 0$. In the first case $\lambda'\underline{\nu\kappa}' = 0$, so the rows of $\underline{\nu\kappa}'$ are linearly dependent and $\det \underline{\nu\kappa}' = 0$. In the second case $\underline{\nu\kappa}'\lambda = 0$, so the columns of $\underline{\nu\kappa}'$ are linearly dependent and $\det \underline{\nu\kappa}' = 0$. In case 2 $\underline{\kappa}'\mu$ is in the null space of $\underline{\nu}$, so $\underline{\nu\kappa}'\mu = 0$. Thus as before, we conclude that $\det \underline{\nu\kappa}' = 0$. The "if" part of the theorem is proved.

To show that 1 or 2 must hold if $\det \underline{\nu\kappa}' = 0$, note that this determinant can vanish only if the columns are linearly dependent. Hence there exists $\mu \in R^n$ such that $\underline{\nu\kappa}'\mu = 0$. Hence either $\underline{\kappa}'\mu = 0$ (case 1), or $\underline{\kappa}'\mu$ is in the null space of $\underline{\nu}$ (case 2). ∎

This theorem may be used to find all sets of species for which $\beta_i(\underline{\gamma}, \mathbf{j}) = 0$. We already know from Theorem V.5 that β vanishes for the species on a simple current cycle, provided there are no entrance reactions and all exit orders equal the corresponding cycle order. We now investigate when $\beta = 0$ for current cycles that are not simple.

Figure 32 shows a number of ways that short circuits can occur in a current cycle. Of the many possibilities, only one feature is important. It is the presence of two or more branches that are not part of the current cycle at the tail of the arrow. Thus the short-circuit reactions *a*, *b*, and *e* do not have a branched tail in Fig. 32. Although reactions *c*, *d*, *f*, and *g* do, one of the branches in the tails of *c* and *g* is part of the current cycle. Hence only one of the tail branches of *c* and *g* forms part of the short circuit; therefore these reactions do not have the important feature. The only reactions with the feature are *d* and *f*. Such reactions will be called *multiple short-circuit tail reactions.* Single short-circuit tail reations sometimes have a second tail that is part of the current cycle. The species that is consumed by this tail is called the *cycle-tail species of the single short-circuit tail reaction.* Examples are X_{11} and X_7, of Fig. 32, which are the cycle-tail species of reactions *c* and *g*, respectively.

Theorem V.7 proves that $\beta = 0$ even when a current cycle has arbitrarily many short circuits that resemble reactions *a*, *b*, or *e* in Fig. 32. Short circuits like *c* or *g* may be present with one minor restriction. The theorem does not apply when short circuits such as *d* or *f* are present.

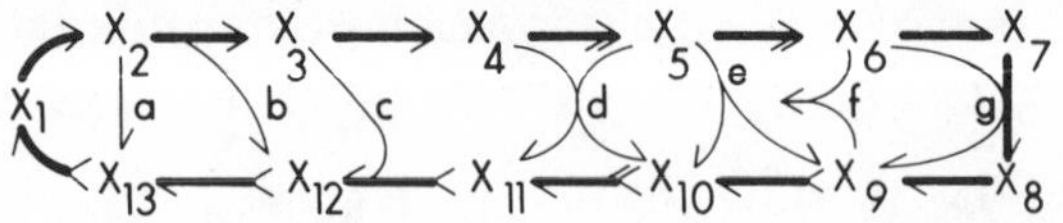

Fig. 32. Some ways in which a current cycle may have short circuits.

Theorem V.7: Critical Current Cycle Theorem. Let $\Sigma \equiv \{X_{\gamma(1)}, \ldots, X_{\gamma(k)}\}$ be the species on a current cycle of $\mathfrak{D}_C$ for the current **Ej**. Then $\beta_k(\gamma, \mathbf{j}) = 0$ if

1. For every $X_i \in \Sigma$, all reactions consuming X_i have the same order of kinetics in X_i.
2. No $X_i \in \Sigma$ has an entrance reaction.
3. The current cycle has no multiple short-circuit tail reactions.
4. The cycle-tail species of every single short-circuit tail reaction is consumed by only that reaction.

Proof. As discussed previously, we may declare all species that are not in Σ to be external species and treat the network containing only the k species Σ. Without loss of generality we may assume that $\underline{\nu}$ is the $n \times r$ current stoichiometric matrix (i.e., $\underline{\nu}\mathbf{e} = 0$ and $\mathbf{e} = \mathbf{Ej}$) of such a network. Thus $n = k$. Then $\beta_k(\gamma, \mathbf{j}) = 0$ if and only if $\det \underline{\nu}\underline{\kappa}^t = 0$. When condition 1 holds, all nonzero elements in each row of $\underline{\kappa}$ are equal. Hence there is a vector $\mathbf{K} \in R^n$ such that

$$\underline{\kappa} = (\operatorname{diag} \mathbf{K})\, \underline{\kappa}^* \tag{V.29}$$

where $\underline{\kappa}^*$ is a matrix whose elements are 0 or 1. Then

$$\det \underline{\nu}\,\underline{\kappa}^t = \det(\underline{\nu}\,\underline{\kappa}^{*t})\det(\operatorname{diag} \mathbf{K})$$

Thus if $\det(\underline{\nu}\mathbf{k}^{*t})$ equals zero, so does $\det(\underline{\nu}\underline{\kappa}^t)$, and so does $\beta_k(\gamma, \mathbf{j})$. Without loss of generality, we may prove $\det(\underline{\nu}\underline{\kappa}^t) = 0$ only when $\underline{\kappa} = \underline{\kappa}^*$ (i.e., $\kappa_{ij} \in \{0, 1\}$).

Let $\mathbf{e}_n$ and $\mathbf{e}_r$ be integer vectors consisting of n and r ones, respectively. Each column of $\underline{\kappa}$ (row of $\underline{\kappa}^t$) has ones to indicate the species that are on the left of the reaction and zeros to indicate the species that are not. (It is a convention to write all species X_i on the left of R_j for which $\kappa_{ij} \neq 0$. In the case of autocatalysis, species must be added to both sides to achieve this. If any other species are on the left for which $\kappa_{ij} = 0$, assumption 1 implies that all reactions consuming this species have zero-order kinetics. Since this gives $\underline{\kappa}$ a row of zeros, we set the appropriate component of $\mathbf{K}$ equal to zero and place ones in $\underline{\kappa}$ to indicate the species on the left.) Define $\mathbf{y} \equiv \boldsymbol{\kappa}^t\mathbf{e}_n$. Then y_j is the number of species on the left of R_j.

Condition 2 says that every reaction has some species on the left. Hence $y_j \neq 0$ for $j = 1, \ldots, n$. The reactions that consume more than one species must have one tail in the current cycle (Fig. 32), for otherwise they would be multiple short-circuit tail reactions, which are forbidden by condition 3. The same condition implies that these reactions cannot consume more than two species. Hence $y_j = 2$ if R_j has two tails, of which one is part of the current cycle and the other is a short circuit (see examples *c* and *g* in Fig. 32). Otherwise $y_j = 1$. Define $\mathbf{z}$ by the equation

$$\mathbf{y} = \underline{\boldsymbol{\kappa}}^t \mathbf{e}_n = \mathbf{e}_r + \mathbf{z} \tag{V.30}$$

Then $z_j = 0$ for all reactions R_j except those having a second tail. These are the single short-circuit tail reactions and for them $z_j = 1$. Let $\mathbf{u}$ be an n-vector such that $u_i \equiv 1$ if X_i is the cycle-tail species of any single short-circuit tail reaction, and $u_i \equiv 0$ otherwise. Condition 4 implies $\underline{\boldsymbol{\kappa}}^t \mathbf{u} = \mathbf{z}$. Substituting for $\mathbf{z}$ in (V.30) and rearranging gives

$$\underline{\boldsymbol{\kappa}}^t (\mathbf{e}_n - \mathbf{u}) = \mathbf{e}_r$$

If we define $\boldsymbol{\mu} = \mathbf{e}_n - \mathbf{u}$, then $\underline{\boldsymbol{\kappa}}^t \boldsymbol{\mu}$ is the vector $\mathbf{e}_r$, which is in the null space of $\underline{\boldsymbol{\nu}}$, because $\underline{\boldsymbol{\nu}} \mathbf{e}_r = 0$. Hence by part 2 of Theorem V.6 $\det \nu \kappa^t = 0$. Therefore $\beta_k(\gamma, \mathbf{j}) = 0$. ■

Example V.8. M. Eigen's theory[68] of the origin of life is based on "autocatalytic hypercycles.". The network consists of a set of amino acid polymers (proteins) X_i and a set of ribonucleotide polymers (e.g., RNA) Y_i. It is assumed that each X_i catalyzes the formation of some Y_k in reactions of the form

$$X_i \rightarrow X_i + Y_k$$

An external species provides the atoms to form Y_i. Similar reactions occur in which Y_k catalyzes X_l, X_l catalyzes Y_m, Y_m catalyzes $X_n, \ldots,$ until eventually an earlier species such as X_i is produced. This network can have a steady state only if we append reactions that remove all species. First-order degradation reactions are the most reasonable. A possible resulting current diagram is shown in Fig. 33. The hypercycle is clearly a current cycle that contains numerous short circuits of the same type as reaction *b* in Fig. 32. Theorem V.7 implies that $\beta_n = 0$ for the *n* species on the hypercycle. From the diagram, the stabilizer at

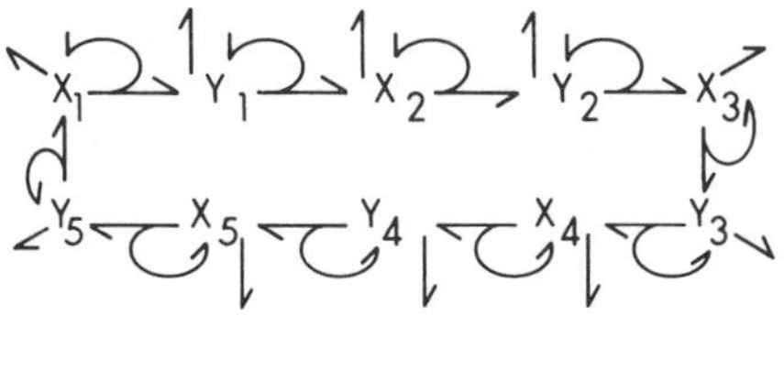

Fig. 33. Current diagram of an "autocatalytic hypercycle."

each species is $+1$, and no other matrix diagram cycles occur. Hence from the SPD-expansion we get for $i = 1, \ldots, n-1$

$$\alpha_i = \sum_{\gamma \in A(n,i)} h_{\gamma(1)} \cdots h_{\gamma(i)}$$

and $\alpha_n = 0$. Hence for all $\mathbf{h} \in R_+^n$, $\alpha_i(\mathbf{h}, \mathbf{j}) > 0$ for $i = 1, \ldots, n-1$. I conjecture that $\Delta_i(\mathbf{h}, \mathbf{j}) \geqslant 0$ for $i = 1, \ldots, n-1$. This has been verified by computer for $n = 6$. Thus one eigenvalue is zero and the rest have negative real parts. From Table II we conclude the current is linearly marginally stable.

With the aid of the theorem, one can construct infinitely many other current cycles that have $\beta_n = 0$. Most of these will also be linearly marginally stable. Thus the autocatalytic hypercycle does not seem to have any special mathematical properties that these other cycles do not also possess. One would expect that biochemical evolution would make use of current cycles with other kinds of short circuit if the chemical reactions were as chemically plausible as those in the hypercycle.

Theorem V.7 generalized part 1 of Theorem V.5 from simple current cycles to the general current cycle. We now generalize the remainder of Theorem V.5; however, for brevity, the arguments are not given in detail.

Consider a network that satisfies conditions 1 to 4 of Theorem V.7. Since $\beta_k = 0$, the SPD-expansion of β_k has an exact cancellation of terms. Select any species X_i and imagine the complete SPD-expansion of β_k. Among the terms in the expansion that contain the stabilizer at X_i, we now make all possible cancellations. Let the numerical value of this sum be S and the numeric value of all the polygon terms through X_i be P. Since $\beta_k = S + P = 0$, we must have $S = -P$. Now, negative terms come from positive feedback cycles. Hence we expect most negative terms to be in P; thus $P < 0$, $S > 0$. The conclusion is not rigorous; however it is reasonable. A rigorous argument would be too long. If $S > 0$, any increase in the stabilizer at X_i that does not affect the arrows through X_i will increase S without affecting P, so β_k increases. Next modify the network by adding an entrance reaction at X_i and satisfy the current (steady-state) condition by increasing the number of barbs on the outgoing reaction. The increase in the number of these barbs causes the stabilizer to increase. The incoming arrows to X_i on $\mathfrak{D}_{CM}$ are not affected. The outgoing arrows depend only on the kinetic order and are not affected. Hence S increases and P remains unchanged, so β_k increases. Making the order of kinetics of an exit reaction larger or smaller than the cycle order increases or decreases S while leaving P unchanged. Thus β_k increases or decreases accordingly. Thus we have proved that parts 2 and 3 (except for the word "only") are true in general, provided $S > 0$.

The only features that have not been discussed are multiple short-circuit tail reactions such as d and f in Fig. 32. What do these do to β? The following example illustrates the general case.

Example V.9. Consider the current of Fig. 34a; $\mathscr{D}_{CM}$ and the SP-diagram of expansion of β_4 appear in 34b and 34c, respectively. Note that if the short-circuit reaction is replaced by two exit reactions at W and Y, the only change in Fig. 34b is the removal of the two dashed arrows, and then only the first two terms in β_4 remain. These cancel (in general) by Theorem V.7. We see that the short-circuit reaction produces an extra negative term in β_4 from the 2-cycle and stabilizers, and *two* positive terms consisting of negative feedback cycles and stabilizers. Thus the short-circuit reaction increases β_4.

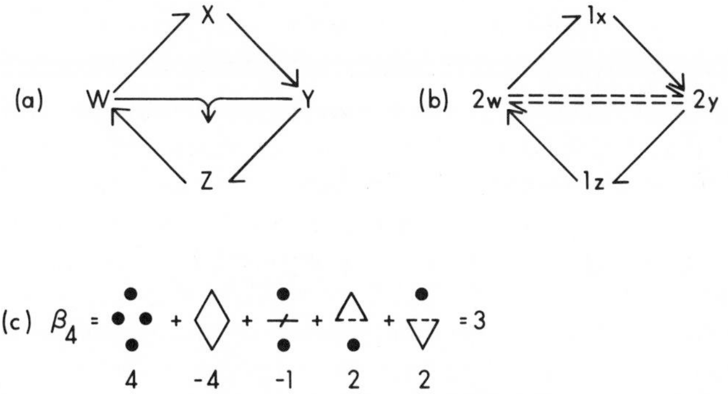

Fig. 34. (a) A network, (b) $\mathscr{D}_{CM'}$ and (c) the SP-diagram expansion of β_4.

It should be clear that in general two or more exit reactions can be converted into a multiple short-circuit tail reaction. The effect will always be to add a pair of negative feedback cycles and one positive feedback cycle for every pair of reactants that are not cycle-tail species. In some circumstances the addition of arrows of $\mathscr{D}_{CM}$ will obscure this conclusion; nevertheless, β_k increases as long as the introduction of short circuits does not make the rows of $\underline{\boldsymbol{\nu}}$ or $\underline{\boldsymbol{\kappa}}$ linearly dependent. See Fig. 35.

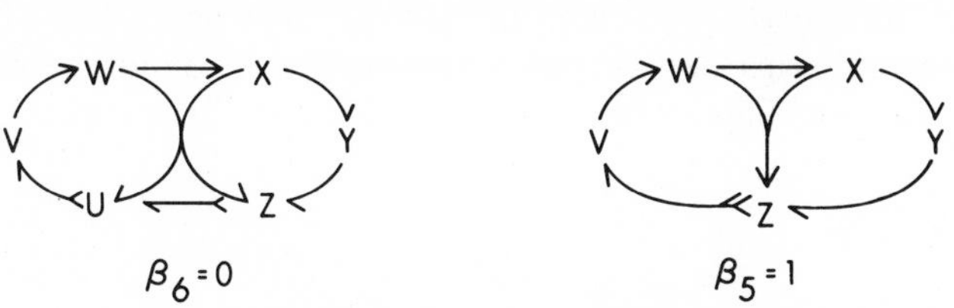

Fig. 35. For the current at the left $\beta_6 = 0$ because the shortcircuit makes the rows of $\boldsymbol{\nu}$ linearly dependent. When this does not happen, as in the example at the right, the current cycle is weakened.

Incidentally, to prove that $\beta_6 = 0$ for the current on the left of Fig. 35, it is sufficient to observe that the short-circuit reaction can be suppressed by changing the stoichiometries to $2W \to 4X \to 4Y \to 4Z \to 2U$; hence the current is not extreme. Thus **e** is not the only vector in S_v, and $\dim S_v > 1$. When there are seven reactions, $\dim S_v > 1$ implies that $\operatorname{rank} \underline{\boldsymbol{\nu}} < 6$. Hence the six rows are linearly dependent.

This section has established (more or less) that $\beta_k(\gamma, \mathbf{j})$ can be negative on a current cycle *only* if there is a species having an exit reaction with a lower order of kinetics than the order of the cycle reaction. If this feature is not present, the minimum possible value of β_k is zero, and even this value occurs only in unusual circumstances.

G. Instability in Realistic Chemical Networks

Any network having a set of species $\Sigma = \{X_{\gamma(1)}, \ldots, X_{\gamma(i)}\}$ and a current **Ej** such that $\beta_i(\gamma, \mathbf{j}) < 0$ is exponentially unstable. Section V.F concluded that if Σ is the set of species on a current cycle, $\beta_i(\gamma, \mathbf{j})$ cannot be negative unless a species of Σ has an exit reaction whose order is less than the cycle order. We now show that this is very unlikely in realistic chemical networks. Hence $\beta_i(\gamma, \mathbf{j}) \geqslant 0$ whenever Σ is the set of species on a current cycle of a realistic network. Then we show that if there is a critical current cycle ($\beta = 0$), it is sometimes possible to find a larger set of species $\Sigma' \supset \Sigma$ such that $\beta' < 0$ for the larger set. This is how instability occurs in all the realistic networks I have examined.

Realistic elementary chemical reaction mechanisms have integer-order kinetics and the order is usually 1 or 2. Termolecular (order 3) elementary reactions can occur, although they are extremely rare. Zero-order reactions occur phenomenologically in a few cases; however the elementary mechanisms for these reactions are never zero order. Hence when dealing with elementary mechanisms, we usually may assume that every species consumed by a reaction has first- or second-order kinetics. ($\kappa_{ij} \in \{1,2\}$, when $\nu_{ij} < 0$.)

A strong current cycle ($\beta < 0$) can occur only if a species on the cycle has an exit order that is less than the cycle order. Since κ_{ij} must be 1 or 2, the exit reaction must be first order and the cycle reaction second order. Hence the current diagram must have the features shown in Fig. 36 with

Fig. 36. The simplest strong current cycle.

Fig. 37. Consequences of a premature breakup of the species V, which becomes 3X. In both cases $\beta = -1$ if $2X \rightarrow V$.

possibly some additional reactions (short circuits) feeding into X, *provided* the stoichiometry of the exit reaction is changed to satisfy the steady-state condition.

The only elementary reaction occurring in realistic networks that has the required kinetics for the cycle reaction is called a *disproportionation*. The general form is

$$2X \rightarrow Y + Z \tag{V.31}$$

Usually Y is X plus a fragment of X, and Z is X less this fragment. From Fig. 36 we see that one of the product species (say Y) must increase in size by attaching more atoms until it can break up into *three* X. It is extremely unlikely in chemistry that any molecule would break up simultaneously into three identical pieces. One piece would break off first, then the remainder would break in two. When this happens, short circuits are introduced as shown in Fig. 37 on the right. A simple calculation shows that β is the same in both cases. In general, a premature breakup of V does not change β.

In the current at the right of Fig. 37, V can be broken up into 3X, and W can be broken up into 2X. It is not very plausible chemically that the trimer V could form except from the dimer W. If W is a precursor in the formation of V, the current is modified to that shown at the left of Fig. 38; however for this current $\beta = 6 > 0$ (assuming $2X \rightarrow W$).This result is surprising because the current diagram has the current cycle $X \rightarrow W \rightarrow V \rightarrow X$ with two simple short circuits. Theorem V.7 would apply if the two

Fig. 38. Currents where the dimer W is a precursor of the trimer V. If W is formed from X in a single step, $\beta_3 = 6, 0, 8$, respectively.

reactions consuming X did not have different orders of kinetics. The three currents in Fig. 38 show that $\beta = 0$ when the kinetic orders of the reactions at X are equal, and that $\beta > 0$ when the exit order is *less than* or greater than the cycle order! Hence the current cycle cannot be strong if the dimer is a precursor of the trimer.

Example V.10. One way to achieve a strong current cycle is shown in the following scheme, which is illustrated in Fig. 39. Here A and B are atoms or molecular fragments, and the formulas are chemical structural formulas. Parentheses mean that a species is to be considered external.

$$2\text{AB} \xrightarrow{1} \text{AA} + (\text{BB})$$

$$(\text{B}) + \text{AA} \xrightarrow{2} \text{ABA}$$

$$\text{ABA} + (\text{BAB}) \xrightarrow{3} \text{ABABAB} \xrightarrow{4} \text{AB} + \text{ABAB}$$

$$\text{ABAB} \xrightarrow{5} 2\text{AB}$$

$$(\text{B}) + \text{AB} \xrightarrow{6} (\text{BAB})$$

This mechanism is just barely plausible chemically. Since AA is the only molecule where two like fragments are bound, AA must be very much less stable than AB (quantum mechanically). Thus R_1 would be slow and would have a significant forward rate only if a very good scavenger for BB were present. Reaction R_2 is reasonable because BAA is unlikely, given that AA is not as stable as AB. The reverse of R_3 (R_{-3}) would occur and would increase β_5. To have an instability, R_{-3} would have to be slower than R_4 because the equilibrium extreme current consisting of R_3 and R_{-3} would contribute a positive term to β_5, which could destroy the source of instability. Similarly, if ABAB broke up into A + (BAB) instead of reacting as in R_5, this sink for ABAB would produce another extreme current with a positive β that could make β_5 positive. We conclude that such a mechanism is unlikely to have suitable rate constants to produce an instability in a real system.

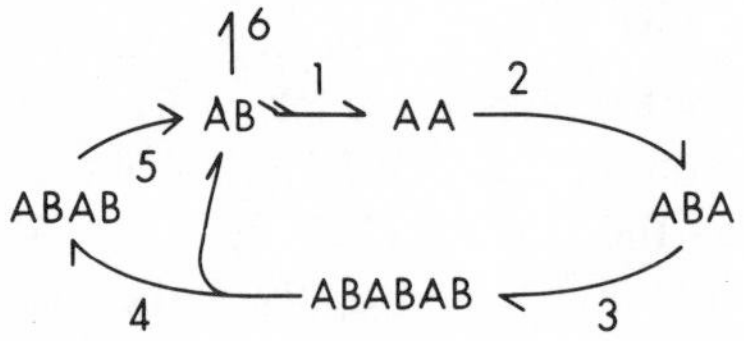

Fig. 39. An attempt to construct a "realistic" strong current cycle: $\beta_5 = -1$.

Let $\mathfrak{N}_{\text{REAL}\,I}$ be the set of networks that can produce instabilities in real chemical systems, and let $\mathfrak{N}_{NSCC}$ be the set of networks that do not have a strong current cycle ($\beta < 0$) and do not have a current cycle with a species having an exit reaction with lower order of kinetics than the cycle order. Because of the difficulty we have had constructing a realistic network containing a strong current cycle, let us conjecture that

$$\mathfrak{N}_{\text{REAL}\,I} \subset \mathfrak{N}_{NSCC} \tag{V.32}$$

If this is so, it is important to understand how instabilities can occur in the networks of $\mathfrak{N}_{NSCC}$. The theorems that follow develop a theory of instabilities in these networks. Note that the nonexistence of strong current cycles and the lack of lower order exit reactions are the main assumptions. The orders of kinetics are considered to be positive real numbers.

Theorem V.8 says roughly that if $\beta_i = 0$ for some current, there must be an extreme current where $\beta_i = 0$. The theorem has great practical importance because it enables us to discover most of the cases where some $\beta_i = 0$ by examining the f extreme currents. Without this theorem we would have to examine infinitely many currents to see if it were possible for β_i to vanish.

Theorem V.8. Let $\Sigma = \{X_{\gamma(1)}, \ldots, X_{\gamma(i)}\}$ be the species of a current i-cycle of a network $N \in \mathfrak{N}_{NSCC}$. If either (1) the current cycle is simple and critical, or (2) conditions 1 through 4 of Theorem V.7 hold; N has an extreme current **Ej** such that $\beta_i(\gamma, \mathbf{j}) = 0$.

Proof. Consider the case where condition 1 holds. Since $N \in \mathfrak{N}_{NSCC}$, there cannot be species satisfying condition 3.e of Theorem V.5: since the cycle is critical, the cycle cannot have properties 2.c and 2.d of that theorem. Hence conditions 1.a and 1.b of Theorem V.5 must hold. If N is not already an extreme network, we construct any extreme subnetwork by deleting reactions. The deletion of reactions cannot add entrance reactions to any X_i on the current cycle; neither are the kinetic orders changed. Hence conditions 1.a and 1.b of Theorem V.5 hold for the extreme subnetwork, and thus, by that theorem, $\beta_i(\gamma, \mathbf{j}) = 0$.

Consider the case where conditions 1 through 4 of Theorem V.7 hold. Let **Ej** be any extreme current of N that has a nonzero reaction velocity for some reaction in the current cycle. The corresponding extreme subnetwork is obtained by deleting reactions from N. Hence all reactions consuming X_i still have the same order of kinetics, no entrance reactions are introduced, no multiple short-circuit tail reactions are added, and no increase in the number of reactions consuming the cycle-tail species of single short-circuit tail reactions is possible. Hence conditions 1 through 4 of Theorem V.7 hold for the subnetwork and $\beta_k(\gamma, \mathbf{j}) = 0$.

The current cycle may not appear in some extreme currents because a reaction on the cycle vanishes. Hence there is a cycle species that is not produced by any reaction in the current (recall that there are no entrance reactions) and hence is not consumed. In such cases the SPD-expansion of $\beta_i(\gamma, \mathbf{j})$ has no terms, so $\beta_i = 0$. ■

As explained previously, in practice, this theorem should be turned around to read "If N does *not* have an extreme current **Ej** with a cycle on which $\beta_i(\gamma, \mathbf{j}) = 0$, there cannot be *any* current where conditions 1 through 4 of Theorem V.7 all hold for the species on the cycle."

If one believes the loose arguments at the end of the previous section that $\beta_i \geqslant 0$ for any current cycle when $N \in \mathfrak{N}_{NSCC}$, and that $\beta > 0$ when the conditions 1 through 4 of Theorem V.7 are not met, it would follow from Theorem V.8 that *whenever* β_i attains the minimum possible value $\beta_i = 0$, it attains this value at an extreme current.

Example V.11. The two extreme current diagrams of the Oregonator are given in Fig. 15. The simple current cycle $X \to X$ satisfies part 1 of Theorem V.4 for E^1, but not for E^2. Hence at X, $\beta_1^{(1)} = 0$ for E^1 and $\beta_1^{(2)} > 0$ for E^2. In general, $\beta_1(\gamma, \mathbf{j}) = 2j_2$. Thus the minimum of β_1 as a function of $\mathbf{j}$ occurs when $\mathbf{Ej}$ is an extreme current.

The theorem cannot be generalized to all current cycles because of Example V.7, where $\beta_4(\gamma, \mathbf{j})$ has an absolute minimum of zero at the nonextreme current shown in the middle diagram of Fig. 31. This is the simplest example I have been able to construct to prove that the theorem cannot be generalized beyond certain limits. This network is not in $\mathfrak{N}_{NSCC}$. A generalization of the theorem that probably can be proved for $N \in \mathfrak{N}_{NSCC}$ replaces conditions 1 and 2 with the much weaker assumption that the current cycle is critical. This generalization would be easy to prove if the discussion at the end of the preceding section could be backed up by theorems.

Example V.12. The network diagram of the original Field-Körös-Noyes mechanism[69] of a Belousov-Zhabotinski system is shown in Fig. 40. The diagram omits BrO_3^-, $CH_2(COOH)_2$, and H_2O, because these species have such high concentrations that they must be considered to be external species. It omits HCOOH and CO_2 because these species occur only on the right-hand side of reactions and hence must be products of the overall reaction; thus they are external species. It also omits H^+ because this species plays a role in so many reactions that every important extreme current probably produces H^+ in more than one reaction. Thus any

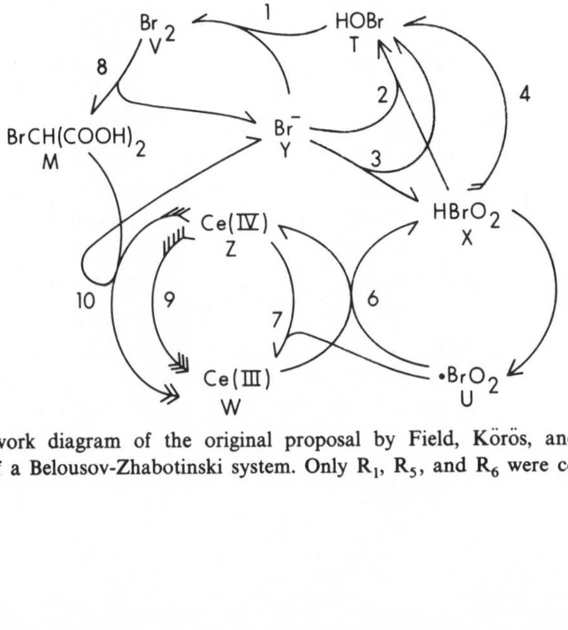

Fig. 40. Network diagram of the original proposal by Field, Körös, and Noyes for a mechanism of a Belousov-Zhabotinski system. Only R_1, R_5, and R_6 were considered to be reversible.

current cycle through H^+ must have an entrance reaction at H^+. The cycle is therefore weak and of no importance in producing instabilities.

Let us find all critical current cycles by finding all *network cycles* on $\mathcal{D}_N$. We must check that a network cycle appears on $\mathcal{D}_C$ of some extreme current and show that β vanishes. The network cycle from X via R_5 to U via R_6 to X is abbreviated X5U6. Since this could be a simple current 2-cycle, Theorem V.8 says that $\beta_2 \neq 0$ for any current unless there is an extreme current where $\beta_2 = 0$. Theorem V.5 says that such an extreme current must not involve the entrance reaction R_3 at X, or the higher order exit reaction R_4 at X. Thus X can leave the cycle only by the exit reaction R_2. We now work out the reaction stoichiometries in such an extreme current, starting with the assumption that *one* Y is consumed by R_2. Then one X is consumed by R_2. Note that $R_5 + R_6 + R_7$ constitutes an extreme current. This is the only current involving R_7 and does not produce the required X. Hence we must delete R_7; then the steady-state condition at U forces us to double the stoichiometry of R_6. Thus $R_5 + 2R_6$ gives steady-state U and the one X needed to consume one Y by R_2. It follows that two Z are formed by $2R_6$. We now must satisfy the steady-state condition at Y. Note that the steady-state condition at V makes R_8 produce the same amount of Y as R_1 consumes. Hence to satisfy the steady-state condition at Y, R_{10} must produce one Y from *four* Z. However only *two* Z are formed by $2R_6$. There is no way to obtain one Y from the two Z that are available. Thus a current having the necessary entrance and exit reaction properties at X is not present. Hence every current of this network that involves X and U has $\beta_2 > 0$ on the cycle X5U6.

There are several network cycles involving both W and Z (W6Z9, W6Z10, W6Z7). Since W and Z are the only forms of cerium in the system, W + Z is constant (a conservation law). The corresponding rows of $\underline{\nu}$ sum to zero and by condition 1 of Theorem V.6 $\beta = 0$ on every set of species containing W and Z. The vanishing of β due to a conservation condition is not significant for the production of instabilities.

The network cycle V8Y1 is also a simple current cycle. It is critical. All other cycles involving the step V8Y must use R_2 or R_3 to leave Y. Hence to meet the steady-state condition at Y, R_{10} must produce Y. Thus Y has the entrance reaction R_{10} and no such cycle can be critical. Similarly, cycles involving Z10Y can only be part of the current that produces Z and X. However X cannot be removed at steady state without ultimately involving R_8, which is an entrance reaction to Y in the cycles we are considering. Thus all such cycles have $\beta > 0$. None of the network cycles through M can satisfy the steady-state conditions; therefore they are not current cycles. The only critical cycles we have found in this example are V8Y1 and the unimportant critical cycles caused by the conservation of W + Z.

This network is probably stable. I became convinced of this shortly after it was first published, but could not prove it by constructing the Hurwitz determinant polynomials because too many terms were present. Stability can probably be proved by using the constrained mixing stable condition in Section IV.A. The crux of the problem is that the simple current cycle X5U6 cannot be critical unless 2Y are produced for every 4Z consumed by R_{10}. The stoichiometry only gives half this number of Y's. The problem can be alleviated by postulating the existence of other reactions that liberate Br^- from $BrCH(COOH)_2$ with the consumption of only two Ce(IV). Then an extreme current is possible where X5U6 is a critical cycle. At present the exact mechanism for the liberation of the required Br^- is still uncertain.

A practical method of locating critical current cycles in very complicated networks for which a current diagram is known requires five simple steps:

1. Circle all species that are produced by only one reaction of $\mathcal{D}_C$.

2. Of these, cross off any species whose consuming reactions do not all have the same order of kinetics.
3. Of the circled species remaining, darken the current pathway between pairs of them when the two species of the pair occur on opposite sides of a reaction.
4. Locate all current cycles along the darkened pathways.
5. Eliminate current cycles that contain multiple tail short-circuit reactions, or single tail short-circuit reactions whose cycle-tail species is consumed by other reactions. Current cycles selected by this procedure satisfy the conditions of Theorem V.7, hence are critical.

Example V.13. Noyes and Sharma[70, 71] have given a mechanism for the system $2H_2O_2 \rightarrow 2H_2O + O_2$, when catalyzed by oxyiodate species. This *Bray-Liebhafsky* system oscillates experimentally. The mechanism given is an extreme network whose current diagram appears in Fig. 41. We treat H_2O_2, H_2O, and $O_2(g)$ as external species; H^+ has been omitted because all current cycles through H^+ have $\beta > 0$. To find the critical current cycles, by step 1, we circle the species $I\cdot$, $IOO\cdot$, I_2, $\cdot IO$, $\cdot IO_2$, and IO_3^-. Step 2 does not eliminate any of these. The figure shows the darkened pathways between the pairs according to step 3. The only cycle that can be drawn along these pathways is the 2-cycle between $\cdot IO_2$ and IO_3^-. This is the only critical current cycle in this network.

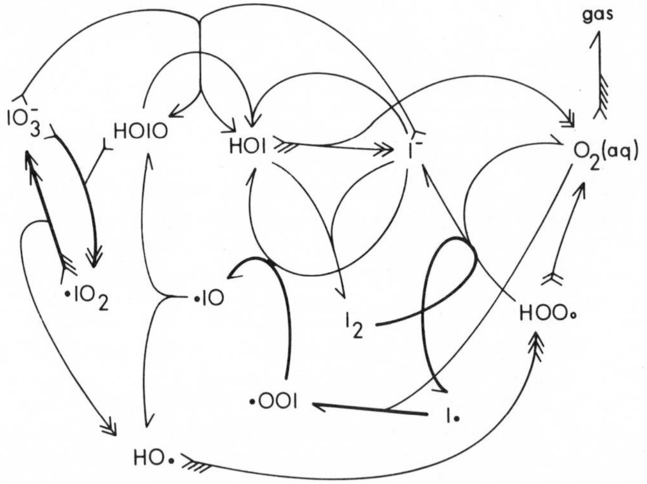

Fig. 41. $\mathfrak{D}_{EC}$ for the Noyes-Sharma mechanism of the Bray-Liebhafsky reaction.

Theorem V.9. Let $\mathfrak{D}_C$ be the current diagram of any network, let $\Sigma = \{X_{\gamma(1)}, \ldots, X_{\gamma(i)}\}$ be any set of species for which $\beta_i(\gamma, \mathbf{j}) = 0$, let Σ' be any set of j species such that $\Sigma \cap \Sigma' = \emptyset$, and consider the SPD-expansion of β_{i+j} on the set $\Sigma \cup \Sigma'$. Then all terms of the expansion that do not contain a polygon spanning Σ and Σ' may be omitted because they cancel identically.

Proof. Divide the SP-diagram terms of β_{i+j} into those having a polygon spanning Σ and Σ' and those not having a spanning polygon. Each term of the latter set is the product of an SP-diagram on Σ and an SP-diagram on Σ'. The set of all such terms therefore factors into

(sum of all SP diagrams on Σ)(sum of all SP diagrams on Σ')

The left factor is $\beta_i(\boldsymbol{\gamma}, \mathbf{j})$, which vanishes by hypothesis. Hence all terms in β_{i+j} that do not contain a polygon spanning Σ and Σ' cancel identically. ■

Theorem V.10. Let $\mathfrak{D}_C$ be a current diagram of any network. Suppose $\Sigma \equiv \{X_{\gamma(1)}, \ldots, X_{\gamma(i)}\}$ is any set of species for which $\beta_i(\boldsymbol{\gamma}, \mathbf{j}) = 0$, suppose a species $Y \notin \Sigma$ reacts directly with the species $X_1 \in \Sigma$ by a reaction of the general form

$$R : pX_1 + qY \to \text{species not in } \Sigma$$

for any $p, q > 0$, suppose $\beta_{i-1}(\gamma_2, \ldots, \gamma_i, \mathbf{j}) > 0$, and suppose no other reaction of $\mathfrak{D}_C$ involves both Y and species of Σ. Then $\beta_{i+1} < 0$ on the set $\{Y\} \cup \Sigma$, and the current is exponentially unstable.

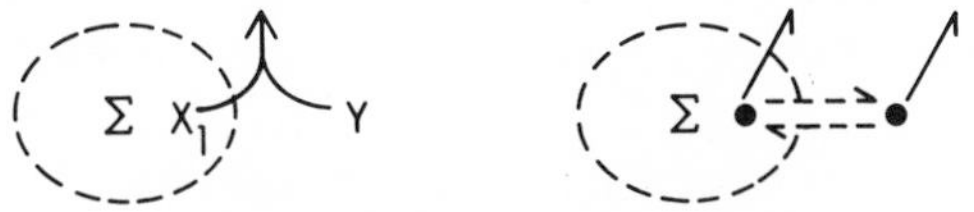

Fig. 42. A source of instability.

Proof. The situation is diagrammed in Fig. 42. Since $\beta_i(\boldsymbol{\gamma}, \mathbf{j}) = 0$, Theorem V.9 implies that β_{i+1} contains only SP-diagrams having a polygon spanning $\{Y\}$ and Σ. The only such polygon corresponds to the positive feedback 2-cycle between X_1 and Y shown in $\mathfrak{D}_{CM}$ in the right-hand diagram. Another polygon could occur only if there were another arrow of $\mathfrak{D}_{CM}$ connecting Y to Σ, but this is impossible because none of the product species of R are in Σ and no other reaction involves Y and a species of Σ. Hence $\beta_{i+1} =$ (the polygon of the positive feedback 2-cycle) β_{i-1}. Since $\beta_{i-1} > 0$, and since positive feedback cycles are negative polygons, $\beta_{i+1} < 0$. ■

Example V.14. The simplest possible example to illustrate this theorem is the current E^1 of the Oregonator, which appears in Fig. 27. If $\Sigma = \{X\}$, $\beta_1 = 0$ because the autocatalytic reproduction of X is a critical current cycle. The reaction $X + Y \to \square$ satisfies the requirements of R. Hence $\beta_2 < 0$ for the species set $\{X, Y\}$ of E^1.

Several remarks on Theorem V.10 should be made.

1. The hypothesis of this theorem cannot be satisfied if β_i vanishes because of a conservation condition involving only species in Σ. This would mean that the corresponding rows of $\underline{\boldsymbol{\nu}}$ were linearly dependent. If X_1 were not involved in the linear dependence, $\beta_{i-1}(\gamma_2, \ldots, \gamma_i, \mathbf{j}) = 0$. Otherwise, X_1 would have to be involved in the linear dependence caused by the conservation condition. However reaction R removes X_1 without producing another species in Σ, so there cannot be a conservation condition involving X and only other species in Σ.
2. The vanishing of β_i and not β_{i-1} is possible only when there are polygon terms corresponding to positive feedback cycles through X_1. (The proof should be obvious.) The simplest case is when X_1 lies on a current cycle through species in Σ. Since R is an exit reaction from this current cycle, the cycle must produce more X_1 than it consumes. Hence the cycle is equivalent to the *autocatalytic reproduction* of X via a series of intermediates in Σ.
3. If a product X_2 (say) of reaction R is in Σ, a negative feedback cycle usually exists of the form $X_1 \rightarrow Y \rightarrow X_2 \rightarrow \cdots \rightarrow X_1$. The part of the cycle from X_2 to X_1 makes use of the same arrows as occur in the positive feedback cycle discussed in remark 2. I have not been able to construct examples where this negative feedback cycle does not contribute a positive term to β_{i+1} which makes $\beta_{i+1} \geqslant 0$.
4. Figure 34 illustrated what appears to be a general principle—that when a reaction such as $X_i + X_j \rightarrow \square$ short-circuits a current cycle, β increases. Theorem V.10 says that when the same reaction is an exit from a current cycle, β decreases.
5. The theorem treats a situation where $\beta_{i-1} > \beta_i > \beta_{i+1}$ and $\beta_i = 0$. Similar reactions with different stoichiometries can give a situation where $\beta_{i-1} > \beta_i > \beta_{i+1}$ and $\beta_{i+1} = 0$. See Example V.15. Then β_{i+1} vanishes from a negative polygon term that is not (the simplest case mentioned in remark 2) caused by a current cycle through X_1. An instability can still result via the mechanism of the theorem through β_{i+2}

Example V.15. The network (fragment) diagrammed in Fig. 43 has the kinetics indicated by the usual convention and the stoichiometries indicated by the parameters e (entrance), d (disproportionation), and a (autocatalysis). From $\mathfrak{D}_{CM}$ one obtains for β_2 on X and Y:

$$\beta_2 = (d + e)(1 + a - d + e) - (a - d + e)^2$$

which may vanish for certain positive values of a, d, and e satisfying the restriction that the flow in the reaction $X + Y \rightarrow \square$ is nonnegative. This condition is $a - d + e > 0$. For such parameter values, β_3 on X, Y, and Z is given by only the stabilizer at X and the 2-cycle on Y and Z,

$$\beta_3 = -(d + e)$$

Thus this current is unstable.

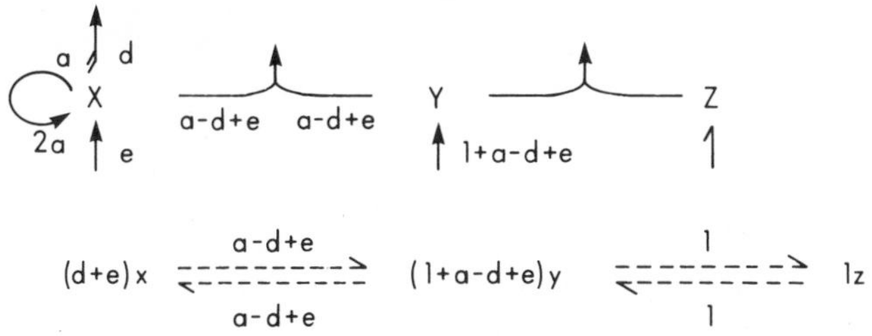

Fig. 43. The stoichiometries at the arrow heads and tails of this current are the nonnegative real numbers a, d, and e. The diagrams $\mathfrak{D}_C$ and $\mathfrak{D}_{CM}$ are shown.

This example proves that a current may be unstable without having a critical cycle. It also proves that a network may be unstable without having an extreme current with a critical cycle. To see this, note that this network fragment can be extended into an extreme network by adding more species and reactions with stoichiometries that allow no flexibility in the steady states, and by fixing a, d, and e such that $\beta_2(a,d,e)=0$. Note that the original network (fragment) has the extreme current $X \to 2X$, $\square \to Y$, and $X+Y \to \square$, which has a critical cycle. These observations lead to the following conjecture. I suspect that β_i cannot be negative on a set of species Σ of any network $N \in \mathfrak{N}_{NSCC}$ if the network fragment that is obtained by restricting N to the species Σ does not have an extreme current with a critical cycle. Since such a cycle cannot occur in a network where $\mathfrak{D}_N$ has no irreversible cycles, this conjecture is consistent with Theorem V.3.

Example V.16. We found all critical cycles of the original Field-Körös-Noyes mechanism for a Belousov-Zhabotinski system in Example V.12. The critical cycles W6Z9, W6Z10, and W6Z7 are caused by the conservation of $W+Z$ and cannot make β negative by the mechanism of Theorem V.10. The critical cycle V8Y1 cannot have an exit reaction at Y unless the Y that leaves is produced by an entrance reaction (i.e., R_{10}). Thus although it looks like $R_2: X+Y \to 2T$ is suitably situated with respect to the critical cycle V8Y1, R_2 does not produce an instability because the network fragment consisting of V, Y, and X has R_1 and R_8 linked together in one extreme current by the steady-state condition at V, while R_2, R_{10} (and possibly R_3) are in an independent current. Neither extreme current of the network fragment has a negative β, nor does any mixture of them. Note that in the preceding sentence "nor" could be replaced by "therefore, nor" if Theorem V.8 were suitably generalized.

We established earlier that the cycle X5U6 is not critical because of the presence of the higher-order exit reaction R_4. This reaction is necessary because the maximum number of Y's that can be produced at steady state is only half that needed to remove all the autocatalytically generated X. However reaction $R_2: X+Y \to 2T$ is ideally situated with respect to this cycle; hence we investigate whether an instability could occur even though X5U6 is not critical. The current fragment in Fig. 44 has the maximum possible ratio v_2/v_4, and the calculation shows that on $\{U,X,Y\}$, $\beta_3=0$. We have not found an instability, but we have found a new critical set of species that might produce an instability as in Example V.15. All

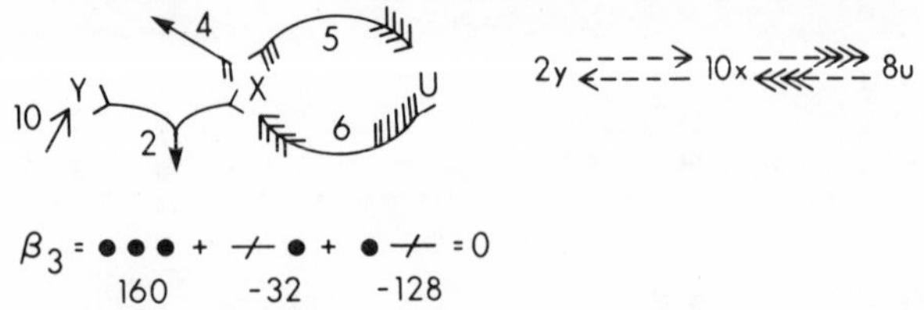

Fig. 44. A current fragment from the Field-Körös-Noyes Belousov-Zhabotinski mechanism. The reaction numbers are shown.

that is needed are reactions such as Y + S→□, U + S→□, or X + S→□, for some species S satisfying the conditions of Theorem V.10. The required reactions are not present; hence an instability has not been found.

Example V.17. We found only one critical current cycle in the Noyes-Sharma mechanism for the Bray-Leibhafsky reaction of Example V.13. This cycle is between $\cdot IO_2$ and IO_3^-. The simplest way to prove instability is to look at Fig. 42 for an exit reaction from either of these species meeting the conditions of Theorem V.10. The only exit reaction is

$$IO_3^- + I^- \rightarrow HOIO + HOI$$

and it satisfies the requirements to make $\beta_3 < 0$ on $\cdot IO_2$, IO_3^-, and I^-.

This current has more negative β's, which will enable us to illustrate Theorem V.9. In Fig. 45, showing $\mathfrak{D}_{CM}$, a positive feedback cycle through $\cdot IO_2$, IO_3^-, HOI, O_2, $\cdot$OOI, $\cdot$IO, and HOIO has been drawn with a heavy line. Since this cycle passes through the set $\{\cdot IO_2, IO_3^-\}$, where $\beta_2 = 0$, the only diagrams that need to be evaluated in the SPD-expansion of β_7 on these species are the diagrams containing polygons spanning $\{\cdot IO_2, IO_3^-\}$ and the remaining

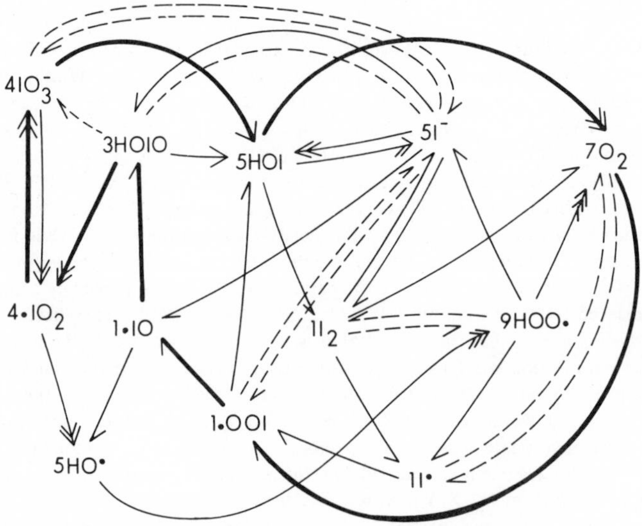

Fig. 45. $\mathfrak{D}_{CM}$ of the Bray-Liebhafsky mechanism.

five species. The dark positive feedback cycle is one such polygon. It makes the contribution -128 to β_7. The only other polygon spanning these sets passes from HOIO to IO_3^- along an inhibition, and then along the darkened arrows through HOI, O_2, $\cdot$OOI, $\cdot$IO, to HOIO. This polygon has the value 8. Since it appears in a diagram along with the stabilizer at $\cdot IO_2$, which is 4, the contribution to β_7 is 32. Hence $\beta_7 = -128 + 32 = -96 < 0$.

The feedback cycle that makes $\beta_7 < 0$ passes through dissolved oxygen, whose concentration at steady state can be controlled almost directly by varying the oxygen partial pressure (an external variable) in the gas above the solution. Hence by allowing gaseous oxygen to escape more easily from the reaction vessel, the parameter associated with O_2 (which is $h_{O_2} \equiv 1/[O_2(aq)]$ at steady state) increases. From (V.15) we see that the terms in $\alpha_7(\mathbf{h},\mathbf{j})$ involving β_7 on sets of species that include O_2 become larger relative to the other terms in α_7. If the concentrations of the other six species ($\cdot$OOI, $\cdot$IO, HOIO, $\cdot IO_2$, IO_3^-, and HOI) are sufficiently small, the negative β_7 examined above would be the dominant term in $\alpha_7(\mathbf{h},\mathbf{j})$. It could make $\alpha_7 < 0$, and the steady state $\mathbf{h}$ of this current would be exponentially unstable. Since half of the other six species are "free radicals," which must have very low concentrations it is quite possible for this negative term to dominate $\alpha_7(\mathbf{h}:\mathbf{j})$. If this term were the only reason for instability at these parameter values, preventing the gaseous oxygen to escape so easily would increase O_2(aq) and decrease the magnitude of the negative β_7, so that α_7 would become positive and the steady state would become stable. The oscillations in the experimental system can be stopped and started by altering the oxygen partial pressure; since other sources of instability can explain the dependence of the oscillation on the oxygen partial pressure as well, however, that does not mean that the explanation above is the source of instability.

H. A Ubiquitous Source of Instability in Oxidation-Reduction Networks

R. M. Noyes[69] has expressed the idea that the instability in Belousov-Zhabotinski systems is a consequence of the interaction between a one-electron oxidizing agent and a two-electron oxidizing agent. This section explains why the interaction between two such oxidizing (or reducing) agents is likely in many instances to give a set of chemical reactions that form an unstable network. Since some readers may be mathematicians or theoreticians who have never studied the chemistry of oxidation-reduction (redox) reactions, I discuss this idea beginning with first principles.

Electrons and atomic nuclei are conserved in every chemical reaction; therefore chemists have devised a sophisticated method for keeping track of the electrons in complicated reaction systems. To use this method, the chemist thinks of a chemical symbol, such as H or O in H_2O, as representing an atomic nucleus plus a certain number of associated electrons. The combined electric charge of the nucleus and associated electrons, when expressed in units such that the electron charge is -1, is called the *oxidation number* of the nucleus.

The electrons associated with a nucleus are assigned partially by convention. Oxidation numbers must be assigned so that the number of electrons associated with all the nuclei of any molecule or ion equals the number of electrons in that molecule or ion. Hence for any species, the sum of the oxidation numbers of all the nuclei equals the electric charge on the species. This condition is called the *principal constraint*.

If oxidation numbers are assigned such that each atomic nucleus has the same oxidation number everywhere in a chemical reaction equation, balancing the nuclei in the equation will balance the electrons; hence the chemist need not keep track of the electrons. For example, in the reaction

$$2H_2 + O_2 \rightarrow 2H_2O \tag{V.33}$$

one may assign zero as the oxidation number of both H and O, because then the principal constraint is satisfied. Since H and O have the same oxidation numbers everywhere, and since the nuclei balance, we know the electrons balance. On the other hand, in the reaction

$$2H_2O \rightarrow H_3O^+ + OH^- \tag{V.34}$$

the principal constraint is not satisfied with the oxidation numbers above. A good assignment is $+1$ for H and -2 for O. Then since the nuclei balance, we know that the electrons balance.

The preceding two examples illustrate that it is not always possible to make one universal assignment of oxidation numbers so that all reactions in a complicated network can be balanced with respect to electrons by balancing the equations with respect to the nuclei only. Chemists have adopted a systematic set of conventions for assigning oxidation numbers. The convention is intended to maximize the number of reactions that can be balanced without considering the electrons. Reactions where the (conventional) oxidation numbers change are called *redox reactions*. Conventionally, H is $+1$ and O is -2; thus reaction (V.34) is not a redox reaction. These oxidation numbers are also satisfactory on the right of reaction (V.33), but on the left they violate the principal constraint. The only satisfactory numbers for the left are $H = 0$, $O = 0$. Hence a change of oxidation numbers occurs, and reaction (V.33) is a redox reaction. Clearly, which reactions are redox reactions is a matter of convention.

A change in oxidation numbers represents a flow of electrons relative to the flow of electrons that belong to the nuclei by convention. *Oxidation* is an increase in oxidation number and represents the departure of electrons; *reduction* is a gain in electrons. In reaction (V.33) the oxidation number of H changes from 0 to $+1$, which indicates a loss of electrons. Oxygen gains these electrons. Hence a flow of electrons occurs from H to O in reaction (V.33).

This flow of electrons is independent of the arbitrary aspects of the convention for assigning oxidation numbers and is therefore an objective mathematical property of a set of chemical reactions. For example, suppose we adopted the convention that the oxidation numbers are $H = 0$, $O = 0$. Then (V.33) would not be a redox reaction, but (V.34) would be. The principal constraint would be violated on the right if these numbers were used. A reasonable choice here is to let $H = +1$, $O = -2$, such that reaction (V.34) causes the oxidation $(0 \to +1)$ of H and the reduction $(0 \to -2)$ of O. Electrons have flowed from H to O, which is the same conclusion we arrived at with the other convention. The arbitrary choices made in assigning oxidation numbers affect only the reaction in which the flow of electrons appears. The choices do not alter the atoms between which the electrons flow, or the relative numbers of electrons in the flow.

A basic principle of chemical bonding and stability is that electrons form pairs whenever possible. A molecule or ion with an even number of electrons almost always has the electrons paired. The molecule is then much more stable and less likely to react chemically than if it had unpaired electrons. Molecules with an odd number of electrons must have an unpaired electron, which is usually shown as a dot (e.g., $\cdot BrO_2$). These highly reactive species are called *free radicals.*

In Belousov-Zhabotinski systems, the bromine atom is the sink for electrons. It begins in the +5 state as BrO_3^- (i.e., oxidation number = 5). If electrons could be added one at a time, the oxidation state would decrease by one for each electron added. Bromine would pass through the following sequence of species, which are the most stable species with the required oxidation number in aqueous solution: BrO_3^-, $\cdot BrO_2$, $HBrO_2$, $\cdot BrO$, $HOBr$, Br_2, Br^-. Note that the number of electrons per bromine atom is alternately even and odd along the sequence. Hence every other species must be either a free radical ($\cdot BrO_2$, $\cdot BrO$) or a dimer (Br_2). Every redox system has sequences of states such as this for the atoms that act as sinks and sources for electrons.

The general situation is diagrammed in Fig. 46, where the species X_i of the first row have an even number of electrons and the species Y_i of the second row have an odd number of electrons. The first row species are therefore stable molecules or ions and the second row species are very

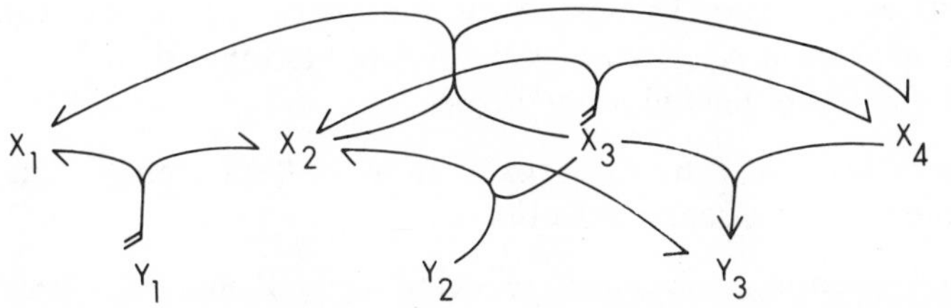

Fig. 46. Reactions among stable species (X_i) and free radicals (Y_i) that conserve electrons.

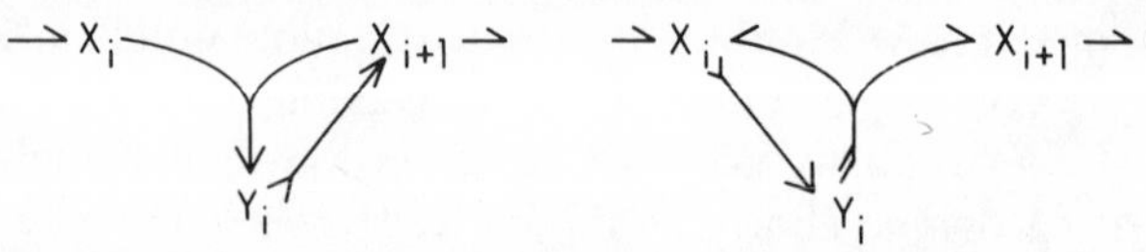

Fig. 47. $\mathfrak{D}_C$ fragment for a redox network: *left*, radical species attack; *right*, paired electron species attack.

reactive free radicals. The oxidation numbers increase or decrease in steps of one from left to right, in the sequence $X_1, Y_1, X_2, Y_2, \ldots$. The reactions that can occur when two species collide are symmetrical on this diagram because of the requirement that electrons be conserved. Some of the most common possibilities are shown. Reactions such as these would be the only reactions involved as the set of species $\{X_i, Y_i\}$ came to thermodynamic equilibrium when the number of electrons is fixed.

Now suppose such a system is acting as a source or sink of electrons and electrons are transferred to or from a species Z, which can take only one electron at a time. Then a reaction between Z and X_i or Y_i results in a species one step to the right (say) on the diagram. If Z attacks a free radical, the reaction is

$$Z + Y_i \rightarrow Z' + X_{i+1} \tag{V.35}$$

If Z attacks a stable molecule or ion with paired electrons, the reaction is

$$Z + X_i \rightarrow Z' + Y_i \tag{V.36}$$

The free radicals involved are likely to be produced or removed at steady state by the disproportionation reaction

$$2Y_i \rightarrow X_i + X_{i+1}$$

or its reverse. Figure 47 shows the current diagrams for "radical attack" and "paired electron species" attack; Z and Z′ have been considered to be species external to this network fragment. In both cases a simple current cycle is formed. Theorem V.5 implies that the cycle caused by "radical attack" is critical, whereas the cycle caused by "paired electron species attack" is weak. From Theorem V.10, the radical attack case will be an unstable network if X_{i+1} reacts with any species other than X_i, Y_i, Z, or Z′. This result gives the following theorem.

Theorem V.11. An extreme redox network is exponentially unstable if one of the following conditions obtains:

1. A one-electron oxidizing or reducing agent Z attacks a free radical species Y_i that is formed by the reverse disproportionation of two

species (X_i and X_{i+1}, say) with oxidation states one greater and one less than that of Y_1.

2. The species produced by the oxidation or reduction X_{i+1} (say) is removed by reacting with any species W, where $W \neq X_i, X_{i+1}, Y_i, Z'$, or Z.
3. The reaction between X_{i+1} and W does not produce X_i, X_{i+1}, Y_i, Z', or Z.

Example V.18. When a sufficient supply of Br^- is produced by reactions not appearing in Fig. 40, this Belousov-Zhabotinski network has an extreme subnetwork fragment resembling the diagram at the left in Fig. 47, where $X_i = BrO_3^-$, $X_{i+1} = HBrO_2$, and $Y_i = \cdot BrO_2$. The one-electron reducing agent is Ce(III). Since the exit from X_{i+1} occurs by a reaction with Br^-, the theorem implies the network is exponentially unstable.

Example V.19. The Bray-Liebhafsky network of Fig. 41 also satisfies Theorem V.11. Now $X_i = HOIO$, $X_{i+1} = IO_3^-$, and $Y_i = \cdot IO_2$. The one-electron oxidizing agent is the external species HOOH. Note that condition 2 is satisfied because IO_3^- reacts with I^-. Hence this extreme network is exponentially unstable.

Example V.20. The source of electrons in Belousov-Zhabotinski systems is usually malonic acid. Figure 48 shows a sequence of oxidation states for malonic acid and its derivatives according to the same plan as Fig. 46. This detailed network has been proposed by Jwo and Noyes[72] as an alternative to R_9 and R_{10} of Fig. 40. Bromomalonic acid is produced by the overall reaction and accumulates. Oscillations are not possible until enough BrMA has

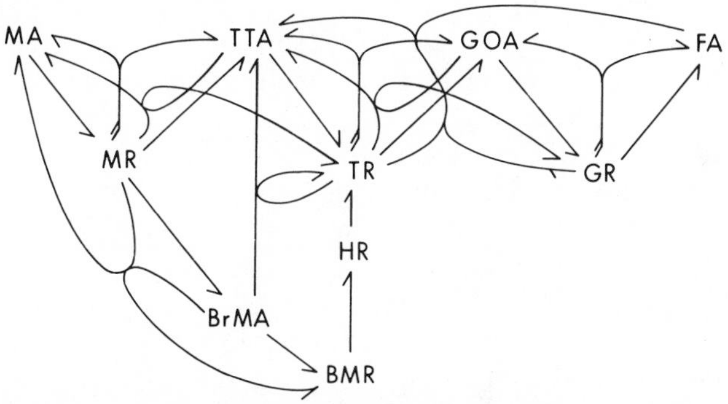

Fig. 48. Proposed reactions for the source of electrons in Belousov-Zhabotinski systems arranged according to oxidation states and free radicals according to the pattern of Fig. 46. The first two rows are species that do not contain bromine, and the last two rows are species containing bromine: MA = $CH_2(COOH)_2$ = malonic acid, TTA = $HOCH(COOH)_2$ = tartronic acid, GOA = OHCCOOH = glyoxylic acid, FA = HCOOH = formic acid, BrMa = $CHBr(COOH)_2$ = bromomalonic acid, MR = $H\dot{C}(COOH)_2$ = malonyl radical, TR = $\cdot OCH(COOH)_2$ = tartronyl radical, HMR = $HO\dot{C}(COOH)_2$ = hydroxymalonyl radical, BMR = $Br\dot{C}(COOH)_2$ = bromomalonyl radical, GR = $O\dot{C}(COOH)_2$. The liberation or addition of Br^- (not shown) can be deduced from the arrangement of species in the figure. All reactions that do not have the symmetry imposed by the conservation of electrons are redox reactions involving Ce(IV)/Ce(III) (not shown).

accumulated for the required bromide ion (Br^-) to be released from it. Note that bromide ion can be produced by only two reactions: BrMa → TTA and BMR → HR. Since the unstable free radical TR catalyzes the first reaction, and since TR is produced from HR, which is produced by the second reaction, TR and the reactions on the right of the figure are important. The amount of Br^- produced for each Ce(IV) consumed can vary between zero and infinity according to the relative rates of these reactions. Since this ratio determines the stability of the network, one cannot predict when the network will oscillate from first principles without a better understanding of these reactions.

During the experiments of Jwo and Noyes, breakpoints in the time dependence of the concentrations of certain species were observed. The systems where the breakpoints occurred contained MA, BrMA, and Ce-(IV). There were no oxybromine species. The important reactions in these systems should be those shown in Fig. 48. A breakpoint could be caused by an instability or by a critical cycle. The next section explains how a critical cycle produces a breakpoint. In either case the most probable cause of the critical cycle (required for an instability) is the oxidation by cerium of a radical species such as MR or TR.

One very reasonable possibility is the following. Initially, there is a large excess of MA that reacts with Ce(IV) and prevents the concentration of Ce(IV) from becoming too large. Since MA and TTA react approximately equally rapidly with Ce(IV), TTA cannot reach steady state until its concentration reaches the MA concentration. As a result, TTA should be a product of the overall reaction and should accumulate. A small amount of this TTA will be oxidized to TR. Now, suppose Ce(IV) is more likely to oxidize TR to GOA than TR is likely to disproportionate into TTA and GOA. Then the critical 2-cycle shown in Fig. 49 will occur. Here TTA is treated like an external species because it should have accumulated to a high concentration. An instability will be caused by this 2-cycle if the reaction removing GOA is bimolecular and does not involve TR or Ce(IV). More reactions than those shown in Fig. 58 must occur because it is known that GOA is eventually oxidized to carbon dioxide. Hence there will be a removal reaction, although we cannot say what it is. Furthermore, there are extremely complicated reasons making it improbable the removal reaction is oxidation by cerium. An instability is very probable. Even if the removal reaction does not produce an instability, an explosive growth of GOA and TTA can still occur because the 2-cycle is critical. These cold explosions are discussed in the next section. Hence the concentration of

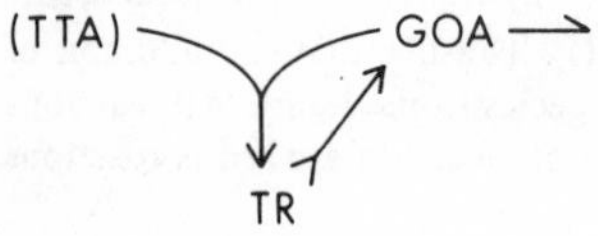

Fig. 49. Conjectured reason for breakpoints in malonic acid, bromomalonic acid, Ce(IV) systems.

TR should increase dramatically because of the explosion or instability when sufficient Ce(IV) is present. The increase in TR would catalyze the liberation of bromide ion via the reaction BrMA → TTA. It is possible that some of the spectacular dynamical phenomena observed in Belousov-Zhabotinski systems involve both this instability and the previous instability (Example V.18) working in tandem.

A tandem oscillator that might occur in Belousov-Zhabotinski systems is shown in Fig. 50. This current diagram has two critical 2-cycles that are coupled by Ce(III)/Ce(IV). The reproduction of $HBrO_2$ by a critical 2-cycle makes Ce(IV) strongly oxidizing and initiates the reproduction of TR by the other critical 2-cycle. However an upsurge in TR liberates Br^- from BrMA, thereby turning off the reproduction of $HBrO_2$ by drawing more of it out of the first critical 2-cycle than is produced.

Although this role for TA is highly speculative, it is appealing because it can explain why the experimental traces of Br^- versus time have much more rapid changes in concentration than are obtained from simulating the Oregonator on a computer. One would expect that a more complicated network which contains only the critical cycle of the Oregonator would have more slowly varying changes in Br^- because of the time delays. This speculation also could also explain the observation that Belousov-Zhabotinski systems have much more spectacular dynamics than any other system known at present. If BZ systems resulted merely from a single redox-type instability, one would expect to find many similar systems. The

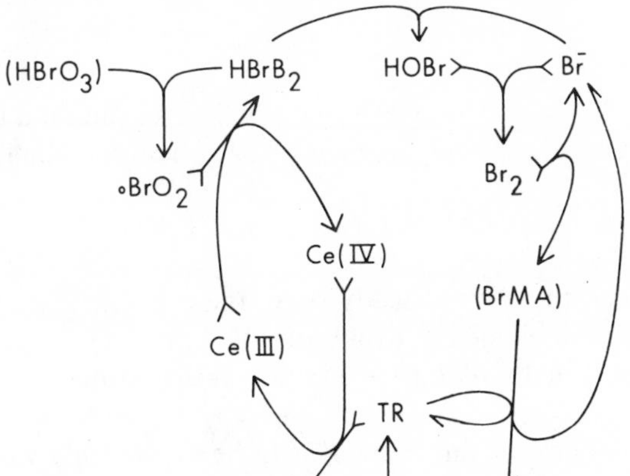

Fig. 50. A current diagram showing how two redox instabilities might be connected in tandem in Belousov-Zhabotinski systems. The species in parentheses should be considered to be external.

uniqueness of the BZ systems could be due to this improbable tandem coupling of two redox instabilities.

Finally, this conjecture about a tandem oscillation is not contrived, but is an almost obvious consequence of the reactions that Jwo and Noyes have proposed as the best explanation for their experimental results on MA, BrMA, Ce(IV) systems. Noyes has assured me that the reactions in the figure are the most likely reactions to occur of all those discussed.

I. Critical Current Cycles, Extinction, and Stoichiometric Explosions

This section explains how current diagrams can be used to recognize networks that are capable of stoichiometric explosions or the extinction of species. The species X_i becomes *extinct* if there is a trajectory $\mathbf{X}(t)$, satisfying (II.2) such that

$$\lim_{t\to\infty} X_i(t) = 0 \tag{V.37}$$

A *stoichiometric explosion* of X_i occurs when $X_i(t)$ becomes larger and larger without limit.

A simple example is a nuclear explosion. This may be regarded as a stoichiometric network in which neutrons reproduce themselves by the reaction

$$R_1 : n + {}^{235}U \to A + B + \alpha n$$

and escape by diffusion or by being absorbed as described by the pseudoreaction

$$R_2 : n \to \square$$

If we regard ^{235}U, A, and B as external species, we obtain a stoichiometric network with a single species (neutrons) that obeys the equation

$$\frac{dX}{dt} = k_1(\alpha - 1)X - k_2X \tag{V.38}$$

If $k_1(\alpha - 1) > k_2$, there is no steady state; the concentration of neutrons X increases in a stoichiometric explosion. If $k_1(\alpha - 1) < k_2$, the neutrons become extinct. Finally, if $k_1(\alpha - 1) = k_2$, this network is at steady state for all X.

A current diagram for this network can be drawn only when the steady state exists; that is, when $k_1(\alpha - 1) = k_2$. The diagram for this case when $\alpha = 2$ appears in Fig. 51*a*. Note that the diagram has a critical current cycle; thus $\alpha_1 = 0$. Hence this current is stable but not asymptotically stable. We emphasize that *explosions and extinction can occur in networks that do not have unstable steady states*.

(a) (b)

Fig. 51. Current *a* is stable and occurs in a network that is prone to a stoichiometric explosion or extinction. Current *b* is unstable.

To understand the connection between instability and explosions, consider the closely related network in Fig. 51*b*. This network has an unstable steady state by Theorem V.10. The equations of motion are

$$\frac{dX}{dt} = k_1X - k_2XY \tag{V.39}$$

$$\frac{dY}{dt} = -k_2XY + k_3$$

and the steady state is

$$X^0 = \frac{k_3}{k_2}, \qquad Y^0 = \frac{k_1}{k_2}$$

If Y is now considered an external species and "frozen" at the value Y^*, we get the former network back again, and its equation is

$$\frac{dX}{dt} = X(k_1 - k_2Y^*)$$

where Y^* is a constant. The former steady state value of Y plays an important role. If $Y^* > Y^0$, extinction occurs; if $Y^* < Y^0$ an explosion occurs. Thus Y^0 is the value of Y^* that separates the extinction case from the explosion case. When the dynamics are given by (V.39) an increase in Y above steady state causes the extinction of X and a decrease in Y below steady state causes an explosion of X. Thus the instabilities we can locate using Theorem V.10 are unstable steady states that represent the knife edge between extinction dynamics and explosion dynamics. If an explosion is analogous to the "on" state of a switch and an extinction is analagous to the "off" state, the instability is analogous to a switch being balanced on the knife edge between "on" and "off." When explosion and extinction dynamics occur in a network that does not have a "switch," no instability is associated with this dynamics. However we will see that networks having such dynamics often can be recognized by the presence of a critical current cycle.

The nuclear explosion example has many features that one might suspect are to be found together in the general case. Let us list these features.

1. The network has no steady state(s) unless a relation is satisfied among the rate constants **k** (and amounts of the components **C**).
2. When the network has a steady state, it has infinitely many steady states with the same **k** and **C**.
3. The relation among the rate constants divides parameter space into two regions. In one region explosion dynamics prevails, and in the other extinction dynamics prevails.
4. The current diagram has a critical cycle that is not caused by a conservation condition.

The discussion that follows establishes the equivalence of the first three properties. The fourth observation holds frequently, but not always.

The discussion in Section II.E made the assumption that the number of steady states available for fixed **k** and **C** was an integer $\rho(\mathbf{k}, \mathbf{C})$. We must now consider what happens when infinitely many steady states occur. Even when there are infinitely many steady states, every steady state of the network has a reaction velocity vector $\mathbf{v} \in \mathcal{C}_v$. Since every reaction R_i has an adjustable parameter k_i that can be used to vary the velocity from zero to infinity for any given concentrations, it is clear that for any steady-state concentrations, we may choose $\mathbf{k} \in R_+^r$ to make $\mathbf{v}$ become any desired point in $\mathcal{C}_v$. That is, every $\mathbf{v} \in \mathcal{C}_v$ is accessible for every set of steady-state concentrations. Also, every steady state has an inverse steady-state concentration vector $\mathbf{h} \in R_+^n$, and every $\mathbf{h} \in R_+^n$ determines a concentration vector. Hence the set of steady states must be in one-to-one correspondence with the set

$$D_C^* \equiv \{(\mathbf{h}, \mathbf{v}) \mid \mathbf{h} \in R_+^n, \mathbf{v} \in \mathcal{C}_v\}$$

This set is equivalent to the parameter domain D_C defined by (II.34); and hence all the theory that has been discussed so far in this chapter is valid when there are infinitely many steady states for any **k** and **C**. Recall that $\Pi_X(\mathbf{C})$ is the set of all states (not necessarily steady) that are accessible for given **C**. If there are an infinite number of steady states with a given **k** and **C**, there must be an infinite number of steady states in $\Pi_X(\mathbf{C})$.

Theorem V.12. Suppose that for a given steady state $(\mathbf{h}, \mathbf{j})$, $\Pi_X(\mathbf{C})$ has an s-dimensional hypersurface H of steady states; then $\alpha_d = \alpha_{d-1} = \cdots = \alpha_{d-s+1} = 0$ identically. That is, all terms in these polynomials in **h** and **j** cancel identically.

Proof. Let $\mathbf{X}^0 = 1/\mathbf{h} \in H \subset \Pi_X(\mathbf{C})$ be the steady-state concentration and let $\mathbf{X}^* \in H$ be any nearby steady state. Regarding $\mathbf{X}^*$ as a perturbation $\zeta = \mathbf{X}^* - \mathbf{X}^0$ from the original steady state, we write down the equation of motion for this perturbation and get

$$\frac{d\zeta}{dt} = \mathsf{M}\zeta = 0$$

The right-hand side vanishes because $\mathbf{X}^*$ is a steady state. Thus $\mathsf{M}\zeta = 0$ for all $\zeta \in T$, which is the linear subspace parallel to the tangent affine subspace to H at $\mathbf{X}^0$. Section III.B proved that $\alpha_{d+1} = \alpha_{d+2} = \cdots = \alpha_n = 0$. This is a consequence of the fact that M has $n - d$ "irrelevant" eigenvalues $\lambda = 0$ because of the $n - d$ independent conservation conditions. These eigenvalues were called irrelevant because they correspond to physically impossible perturbations orthogonal to $\Pi_X(\mathbf{C})$. The irrelevant eigenvalues tell us that $n - d$ columns of M are dependent on the remaining d columns. Now that we have proved $\mathsf{M}\zeta = 0$ for all $\zeta \in T$, we conclude that s of these d columns of M must be dependent on the remaining $d - s$. The reason is as follows. These s dependent columns are new because T is parallel to $\Pi_X(\mathbf{C})$. Hence there are at least $(n - d) + s$ dependent columns of M, therefore at most $d - s$ independent columns. Therefore all minors of M containing more than $d - s$ columns must vanish. We conclude that $\alpha_i = 0$ for $i > d - s$. ■

Theorem V.13. $\alpha_d = \alpha_{d-1} = \cdots = \alpha_{d-s+1} = 0$ and $\alpha_{d-s} \neq 0$ identically, if and only if every concentration polytope $\Pi_X(\mathbf{C})$, $\mathbf{C} \in R_+^{n-d}$, has an s-dimensional hypersurface of steady states.

Proof. If $\alpha_d = \cdots = \alpha_{d-s+1} = 0$ and $\alpha_{d-s} \neq 0$ identically, then $\mathsf{M}(\mathbf{h},\mathbf{j})$ has $n - (d - s)$ eigenvalues $\lambda = 0$ for all $(\mathbf{h},\mathbf{j})$. It does not have more than $n - d + s$ zero eigenvalues except possibly for certain $(\mathbf{h},\mathbf{j}) \in D_R$. The eigenspace for $\lambda = 0$ is $n - d + s$ dimensional almost everywhere. Thus near to any steady state are many other steady states that can be reached by making a small perturbation in the eigenspace of $\lambda = 0$. This eigenspace separates into an $(n - d)$-dimensional space orthogonal to $\Pi_X(\mathbf{C})$ and an s-dimensional space parallel to $\Pi_X(\mathbf{C})$. Hence the nearby set of steady states in $\Pi_X(\mathbf{C})$ has dimension s. This argument is valid for almost all $(\mathbf{h},\mathbf{j})$, in any $\Pi_X(\mathbf{C})$. Thus every $\Pi_X(\mathbf{C})$ must have an s-dimensional hypersurface of steady states.

If any $\Pi_x(\mathbf{C})$ has an s-dimensional hypersurface of steady states, then $\alpha_d = \cdots = \alpha_{d-s+1} = 0$ identically by Theorem V.12. To prove that $\alpha_{d-s} \neq 0$ identically, note that if this were not so, we could deduce from the first half of this theorem that $\Pi(\mathbf{C})$ has an $(s + 1)$-dimensional hypersurface of steady states. This contradicts our assumption. Hence $\alpha_{d-s} \neq 0$ identically. ■

We now look at the situation in terms of the kinetic parameters. When there is an s-dimensional hypersurface of steady states in every $\Pi_X(\mathbf{C})$, we can set up a coordinate system $\mathbf{q} \in R^s$ on the hypersurface. One can choose the coordinates so that the domain $\mathbf{q} \in D_q \subset R^s$ corresponds to the entire hypersurface for any $\Pi_X(\mathbf{C})$. (There may be trouble here if the topology of the hypersurface changes. I shall omit these possible complications.) To specify any steady state using the kinetic parameters, one must specify $\mathbf{k}$, $\mathbf{C}$, and $\mathbf{q}$. Hence the kinetic parameter domain is

$$D_K = \left\{(\mathbf{k}, \mathbf{C}, \mathbf{q}) \,|\, \mathbf{k} \in R_+^r, \mathbf{C} \in R_+^{n-d}, \mathbf{q} \in D_q\right\}$$

This set has dimension $n + r - d + s$, and D_C has dimension $n + r - d$. Hence this set has s more dimensions than the set of steady states. Either more than one point in D_K refers to the same steady state, or steady states exist only for an $(n - r - d)$-dimensional subset of D_K. We now eliminate the first possibility. Suppose two different points $(\mathbf{k}, \mathbf{C}, \mathbf{q})$ and $(\mathbf{k}', \mathbf{C}', \mathbf{q}')$ $\in D_K$ refer to the same steady state $(\mathbf{h}, \mathbf{j})$. The total amount of each component must be the same for both points and this implies $\mathbf{C} = \mathbf{C}'$. If we calculate $\mathbf{k}$ and $\mathbf{k}'$ using (II.50), we obtain the same result; so $\mathbf{k} = \mathbf{k}'$. These two points both lie in the same $\Pi_X(\mathbf{C})$. The coordinate system for the hypersurface allows $\mathbf{X}^0$ to be the same in both cases if and only if $\mathbf{q} = \mathbf{q}'$. Hence these points are identical, contrary to our assumption. We therefore conclude that not more than one point in D_K can correspond to a steady state. It then follows that the points of D_K that correspond to steady states have the same dimension as D_C, namely $n + r - d$. The steady states of D_K therefore lie in the intersection of s independent hypersurfaces. The equations of these hypersurfaces restrict only $\mathbf{k}$ and $\mathbf{C}$, because, by hypothesis, the domain of $\mathbf{q}$ is always D_q. The set of restricted $(\mathbf{k}, \mathbf{C})$ will now be called D_K^0. This result and the previous two theorems may now be stated as a single theorem.

Theorem V.14. The following statements are equivalent.

1. The subset $D_K^0 \times D_q$ of D_K for which steady states exist satisfies s independent relations among $\mathbf{k}$ and $\mathbf{C}$.
2. Every $\Pi_X(\mathbf{C})$ contains an s-dimensional hypersurface of steady states.
3. $\alpha_d = \alpha_{d-1} = \cdots = \alpha_{d-s+1} = 0$ identically. $\alpha_{d-s} \neq 0$ identically (but may vanish for some $\mathbf{h}, \mathbf{j}$).

If $(\mathbf{k}, \mathbf{C}) \notin D_K^0$, several dynamical possibilities are still open. There could be explosion-extinction dynamics, a limit cycle oscillation, chaos, or ergodic dynamics. We now argue that explosion-extinction dynamics is the most reasonable possibility. Assume $(\mathbf{k}, \mathbf{C}) \notin D_K^0$. We can increase and decrease the components of $\mathbf{k}$ to obtain other parameters $(\mathbf{k}^*, \mathbf{C}) \in D_K^0$.

Comparing **k** and **k*** tells us that for the first system, certain reactions are faster and others are slower than they are at steady state. Thus certain species are produced more rapidly and others are produced more slowly. Neither oscillations, chaos, nor ergodic behavior seems probable under these conditions. It appears that explosion-extinction dynamics is the most reasonable possibility.

Theorem V.14 and this argument generalize the first three observations we made for the nuclear explosion example. I now give two examples that prove that critical current cycles need not be present in networks satisfying this theorem. The examples show that the fourth observation for the nuclear example is not as general as the first three observations.

Example V.21. Figure 52 shows the current diagrams of two networks that were discussed in Examples IV.6 and V.8. We assume that these are complete networks and that no other reactions are available. Both networks were previously proved to be stable. In both cases the steady-state condition imposes the constraint $v_1 = v_2$; that is, $k_1 = k_2$. Hence a steady state exists only when this relation holds among the rate constants. Theorem V.14 implies $\alpha_4 = 0$ in both cases (proved earlier) and that $\Pi_X(\mathbf{C})$ has a set of steady states whose dimension is at least one.

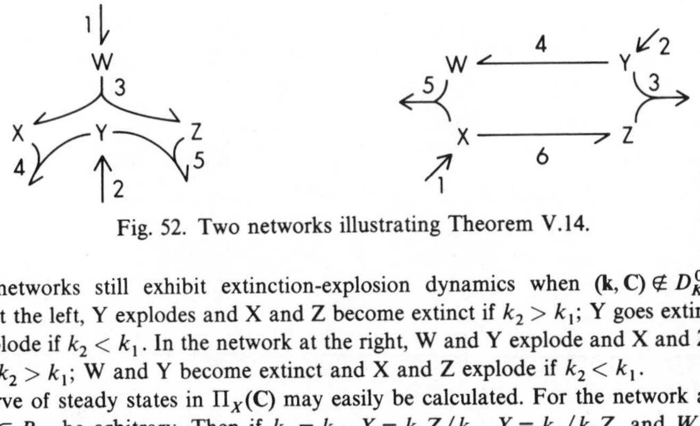

Fig. 52. Two networks illustrating Theorem V.14.

These networks still exhibit extinction-explosion dynamics when $(\mathbf{k}, \mathbf{C}) \notin D_K^0$. In the network at the left, Y explodes and X and Z become extinct if $k_2 > k_1$; Y goes extinct and X and Z explode if $k_2 < k_1$. In the network at the right, W and Y explode and X and Z become extinct if $k_2 > k_1$; W and Y become extinct and X and Z explode if $k_2 < k_1$.

The curve of steady states in $\Pi_X(\mathbf{C})$ may easily be calculated. For the network at the left we let $Z \in R_+$ be arbitrary. Then if $k_1 = k_2$, $X = k_5Z/k_4$, $Y = k_2/k_5Z$, and $W = k_2/k_3$. For the network at the right, when $k_1 = k_2$ the only constraint on X and Y is $2k_1 = k_4X + k_6Y$. Then $W = k_4Y/k_5X$ and $Z = k_6X/k_3Y$.

These examples suggest the following generalization. Whenever an extreme subnetwork has more than one entrance reaction (□ on the left), the steady-state condition forces the entrance reaction rates to have a ratio that is determined by stoichiometry. Hence restrictions apply to the rate constants, and if these are not satisfied, extinction or explosion dynamics results.

We now examine how Theorem V.14 applies to systems with critical current cycles. Let Σ be a subset of species lying on a critical current i-cycle and assume that no conservation condition involves only species in

Σ. Thus the rows of $\underline{\nu}$ corresponding to species in Σ are linearly independent. Since Σ lies on a critical current cycle, $\beta_i = 0$ for the species Σ. We know that $\alpha_i = 0$ for all $i > d$ because of the conservation conditions. Note also that $\beta_i = 0$ whenever the set of species used to calculate β_i contains all the species in a conservation condition. Hence the only terms in α_d that might have a nonvanishing coefficient β_d are the terms that come from sets of species that omit a species corresponding to each of the $n - d$ conservation conditions. Let Σ' be such that $\Sigma \cap \Sigma' = \emptyset$ and assume that $\Sigma \cup \Sigma'$ is such a set. That is, Σ' has been chosen such that if $X_{\gamma_{i1}}, \ldots, X_{\gamma_{in}}$ $(\gamma_{ik} \neq 0)$ are the species involved in any conservation condition, there must be a species $X \in \{X_{\gamma_{i1}}, \ldots, X_{\gamma_{in}}\}$ such that $X \notin \Sigma \cup \Sigma'$. Any other choice of Σ' will make $\beta_d = 0$ on the set $\Sigma \cup \Sigma'$ from the conservation condition. Now suppose that the diagram of all stabilizers and polygons does not have a polygon that spans Σ and Σ' for any acceptable choice of $\Sigma \cup \Sigma'$. Then every remaining β_d vanishes by Theorem V.9. Hence $\alpha_d = 0$ identically.

Suppose that a network has s critical current cycles. Let the ith cycle involve the species set Σ_i, and assume that $\Sigma_1 \cap \Sigma_2 \cap \Sigma_3 \cap \cdots \cap \Sigma_s = \emptyset$. Assume as well that none of these cycles contains a species in a conservation condition. Now β_j will vanish if it involves all species in a conservation condition. It will also vanish if it contains all species in Σ_i and there are no polygons that span Σ_i and the other species used to calculate β_j. It might happen that none of the Σ_i have such polygons. Then β_j would only be nonvanishing if it omitted one (or more) species from every conservation condition (a minimum of $n - d$ species) and one (or more) additional species from every Σ_i (a minimum of s additional species). These species are additional because no conserved species is in a Σ_i. In this case, β_j would vanish unless at least $n - d + s$ species were omitted. The maximum number that can remain is $d - s$. Hence $\beta_j = 0$ for $j > d - s$. Then $\alpha_j = 0$ for $j > d - s$. This is one situation where Theorem V.14 can apply. Hence every $\Pi_X(\mathbf{C})$ has an s-dimensional hypersurface of steady states, and s independent relations among the rate constants must be satisfied.

My intuitive picture in this case is the following. Each of the s independent relations among the $\mathbf{k}$ and $\mathbf{C}$ corresponds to a critical current cycle. The equation describes a hypersurface that divides D_K into two regions. In one region the species on the current cycle explode, and in the other region they approach extinction. The s independent relations among $\mathbf{k}$ and $\mathbf{C}$ therefore divide D_K into 2^s regions. In each region certain current cycles explode and the remaining ones approach extinction.

This picture does not apply to a current cycle if any species on the cycle appears in a conservation condition because the conservation condition limits the total amount of this species. Then an explosion is not possible.

This picture applies only when a polygon does not span the set of species on the current cycle and the set of species not on the current cycle. The picture corresponds to Fig. 51*a*. If there is such a polygon, new terms will appear in α_j with signs that depend on the sign of the feedback. Then α_j may be positive in one domain and negative in another. One gets domains of stability and instability, and the situation described in Theorem V.14 no longer applies. There is no longer a hypersurface D_K^0; instead there exists a steady state of this current cycle, which, however, is unstable. A perturbation will cause the current cycle to explode or go extinct depending on the direction of the perturbation from steady state.

Example V.22. For the "autocatalytic hypercycle" of M. Eigen (Fig. 33), $\alpha_d = 0$ and $\alpha_{d-1} \neq 0$. Hence a steady state exists if and only if one condition on the rate constants is satisfied. Otherwise the species explode or go extinct. Similarly two *independent* hypercycles would have a steady state if and only if two restrictions on their rate constants were satisfied. Figure 53 is a current diagram showing two *interacting* hypercycles. Now $\alpha_d \neq 0$, so a steady state exists for all **k** and **C**. The steady state is unstable ($\alpha_d < 0$). A perturbation will decide which cycle explodes and which goes extinct.

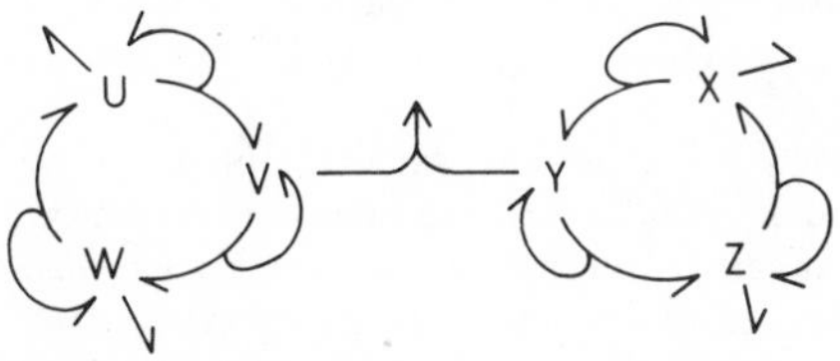

Fig. 53. Two competing autocatalytic hypercycles. An instability comes from a positive feedback 2-cycle between V and Y. This current diagram represents the unstable steady state where both hypercycles are equally likely to survive. A perturbation decides which cycle explodes and which goes extinct.

Example V.22 can be generalized to any pair of competing critical current cycles. In a large system with many competing critical current cycles, one would expect this type of competition to cause many species to go extinct. The current of the network would then contain only critical current cycles that could coexist. As an example of coexistence, consider the steady state of the biochemical network of a bacterium. It has several critical current cycles, such as the cycle by which DNA reproduces itself and the citric acid (Krebs) cycle. At steady steady these cycles do not compete; rather, they are independent. Of course, the rate constants are not such that these cycles are at steady state. Both are slowly exploding, which is why the bacterium must divide and thereby reproduce.

Example V.23. The detailed mechanism for the reaction

$$2H_2 + O_2 \rightarrow 2H_2O$$

is extremely complicated. Hinshelwood[73] has given the following simplified mechanism,

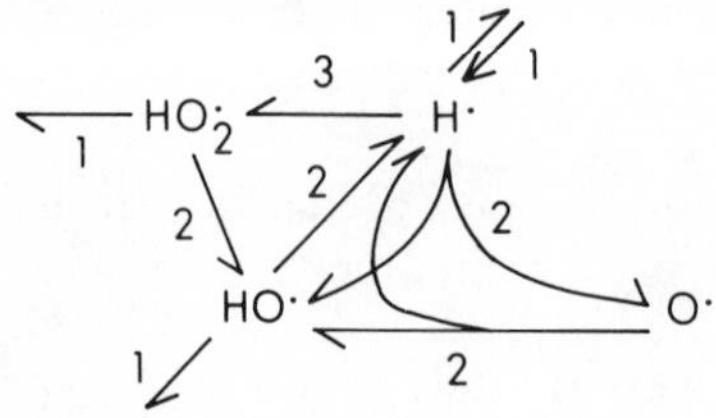

Fig. 54. $\mathfrak{D}_N$ for the reaction between H_2 and O_2. The numbers are the orders of the reactions.

which contains the main features of the mechanism:

$$
\begin{array}{rl}
R_1: & H_2 \rightarrow 2H\cdot \\
R_2: & \cdot H + O_2 \rightarrow \cdot OH + O\cdot \\
R_3: & \cdot O + H_2 \rightarrow \cdot OH + \cdot H \\
R_4: & \cdot H + O_2 + M \rightarrow HO_2^{\cdot} + M \\
R_5: & HO_2^{\cdot} + H_2 \rightarrow H_2O + \cdot OH \\
R_6: & \cdot OH + H_2 \rightarrow H_2O + H\cdot \\
R_7, R_8, R_9: & HO_2^{\cdot}, H\cdot, \cdot OH \rightarrow \text{removal at surface}
\end{array}
$$

The species appearing in the overall reaction and M will be considered external. The rates of the reactions above are proportional to the pressure to the κth power, where κ is the number of species on the left of the reaction. Figure 54, the network diagram of the internal species, indicates the order (κ) of each reaction.

At low pressures only the first order reactions are important. At higher pressures the second-order reactions become dominant, and at even higher pressures the third-order reaction plays a role. Only one extreme current can be constructed from the low pressure (first-order) reactions. It is shown in Fig. 55*a* and is asymptotically stable. At medium pressures the two extreme currents shown in Fig. 55*b* and 55*c* should be dominant. Current *b* has a critical current cycle involving all three species. Hence $\alpha_3 = 0$ for this current by itself. Current *c* has a critical current cycle involving H· and O·. Since no polygons span these species and HO·, $\alpha_3 = 0$ as well. Thus any mixture of currents *b* and *c* has $\alpha_3 = 0$. From Theorem V.14 we conclude that the system can be in a steady state described by a mixture of these two current diagrams alone only if a relation is satisfied by the rate constants; that is, $(\mathbf{k}, \mathbf{C}) \in D_K^0$. Since such a relation is unlikely to be satisfied, we expect either explosion or extinction dynamics to occur.

In this example we encounter for the first time a steady state that is a mixture of currents, some having explosion-extinction dynamics and some being stable. Let D_K^0 be the set of $(\mathbf{k}, \mathbf{C})$ that satisfy the restrictions making currents *b* and *c* possible in the absence of current *a*. If $(\mathbf{k} \in \mathbf{C}) \in D_K^0$ and we now add some current *a*, the presence of this current makes $\alpha_3 > 0$. A term in α_3 comes from the stabilizers of H· in current *a* and O· and ·OH in currents *b* and *c*. The steady state is now asymptotically stable. Note that a steady state described by only *b* and *c* has the property that if the concentrations of H·, O·, and HO· are multiplied by λ, all reactions go λ times faster and the system is still at steady state. Neither an explosion nor an extinction will occur; this marginal stability produces a line of possible steady states in $\Pi(\mathbf{C})$. Now mix in some of current *a*. These reactions of this current determine a steady state of H· that is asymptotically stable. Therefore they determine a particular point on the previous line of steady states in $\Pi(\mathbf{C})$. The system is now at steady state at this one point only. Marginal stability has been replaced by asymptotic stability.

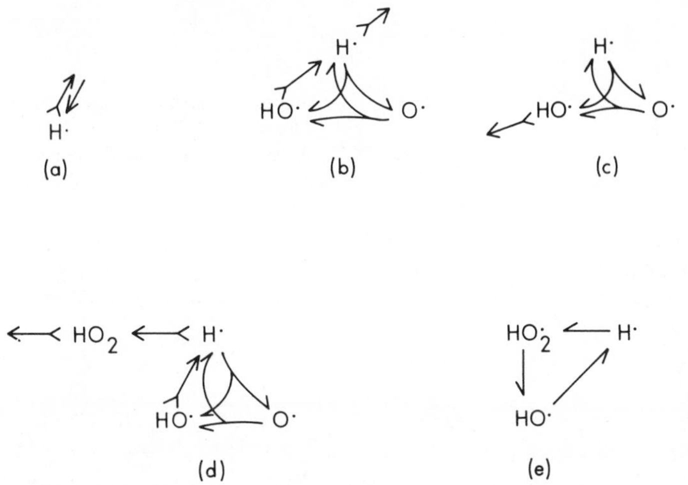

Fig. 55. The extreme currents that are dominant (*a*) at low pressure, (*b*) and (*c*) at medium pressures, and (*d*) and (*e*) at high pressures.

Next, suppose that $(\mathbf{k}, \mathbf{C}) \notin D_K^0$ and that extinction dynamics prevails. In the absence of the reactions of current *a*, the concentrations of H ·, HO ·, and O · would approach zero. However if the reactions of *a* can occur, the replenishment of H · must prevent H · from approaching zero and will bring about a steady state that is a mixture of currents *a*, *b*, and *c*. This current has $\alpha_3 > 0$ and is asymptotically stable.

The remaining case occurs when $(\mathbf{k}, \mathbf{C}) \notin D_\kappa^0$ and explosion dynamics prevails. We now examine how this situation is modified by the presence of current *a*. The explosion dynamics will tend to increase H ·, O ·, and HO ·. As H · increases, the exit reaction (R_8) in current *a* will become faster but the input reaction (R_1) will not. This removal of H · may stabilize the steady state and prevent explosion. However it may happen that currents *b* or *c* produce H · so rapidly that R_8 cannot remove it as fast as it is produced. Then an explosion will occur.

What Example V.23 illustrates can now be stated generally. *If a network has extreme currents that are capable of explosion-extinction dynamics, and if other asymptotically stable extreme currents are present and involve some of the species that may explode or go extinct, complete extinction cannot occur, but explosions can occur.* If there is extinction dynamics, the asymptotically stable extreme currents become dominant and stabilize the steady state so that complete extinction does not occur. Explosions are not prevented by these asymptotically stable extreme currents because an explosion makes the reactions in the explosive extreme current become relatively more important, thereby enhancing the conditions for explosion.

Let us reconsider the example in the case where explosion prevails. If we begin at the stable steady state (current *a*), and then shift to new parameters (steady states) having more of currents *b* and *c* and less of current *a*,

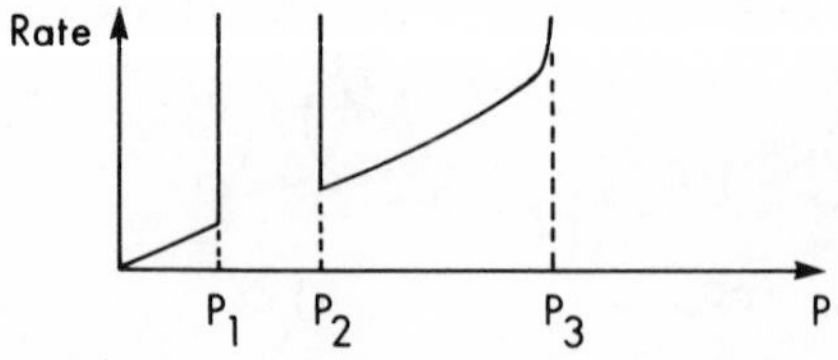

Fig. 56. Explosion limits for the reactions $2H_2 + O_2 \rightarrow 2H_2O$ at approximately 550 °C. The lower explosion limit occurs at P_1, the second explosion limit occurs at P_2. The third limit at P_3 is thought to be a thermal explosion, not a stoichiometric explosion.

we will eventually reach a critical mixture of currents beyond which no steady state exists and an explosion occurs. This steady state is called an *explosion limit*.

The hydrogen-oxygen reaction has a "lower" and an "upper" explosion limit (Fig. 56). Starting at zero pressure, as the pressure increases, currents 55*b* and 55*c* become more significant relative to current 55*a*, and an explosion occurs when the lower explosion limit at pressure P_1 is crossed. Between pressures P_2 and P_3, currents 55*d* and 55*e* are dominant. Note that current 55*c* has reactions that produce HO· and H·. When this current is mixed with currents *b* and *c* the current cycle changes from critical to weak because of these entrance reactions. Thus a steady state exists and is asymptotically stable. As the pressure is lowered, currents *d* and *e* become less important, and currents *b* and *c* become more important. An explosion occurs when the pressure P_2 is crossed. At pressures between P_1 and P_2, the dominant reactions of the network are those of currents *b* and *c*; however there is no steady state, and an explosion occurs.

This section has shown that nuclear explosions, DNA reproduction, and the explosion of H_2 and O_2 are essentially the same phenomena. Stoichiometric explosions can occur whenever a network has an extreme current exhibiting certain features, and rate constants that do not allow the steady-state conditions to be satisfied. These extreme currents often can be recognized easily. Currents containing more than one entrance reaction fall into this category. The category also includes currents having critical current cycles without the accompanying features that produce instability. When such extreme currents occur in a network that also has some asymptotically stable extreme currents, explosions occur only when overlapping asymptotically stable currents are sufficiently small relative to these explosive currents.

VI. ESTIMATING STABILITY AND THE BIFURCATION SET USING EXPONENT POLYTOPES

A. Introduction to the Problem

The basic ideas of the Hurwitz determinant approach to stability analysis were discussed in Section III.C. Previously we defined five sets of stability polynomials, called F_i, $i = 1, \ldots, 5$, and used them to express the domains of linear asymptotic stability and linear exponential instability. Now (III.50), (III.51), (III.28), and (III.29) give

$$D_a \supset \{\mathbf{p} \in D \mid f(\mathbf{p}) > 0 \quad \text{for all } f \in F_i\} \qquad \text{(VI.1)}$$

$$D_u \supset \{\mathbf{p} \in D \mid f(\mathbf{p}) < 0 \quad \text{for some } f \in F_i\} \qquad \text{(VI.2)}$$

Usually no stability polynomial vanishes identically, and usually the polynomials change sign in a neighborhood of every $\mathbf{p} \in D$ where they vanish. In these circumstances, D will be divided into two regions D_a and D_u by the hypersurface

$$B \equiv \{\mathbf{p} \in D \mid f(\mathbf{p}) \geqslant 0 \quad \text{for all } f \in F_i \quad \text{and}$$
$$f(\mathbf{p}) = 0 \quad \text{for some } f \in F_i\} \qquad \text{(VI.3)}$$

where B is called the *bifurcation set for steady states* because as $\mathbf{p}$ crosses B, an asymptotically stable steady state changes into an unstable steady state. The qualitative dynamics must change discontinuously. B is also called the *catastrophe set* or the *boundary of stability*.

In chemical reaction systems, $\mathbf{p}$ can be varied by changing the temperature or the concentrations of the external species. Thus if we can calculate B, we can predict, for example, how much sodium bromate, malonic acid, cerium, and sodium bromide should be mixed together to create a Belousov-Zhabotinski system just on the verge of oscillation. To make this prediction, we would have to know the true rate constants for the significant reactions in the actual mechanism. Usually the mechanism is partly uncertain and the rate constants are known only to a limited accuracy. If we could make a theoretical calculation that shows how B depends on the true rate constants and concentrations for a particular mechanism, then by experimentally determining B we could gain information that would enable us to refine the rate constants and test the mechanism. A method for determining B to the accuracy needed by chemists would therefore be of great practical value.

Let us examine the nature of the mathematical problem of calculating B. The stability polynomials can have very many terms. From (V.15) we see that the number of coefficients $\beta_i(\gamma, \mathbf{j})$ in the expansion of $\alpha_i(\mathbf{h}, \mathbf{j})$ is $\binom{n}{i}$. If each $\beta_i(\gamma, \mathbf{j})$ is expanded into a polynomial in $\mathbf{j} \in R_+^g$ (allowing for the possibility that not all f extreme currents are used), the number of terms that will occur if all coefficients are nonzero is the number of lattice points in a g-dimensional simplex with $i + 1$ points along each edge. The number of terms is then

$$N_\alpha(n, g, i) = \frac{n!}{i!\,(n-1)!} \times \frac{(i+g-1)!}{i!\,(g-1)!} \qquad \text{(VI.4)}$$

The total number of terms in all α_i, $i = 1, \ldots, n$ is given in Table III. The Hurwitz determinant $\Delta_i(\mathbf{h}, \mathbf{j})$ is a homogeneous polynomial of degree $i(i+1)/2$ in both $\mathbf{h} \in R_+^n$ and $\mathbf{j} \in R_+^g$ independently. The number of lattice points in a g-simplex with $i(i+1)/2+1$ points on an edge is an accurate estimate of the number of terms in $\mathbf{j}$ for each $\mathbf{h}$. The number of terms in $\mathbf{h}$ for each $\mathbf{j}$ is slightly overestimated by the number of lattice points in an n-simplex; however we use this formula as the simplest approximation. Then the number of terms is approximately

$$N_\Delta(n, g, i) = \frac{[i(i+1)/2+n-1]!}{(n-1)![i(i+1)/2]!} \times \frac{[i(i+1)/2+g-1]!}{(g-1)![i(i+1)/2]!} \qquad \text{(VI.5)}$$

When there are no conservation conditions ($d = n$), the largest Hurwitz determinant that must be studied is Δ_{n-1}. The approximate number of terms in this polynomial is given as a function of n and g in Table IV. It is best to make a simplicial decomposition of $\mathcal{C}_v$ and then calculate B only in one simplicial subcone at a time. The number of frame vectors required to span each simplicial subcone is $r - d$. Hence for each g and n in Tables III and IV, the maximum number of reactions that will always allow a full treatment of the problem is $r = n + g$ (when $d = n$). Of course when there are many conservation conditions, many terms vanish and the polynomials are usually very much smaller. The enormous number of terms that occur in α_i and Δ_i for medium size networks make it very desirable to prove stability by finding a Lyapunov function. The calculation of B in unstable networks can be accomplished only when the problem can be simplified mathematically.

We saw in Section V.D that each term in α_i is a product of feedback cycles. So is each term in Δ_i. There are very many terms in Δ_i because every possible combination of feedback cycles that might influence stability for every possible value of the parameters must appear. The number of possibilities is enormous. However if one knows that a network has a critical current cycle that is likely to produce an instability by interacting with certain other feedback cycles, one can often pick out the terms in Δ_i that will produce instability easily, even for networks where Δ_i cannot be constructed on the computer. We must therefore study the network interpretation of the terms in Δ_i. When an instability is possible, there must be some negative terms that make Δ_i negative for an appropriate $\mathbf{p} \in D$. Since the parameters are all nonnegative, these negative terms must have a negative algebraic sign. Therefore they are easily recognized. The mathematical problem we must now examine is how terms with negative algebraic signs can make a polynomial zero, when the polynomial consists

TABLE III
Logarithm (base 10) of the Sum of the Numbers of Terms in All Polynomials $\alpha_i(\mathbf{h}, \mathbf{j})$ for $i = 1, \ldots, n$[a]

$n = d$	$g = 1$	$g = 2$	$g = 3$	$g = 4$	$g = 5$	$g = 6$
1	0	0.3	0.5	0.6	0.7	0.8
2	0.3	1.1	1.6	1.9	2.2	2.4
3	1.0	2.3	3.2	3.9	4.4	4.8
4	2.0	4.1	5.4	6.4	7.3	7.9
5	3.4	6.3	8.2	9.6	10.8	11.8
6	5.2	8.9	11.4	13.3	14.9	16.2

[a] The number of species is n, and the number of current parameters being used is g. Data are for the worst case when $n = d$.

TABLE IV
Logarithm (base 10) of the Number of Terms in $\Delta_{n-1}(\mathbf{h}, \mathbf{j})$ as a Function of g and n ($n = d$)

$n = d$	$g = 1$	$g = 2$	$g = 3$	$g = 4$	$g = 5$	$g = 6$
2	0.3	0.6	0.8	0.9	1.0	1.1
3	1.0	1.6	2.0	2.3	2.5	2.8
4	1.9	2.8	3.4	3.9	4.3	4.6
5	3.0	4.0	4.8	5.5	6.0	6.5
6	4.2	5.4	6.3	7.1	7.8	8.4

mainly of positive terms, in the case when all parameters (variables) are positive.

The final important aspect of this problem is that the chemist is interested primarily in solutions in instances when the parameters are very different from each other in order of magnitude. Since **h** is the reciprocal of the steady-state concentration, it has the range of chemical concentrations, which is about 30 orders of magnitude. Similarly, **j** is determined by the relative reaction velocities, and these may range over even more orders of magnitude.

The mathematical problem can now be stated much more precisely. Given a polynomial $f(\mathbf{p})$ with extremely many terms, most of them positive and a few of them negative, construct approximate solutions to the equation $f(\mathbf{p}) = 0$ [i.e., "zeros" of $f(\mathbf{p})$] that are valid when the components of **p** differ widely in order of magnitude.

For any fixed **p** we can easily determine the numerical value of every term in the polynomial. If the components of **p** vary widely in magnitude, most terms should be negligible. We might try throwing away these terms and keeping only the few largest terms. If this set of terms contains terms with both positive and negative signs, we could reexpress them using **p** again to obtain a *subpolynomial* $f^*(\mathbf{p})$ and solve the equation $f^*(\mathbf{p}) = 0$ to obtain a valid approximation to the zeros of $f(\mathbf{p})$ near the original numerical value of **p**. A source of concern with this approach is that when **p** is changed to other numerical values, one has no way of knowing whether some of the dropped terms have become so large that they must be included.

The exponent polytope is the key to choosing the sets of terms in $f(\mathbf{p})$ properly, and it may be used to solve this mathematical problem in any context. In addition to the bifurcation set problem in stability analysis, it can be used to determine the concentrations of species at chemical equilibrium. It can be used to approximate the steady-state concentrations in terms of the rate constants. What these problems have in common is that they involve polynomials in parameters (concentrations, rate constants) that are positive and frequently range over many orders of magnitude.

For many years chemists have been making intuitive approximations of this sort based on their "physical picture" of what is happening. The exponent polytope method yields the same approximations chemists have obtained intuitively in the simple case; however it can handle much more complex problems. If one has a "physical picture," the exponent polytope method assures us of the consistency of this picture. If one does not have a picture, the method enables one to generate by computer all possible self-consistent approximations. These approximations suggest all possible self-consistent "pictures." The method can help build one's intuitive understanding of any system that is so complex that a clear intuitive picture of it is difficult to construct. We now explain the method and use it to develop a picture of how instabilities occur in complex networks.

B. Basic Ideas of the Exponent Polytope Method

A thorough mathematical treatment of the exponent polytope method may be found in Clarke.[42] This section explains the method in a manner that I hope will make the method transparent without giving detailed proofs. The reference paper contains the proofs.

Consider the polynomial

$$X^3 + 4X^7 - 16X^9 + 2X^{12} \qquad \text{(VI.6)}$$

As X becomes large, the dominant term is $2X^{12}$; as X approaches zero, the dominant term is X^3. I want to examine how we arrived at this obvious conclusion and then turn this thought process into an abstract algorithm that can be generalized to huge polynomials in many variables.

As X becomes large (or approaches zero) the dominant term is the one

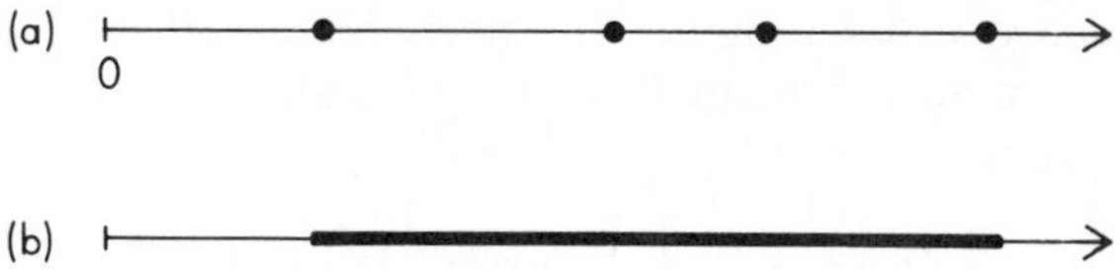

Fig. 57. (*a*) The exponents of polynomial (VI.6); (*b*) their convex hull.

with the largest (or smallest) exponent. Let us plot the exponents on the real line, as shown in Fig. 57*a*. The dominant term when $X \to \infty$ is represented by the point farthest to the right, and the dominant term when $X \to 0$ is represented by the point farthest to the left. We now develop a more abstract way of stating the two dominant terms.

Let Y be the set of exponent points on the line; that is, $Y = \{3, 7, 9, 12\}$. The *convex hull* of any set is defined to be the smallest convex set that contains all the points on the set. We write conv Y for the convex hull of Y. Figure 57*b* shows the line segment conv Y. In general, conv Y is a convex polytope, and since it is very important, we call it the *exponent polytope* and use the symbol $P \equiv \text{conv}\, Y$. The set of vertices of P consists of the points $\{3, 12\}$, which we call vert P. Note that *the two possible dominant terms are the terms whose exponents correspond to the two points in vert P.*

This description is pleasing because it does not make any reference to directions on the line. On the other hand, our method of saying $X \to \infty$ or $X \to 0$ is not symmetrical. To achieve symmetry, let us take natural logarithms of X and define $x = \ln X$. Then as $x \to \infty$, the dominant term is the vertex of P farthest in the direction toward $+\infty$. As $x \to -\infty$, the dominant term is the vertex of P farthest in the direction toward $-\infty$. We now have a symmetrical way of stating which term is dominant.

Since we must speak of "the vertex of P farthest in the direction . . . ," let us develop a better way of saying this. Mathematicians would call the point of P farthest to the right "sup P," the *supremum* of P, and the point of P farthest to the left "inf P," the *infimum* of P. The concept we need is an n-dimensional generalization of the supremum and infinum.

The vertices of any polytope P are considered to be zero-dimensional faces of P. Thus in the example the points 3 and 12 are "0-faces" of P. The face of P that is farthest in any direction is called an *extreme face*. The direction toward $+\infty$ is called $+1$ and the direction toward $-\infty$ is called -1. Let us define $\text{exface}_Y \hat{x}$ to be the extreme face in the direction $\hat{x}$ for the polytope conv Y. Thus we have for the example $12 = \sup P = \text{exface}_Y(+1)$, $3 = \inf P = \text{exface}_Y(-1)$. We now restate our observation about the dominant terms of the polynomial (VI.6) as follows. *As x changes sufficiently far in the direction $\hat{x}$, the dominant term is the term whose exponent is $\text{exface}_Y \hat{x}$.*

This abstract way of stating the intuitively obvious has the advantage of usually being true for polynomials in many variables. Consider the polynomial in the variables X and Y,

$$X^4Y^2 + X^4Y^4 + X^3Y + X^3Y^3 + X^3Y^4 + X^3Y^5 + X^2Y^2 + X^2Y^4 + XY + XY^2 + XY^3 + XY^4 + XY^5 + Y^3 \qquad \text{(VI.7)}$$

in the quadrant, $(X, Y) \in R^2_+$. The term X^mY^n has the exponent point (m, n) which is an element of the exponent set Y. Figure 58 shows the set Y in the left-hand drawing and the exponential polytope $P = \operatorname{conv} Y$ in the middle drawing. As before, we define new variables from the old ones,

$$x = \ln X \qquad \text{and} \qquad y = \ln Y$$

and ask what term (or terms) is dominant as x and y approach $\pm\infty$.

There are many curves along which (x, y) could approach infinity. It turns out that we can understand enough by studying only what happens as (x, y) approaches infinity along straight lines. "Lines" that have one end point and go to infinity only toward the other end are called *rays*. For any two points (x_0, y_0) and $(\hat{x}, \hat{y}) \in R^2$, there is a ray Λ, whose parametric equation is

$$(x(\lambda), y(\lambda)) = (x_0, y_0) + (\lambda\hat{x}, \lambda\hat{y}), \qquad \lambda \geqslant 0 \qquad \text{(VI.8)}$$

whose *base point* is (x_0, y_0), and whose *direction vector* is $(\hat{x}, \hat{y})$. Without loss of generality we may assume

$$(\hat{x}, \hat{y}) \in \Sigma \equiv \{(\hat{x}, \hat{y}) \in R^2 \,|\, \hat{x}^2 + \hat{y}^2 = 1\} \qquad \text{(VI.9)}$$

where Σ is the *unit sphere*.

As (x, y) goes to infinity along the ray Λ, the terms that become dominant are the terms whose exponent points are elements of the set $\operatorname{exface}_Y(\hat{x}, \hat{y})$. This statement is true for this example; however for other

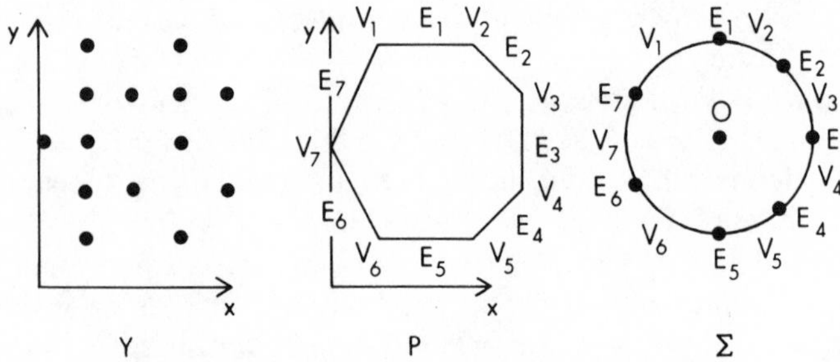

Fig. 58. Y, P, and Σ for the polynomial (VI.7).

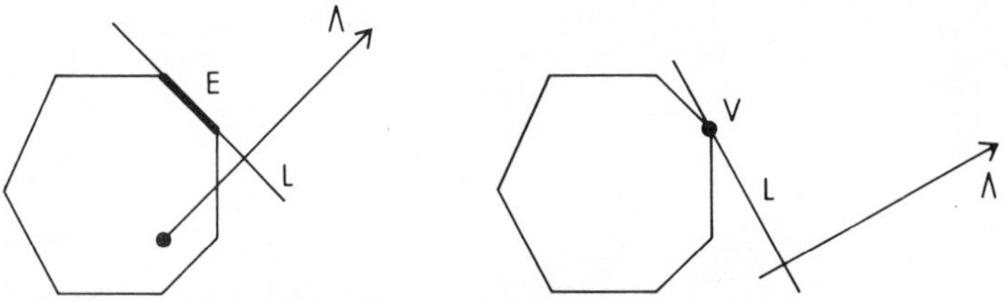

Fig. 59. How to determine $\text{exface}_Y(\hat{x}, \hat{y})$ for the polynomial (VI.7). At the left is a ray Λ whose extreme face is the edge E. At the right is a ray Λ whose extreme face is the vertex V.

examples there are certain values of (x_0, y_0) for which this statement is not true. The proof may be found in Theorem 1 Ref. 42.

I now explain how to determine $\text{exface}_Y(\hat{x}, \hat{y})$ in the two-dimensional case. Figure 59 shows P and two arbitrary rays. In both cases imagine a line L perpendicular to the ray that intersects the ray at the point $(x(\lambda), y(\lambda))$. Assume that λ is very large. If it is large enough, decreasing λ will move the line L toward P. We now decrease λ until L just touches P. Then L is called a *supporting line* of P. Then we define

$$\text{exface}_Y(\hat{x}, \hat{y}) = L \cap P \tag{VI.10}$$

In two dimensions, P has only two types of faces, vertices and edges. Figure 59 shows an example of each case.

The set of dominant terms depends only on the direction of the ray in this example (but not in general). At the right in Fig. 58 the unit sphere Σ has been divided into regions such that each region is the set of directions where a given edge or vertex is dominant. Note that the terms on an edge can be dominant only on rays in one special direction—the direction orthogonal to the edge. On the other hand, the term corresponding to a vertex is dominant for all rays whose directions lie on an open arc of Σ.

Let us now consider the polynomial

$$X^3 - 10X^2Y^2 + 10XY^4 - Y^6 \tag{VI.11}$$

Figure 60 shows the region of (x, y) where this polynomial is positive, zero, and negative. This example illustrates Theorem 3 Ref. 42, which proves that when P has fewer dimensions than the number of variables, the regions where the polynomial has certain signs have translation symmetry in all directions orthogonal to P.

We are now ready to understand how the polynomial curve $f(X, Y) = 0$ approaches the boundary of the quadrant $X > 0$, $Y > 0$. First we convert to the new variables $x = \ln X$, $y = \ln Y$. This mapping distorts the quadrant

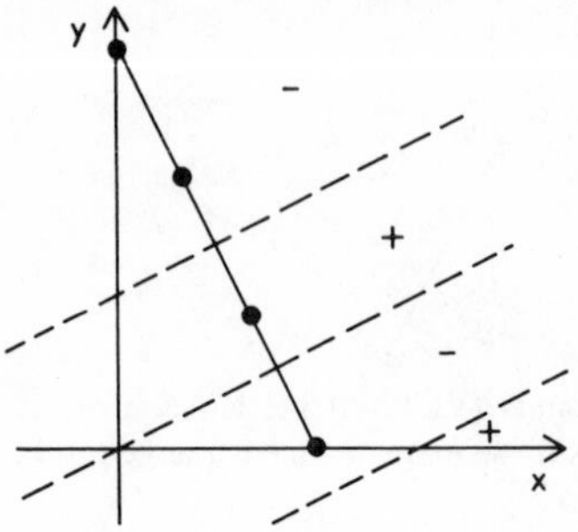

Fig. 60. Signs and zeros of polynomial (VI.11). The drawing shows the points Y, the exponent polytope P, and the regions of (x, y) where the polynomial is positive, zero, and negative.

as shown in Fig. 61. The curve $f(X, Y) = 0$ will be mapped into the curve $g(x, y) = 0$. The finite and infinite boundary of the orthant are all mapped into infinity. Hence the behavior of $f(X, Y) = 0$ near the orthant boundary is equivalent to the behavior of $g(x, y) = 0$ as (x, y) approaches infinity in every direction.

To find this curve at infinity, we combine the ideas of the two preceding examples. As we go out to infinity along any ray whose corresponding extreme face is an edge, the terms on the edge become dominant [see polynomial (VI.7)]. However the edge terms form a polynomial whose exponent polytope has a lower dimension (it is one-dimensional); thus "at infinity" the situation resembles the case of polynomial (VI.11) shown in Fig. 60. That is, the curve $g(x, y) = 0$ lies approximately on a set of lines that are perpendicular to the edge.

Let us modify the signs at some terms in polynomial (VI.7) so that the

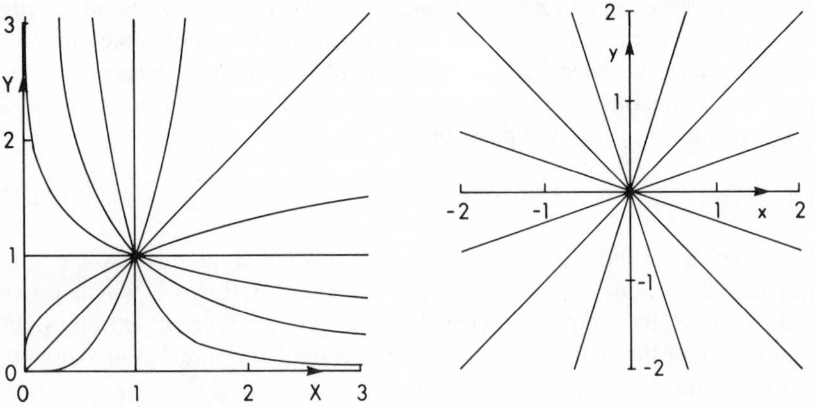

Fig. 61. The logarithmic mapping $(X, Y) \rightarrow (x, y)$. The graphs show curves in the (X, Y)-plane that map into rays in the (x, y)-plane.

subpolynomials corresponding to the edges can change sign. Take as an example

$$-X^4Y^2 + X^4Y^4 + X^3Y + X^3Y^3 + X^3Y^4 - X^3Y^5 + X^2Y^2$$
$$+X^2Y^4 - XY + XY^2 + XY^3 + XY^4 + XY^5 + Y^3 \quad \text{(VI.12)}$$

where P, Y, and Σ are still the same as in Fig. 58. Along all rays in the direction (0, 1), the terms on the edge E_1 become dominant at infinity. Hence we can throw away all the remaining terms to leave only the two terms on E_1. These terms are XY^5 and $-X^3Y^5$. When the polynomial can be approximated only by these terms, we calculate that the polynomial vanishes when $XY^5 = X^3Y^5$. Taking logarithms gives

$$\ln X + 5\ln Y = 3\ln X + 5\ln Y$$

or

$$x = 0$$

This line is plotted in Fig. 62. We see that as $y \to \infty$, the function $g(x, y)$ changes sign as x crosses the line $x = 0$. Most of the other edges give rise to similar lines where $g(x, y)$ changes sign at infinity.

The rays in Fig. 62 that approach the curves $g(x, y) = 0$ at infinity are called *asymptotic rays*. These rays are special cases of *critical rays*, which are defined as follows in the two variable case. For any edge E, let $f_E(X, Y)$ consist of the terms of f whose exponent points lie on the edge E of P. Let X_0 and Y_0 be such that $f_E(X_0, Y_0) = 0$. Let $x_0 = \ln X_0, y_0 = \ln Y_0$,

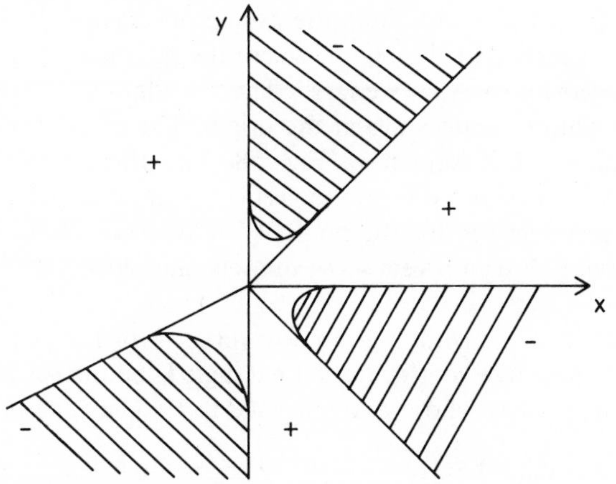

Fig. 62. Regions where the polynomial (VI.12) is positive, zero, and negative (shaded). This plot uses the new variables $x = \ln X$, and $y = \ln Y$.

and let $(\hat{x}, \hat{y})$ be the direction whose extreme face is E. Then the ray

$$(x(\lambda), y(\lambda)) = (x_0, y_0) + \lambda(\hat{x}, \hat{y})$$

is called a *critical ray*. As illustrated in Fig. 60, every point on this ray gives another (X_0', Y_0') such that $f_E(X_0', Y_0') = 0$. An edge may have any finite number of critical rays, but not more than the number of points of Y on the edge minus one (Newton's theorem).

The function of (x, y) that corresponds to $f_E(X, Y)$ will be called $g_E(x, y)$. If $g_E(x, y)$ changes sign near a critical ray, the ray is always an asymptotic ray. When $g_E(x, y) = 0$ on the ray but does not change sign, the ray may or may not be an asymptotic ray. Complications may occur that are probably not of much interest in the stability problem, so they are not discussed here. The reference paper[42] shows that the possibilities in this situation can be extremely complicated. The theorem that is of most interest in stability analysis is now stated for the two variable case. (See Theorem 6 of Clarke[42].)

Theorem VI.1. Given any $\varepsilon > 0$, there exists a real number $\lambda > 0$ such that every real solution of $g(x, y) = 0$, with $\sqrt{x^2 + y^2} > \lambda$, lies within ε of a critical ray.

This theorem implies that the curves $g(x, y) = 0$ cannot go to infinity in any way other than by approaching critical rays asymptotically. The theorem is important because it implies that if we approximate the negative regions at infinity using asymptotic rays, we will not miss any of the negative region at infinity. Even then, the error will be less than ε and will be confined to a small band along the asymptotic rays.

Figure 62 suggests that the region where the polynomial is negative can be approximated by several "wedges". These wedges are two-dimensional convex cones whose vertices are at the origin. The cones can be obtained from the regions of Σ shown in Fig. 58. The three negative terms in polynomial (VI.12) correspond to the vertices of P labeled V_2, V_4, and V_6. The three negative cones are the points lying on rays whose base point is the origin, having direction vectors in the sets marked V_2, V_4, and V_6 on Σ in Fig. 58.

The general principle is this: given any vertex V of P whose corresponding term has a negative coefficient, let extangle V be the set of points in Σ representing directions whose extreme face is V. The *cone of* V is the set

$$\mathcal{C}_0(V) \equiv \{(\lambda\hat{x}, \lambda\hat{y}) \mid (\hat{x}, \hat{y}) \in \text{extangle } V\} \qquad \text{(VI.13)}$$

The negative region in Fig. 62 can be approximated by $\mathcal{C}_0(V_2) \cup \mathcal{C}_0(V_4) \cup$

$\mathcal{C}_0(V_6)$. This approximation is discussed in Section 12 of Clarke[42] and is called the *vertex approximation* to the negative region or to the zeros.

The vertex approximation is completely determined by the shape of P and the signs of the terms corresponding to the vertices of P. All the cones in this vertex approximation have their vertex at the origin. This means that asymptotic rays that cannot be made to pass through the origin by choosing (x_0, y_0) appropriately do not lie along the edges of the cones. Since the curves $g(x, y) = 0$ approach these rays at infinity, the edges of the cones are displaced from the true curves by a fixed distance at infinity. One can partly remedy the situation by moving the common vertex from the origin to a point that optimizes the approximation; however this procedure is not very good because the asymptotic rays do not always intersect in a common point.

The next approximation, called the *edge approximation*, is obtained by modifying the vertex approximation slightly along the edges of $\mathcal{C}_0(V)$. We know that when one goes to infinity along an edge of the cone $\mathcal{C}_0(V)$, the dominant terms of the polynomial correspond to points on a single edge of P. The picture at infinity is like Fig. 60. Hence we simply replace the ray representing the edge of $\mathcal{C}_0(V)$ with a set of parallel rays. Each ray is positioned so that the subpolynomial of terms from the edge of P vanishes on the ray. The edge approximation to a complicated polynomial could give a plot similar to that shown in Fig. 63. Note that the asymptotic rays approximate the curve $g(x, y) = 0$ very well outside of a central region

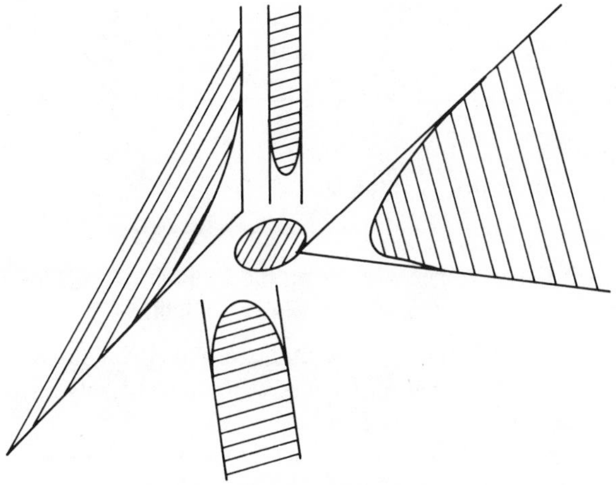

Fig. 63. A typical edge approximation for a very complicated polynomial.

where the curve is much more complicated. The approximate regions where $g(x, y) < 0$ are either cones or cylinders (strips, bands). The cones are still associated with vertices of P. We therefore call the one associated with the vertex V $\mathcal{C}_1^0(V)$. The cylinders are associated with edges, so we call the one associated with the edge E $\mathcal{C}_1^1(E)$.

The exponential polytope method can also determine in detail how the branches of $g(x, y) = 0$ approach the asymptotic rays. It is possible for many branches to approach the same ray. When this happens, the branches approach in groups with roughly the same rate of approach. Within these groups the branches fall into further subgroups with roughly the same rate of approach in each subgroup. How all this occurs can be understood from the exponent polytope. This problem, called "the resolution of singularities," is not discussed further here.

The reader who wishes to get an overview of developments in two-dimensional algebraic geometry during the last several centuries will find the review article by Abhyankar[74] interesting. The exponent polytope enables one to understand the subject from a new perspective. A two-dimensional forerunner of the exponent polytope is called *Newton's polygon* (NP). The NP was used to construct Newton-Puiseux expansions (Ref. 75, p. 91) of functions near their branch points. In effect, it solved the resolution of singularities problem in two-dimensions. The general resolutionof singularities problem has recently been solved by Hironaka[76] using higher-dimensional polyhedra that are associated with the singularity. The relationship between my work and Hironaka's is analogous to the relation between the work of Jean-Paul de Gua de Malves[77] and that of Isaac Newton (Ref. 78, p. 50). Seventy years after Newton discovered the polygon, Gua de Malves discovered that the polygon also determined how two-dimensional curves of any order go to infinity in the quadrant. The exponent polytope resolves singularities and determines how all solutions of $f(\mathbf{X}) = 0$ approach both the finite and infinite part of the orthant boundary in any number of dimensions.

C. Practical Aspects of Using the Exponent Polytope for Polynomials in Many Variables

As a simple introduction to the n-dimensional problem, we derive some formulas that approximate the pH of a weak acid dissolved in water. The ionization reaction for the acid HA is

$$HA + H_2O \rightleftarrows A^- + H_3O^+ \qquad \text{(VI.14)}$$

The law of mass action says that the equilibrium concentrations of HA,

A^-, and H_2O^+ must satisfy the equation

$$\frac{[A^-][H_3O^+]}{[HA]} = K$$

If there is no H_3O^+ present initially, when the reaction produces H_3O^+ with a concentration H (say), it must also increase the concentration of the base A^- from its initial value B (say) to $B + H$, and decrease the concentration of the acid HA from its initial value A (say) to $A - H$. The law of mass action is then satisfied if and only if the polynomial

$$f(A, B, H, K) \equiv H^2 + H(B + K) - KA \tag{VI.15}$$

is zero. We wish to find the real positive zeros of this polynomial in four variables. The real zeros form a hypersurface in R^4.

First we find the exponent polytope by writing

$$f(A, B, H, K) = A^0B^0H^2K^0 + A^0B^1H^1K^0 + A^0B^0H^1K^1 - A^1B^0H^0K^1 \tag{VI.16}$$

The exponents of the four terms give us the set of four points

$$Y = \{(0,0,2,0)', (0,1,1,0)', (0,0,1,1)', (1,0,0,1)'\}$$

whose convex hull is P. One may verify that P is a simplex (tetrahedron) by showing that the determinant whose four columns are the coordinates of the points of Y does not vanish. Hence every vertex of P is connected to every other vertex of P by an edge. Four points in R^4 must lie in a three-dimensional affine subspace, so the tetrahedron P must be in a subspace. Figure 64 shows P at the left. At the right we see how Σ is subdivided according to the face structure of P: $\hat{V}_i$, $\hat{F}_i$, and $\hat{E}_i$ are the

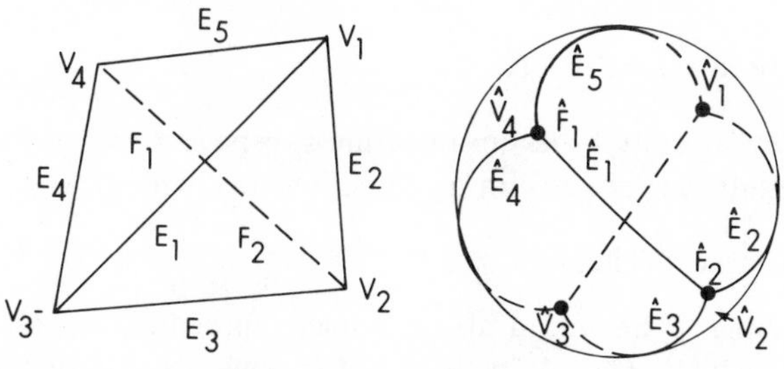

Fig. 64. The vertices, edges, and facets of the polytope P at the left are marked V_i, E_i, and F_i. At the right is the unit sphere Σ, which has been subdivided so that $\hat{V}_i$ (or alternately, $\hat{E}_i$ or $\hat{F}_i$) is the set of $\hat{\mathbf{x}} \in \Sigma$ such that $\text{exface}_Y \hat{x}$ is in V_i (or alternately E_i or F_i) of P. Only the front faces are labeled.

vertices, faces, and edges, respectively, of a spherical polytope that is dual to P.

Next define the logarithmic variables $a = \log A$, $b = \log B$, $h = \log H$, and $k = \log K$. Let $\mathbf{x} = (a, b, h, k)'$, and then define the *exponomial*

$$g(\mathbf{x}) = 10^{(0\,0\,2\,0)\mathbf{x}} + 10^{(0\,1\,1\,0)\mathbf{x}} + 10^{(0\,0\,1\,1)\mathbf{x}} - 10^{(1\,0\,0\,1)\mathbf{x}} \qquad \text{(VI.17)}$$

where the multiplication of a row and column matrices in the exponents is equivalent to a vector inner product. It follows from these definitions that $f(A, B, H, K) = g(\mathbf{x})$. Hence the hypersurface $g(\mathbf{x}) = 0$ in the logarithmic space corresponds to the hypersurface $f(A, B, H, K) = 0$ in the orthant, and this is the set of physical (i.e., real positive) values of A, B, H, and K where reaction (VI.14) is at equilibrium.

What we have done so far in this example can be generalized as follows. Let the polynomial $f(\mathbf{X})$ have m terms each with different exponents. Let $\mathbf{y}_i$ be the column matrix representing the exponents of the ith term. Then $Y = \{\mathbf{y}_i \mid i = 1, \ldots, m\}$ and $P \equiv \operatorname{conv} Y$. From the face structure of P we may always subdivide the unit sphere in n dimensions in analogy with Fig. 64. If we choose any base a for the logarithms ($a = e$ and 10 previously), then we may define

$$g(\mathbf{x}) = \sum_{i=1}^{m} c_i a^{\mathbf{y}_i'\mathbf{x}}$$

where c_i is the coefficient of the ith term in $f(\mathbf{X})$. Then $f(\mathbf{X}) = g(\mathbf{x})$. Equation VI.17 is a special case of this general form. Since the sum is taken over all points in Y, we do not need an index i to label the terms. Instead we write

$$g(\mathbf{x}) = \sum_{\mathbf{y} \in Y} c(\mathbf{y}) a^{\mathbf{y}'\mathbf{x}} \qquad \text{(VI.18)}$$

where c_i has been relabeled $c(\mathbf{y})$.

Theorem VI.2. If conv Y lies in an affine subspace $\mathcal{A}$, the set of zeros of $g(\mathbf{x})$ has translation symmetry in all directions orthogonal to $\mathcal{A}$.

Proof. See Clarke,[42] Theorem 3.

In the example, P lies in an affine 3-space embedded in R^4. If we can visualize the surface $g(\mathbf{x}) = 0$ in three dimensions, the extension to R^4 by translation symmetry is trivial. In the discussion that follows I refer to points, lines, and planes in $\mathcal{A}$ rather than the corresponding surfaces of one higher dimension in R^4.

If we put numbers into (VI.16) for a system in equilibrium, we always find that the negative term is the largest in magnitude. Usually the negative term will be mostly canceled by just one of the other three terms. There are three possibilities, which we state as equalities even though numerically the terms would only be approximately equal. The possibilities are

$$H^2 = AK, \qquad BH = AK, \qquad HK = AK$$

Taking logarithms and solving for h gives

$$h = \frac{a+k}{2}, \qquad h = k + a - b, \qquad h = a \tag{VI.19}$$

Every chemist will recognize these equations when written in the conventional notation, where $\mathrm{pH} \equiv -h$, $\mathrm{p}K_a \equiv -k$, $\log[\mathrm{HA}]_0 \equiv a$, and $\log[\mathrm{A}^-]_0 \equiv b$. Then they become

$$\mathrm{pH} = \frac{\mathrm{p}K_a - \log[\mathrm{HA}]_0}{2}$$

$$\mathrm{pH} = \mathrm{p}K_a + \log\frac{[\mathrm{A}^-]_0}{[\mathrm{HA}]_0}$$

$$\mathrm{pH} = -\log[\mathrm{HA}]_0$$

Only one of these approximations can apply at once. In the first case, BH and HK are small relative to H^2, which equals AK. Thus $B \ll H$ and $H \ll A$. This case applies when the acid is concentrated and mostly un-ionized. In the second case H^2 and HK are small relative to BH, which equals AK. Thus $H \ll B$ and $H \ll A$. This is the situation in a buffer solution. In the third case H^2 and BH are small relative to HK, which equals AK. Thus $H \ll K$ and $B \ll K$. Since $H = A$, we get $A \ll K$. This case applies when the initial amount of acid is much less than K. Chemists rarely have to solve the quadratic equation (VI.15) because the variables A, B, H, and K are usually so different in order of magnitude that one of these limiting situations gives a good approximation.

We now look at these results in terms of the exponent polytope in the affine three-dimensional space $\mathscr{A}$ using the theorems proved in Ref. 42. A face of P can have any dimension from 0 to $\dim P$. For any face F of P we define $g_F(\mathbf{x})$ to be the *subexponomial* consisting of all terms whose exponent matrices (vectors) $\mathbf{y}_i$ lie in F.

$$g_F(\mathbf{x}) \equiv \sum_{\mathbf{y} \in Y \cap F} c(\mathbf{y}) a^{\mathbf{y}'\mathbf{x}} \tag{VI.20}$$

Theorem VI.3. We can find the limiting form of $g(\mathbf{x}(\lambda))$ as $\lambda \to \infty$ along any ray $\mathbf{x}(\lambda) = \mathbf{x}_0 + \lambda\hat{\mathbf{x}}$ as follows. The ray determines $\mathbf{x}_0 \in R^n$ and $\hat{\mathbf{x}} \in \Sigma$. First find the face F defined by $F = \text{exface}_Y \hat{\mathbf{x}}$. Then test to see if $g_F(\mathbf{x}_0) \neq 0$. If $g_F(\mathbf{x}_0) \neq 0$ then

$$\lim_{\lambda \to \infty} \frac{g(\mathbf{x}(\lambda))}{g_F(\mathbf{x}(\lambda))} = 1$$

That is, the limiting form of $g(\mathbf{x})$ is $g_F(\mathbf{x})$. Otherwise, delete all the terms in $g_F(\mathbf{x})$ from $g(\mathbf{x})$ to obtain a new polynomial. Repeat the test for that polynomial. If a limiting form is not found, more terms should be deleted and the test repeated again. Iterate the test as often as necessary to find a limiting form. If no such form is found before all terms have been deleted from $g(\mathbf{x})$, then $g(\mathbf{x}) = 0$ everywhere on the ray.

Proof. See Theorem 1, part a, and Theorem 2 of Clarke[42]

Suppose $\hat{\mathbf{x}}$ in the example lies in one of the regions $\hat{V}_i$, $i = 1, 2, 3, 4$ of Fig. 64. Then $F = \text{exface}_Y \hat{\mathbf{x}} = V_i$. Each of the four possible subexponomials has a single term whose exponents are the coordinates of the vertex V_i. Theorem VI.3 implies that the limiting form of $g(\mathbf{x})$ along rays in these directions (i.e., $\hat{\mathbf{x}} \in \hat{V}_i$) must have the sign of the term whose exponents are the corresponding vertex of P. Since there is only one negative term in the polynomial $(-AK)$, there is only one vertex of P that can be the extreme face of direction vectors $\hat{\mathbf{x}}$ of rays along which $g(\mathbf{x}(\lambda))$ has a negative limit. If we call the vertices $V_1, \ldots, V_4$ in the order listed in Y, this vertex is V_4, and this set of rays has direction vectors $\hat{\mathbf{x}} \in \hat{V}_4$. To visualize these rays, let $\mathbf{x}_0 = 0$; then the rays form the cone

$$\mathcal{C}_0(V_4) \equiv \{\lambda\hat{\mathbf{x}} \mid \lambda \geqslant 0, \quad \hat{\mathbf{x}} \in \hat{V}_4\}$$

When $\mathbf{x}(\lambda)$ goes out a ray whose extreme face is an edge, the limiting form of $g(\mathbf{x})$ contains only the terms on the edge, provided $g_F(\mathbf{x}_0) \neq 0$. Consider the edge E_5 (Fig. 64) in the example. This edge connects vertices V_1 and V_4 so

$$g_{E_5}(\mathbf{x}) = c(\mathbf{y}_1)a^{\mathbf{y}_1'\mathbf{x}} + c(\mathbf{y}_4)a^{\mathbf{y}_4'\mathbf{x}} = H^2 - AK$$

Far out along rays whose extreme face is E_5, the limiting form of $g(\mathbf{x})$ has the sign of $H^2 - AK$, provided this expression does not vanish. We see that the limiting form changes sign when $H^2 - AK$ changes sign. Hence there must be a zero of $g(\mathbf{x})$ near the surface $H^2 = AK$ [or equivalently $h = \frac{1}{2}(a + k)$] at infinity.

Recall from the previous subsection that $\mathbf{x}_0 + \lambda\hat{\mathbf{x}}$ is called a *critical ray* if

$g_F(\mathbf{x}_0) = 0$, where $F = \text{exface}\,\hat{\mathbf{x}}$. Rays having $\hat{\mathbf{x}} \in \text{extangle}\, E_5$ are critical if and only if $g_{E_5}(\mathbf{x}_0) = 0$; that is, if $h = \frac{1}{2}(a + k)$. This linear equation in the logarithmic variables is a hyperplane in R^4. I now show that the edge E_5 is perpendicular to this hyperplane. The *edge vector* of E_5 is the vector between the vertices at the opposite ends of E_5; that is, $\mathbf{y}_1 - \mathbf{y}_4 = (0,0,2,0) - (1,0,0,1) = (-1,0,2,-1)$. The equation of the hyperplane can be rearranged as follows:

$$(-1,0,2,-1)(a,b,h,k)^t = 0$$

Hence the hyperplane is orthogonal to E_5. It also passes through the origin and the set $\hat{E}_5$ marked on Σ in Fig. 64. This hyperplane is a facet of the cone $\mathcal{C}_0(V_4)$ because $\hat{E}_5$ is part of the boundary of $\hat{V}_4$. Similar arguments may be applied to the two other boundaries of $\hat{V}_4$, which are associated with the other two edges of P meeting at the vertex V_4, corresponding to the vertex negative term in $f(\mathbf{X})$. We conclude that the equations (VI.19) are the three hyperplane facets of $\mathcal{C}_0(V_4)$. All rays lying in these hyperplanes and having the corresponding edges as extreme faces in the direction of the ray are critical rays. Viewed from within $\mathcal{Q}$, the hyperplanes become planes and $\mathcal{C}_0(V_4)$ is a triangular cone in three dimensions with three plane facets.

Theorem VI.4. Given any $\varepsilon > 0$, there exists $\lambda > 0$ such that every real solution to the equation $g(\mathbf{x}) = 0$ with $\|\mathbf{x}\| > \lambda$ lies within ε of a critical ray.

Proof. See Theorem 6 of Clarke.[42]

If the three hyperplanes (VI.19) contained all critical rays, this theorem would imply that the hypersurface $g(\mathbf{x}) = 0$ lies within ε of the boundary of $\mathcal{C}_0(V_4)$ for all $\|\mathbf{x}\| > \lambda$. The situation is not quite so simple because there are more critical rays. However we already have the main features of the hypersurface $g(\mathbf{x}) = 0$.

There are no critical rays associated with the vertices because when F is a vertex, $g_F(\mathbf{x}_0)$ is a single term and cannot vanish. Three of the edges have a hyperplane of critical rays, and the other three edges of P have no critical rays. The hyperplanes are given by (VI.19). We now find all critical rays associated with the four facets F_i of P. From Fig. 64 we see that these rays can have only four directions $\hat{F}_i$, $i = 1, \ldots, 4$ when viewed from within $\mathcal{Q}$. As λ increases, $g(\mathbf{x}(\lambda))$ resembles $g_F(\mathbf{x}(\lambda))$, provided $g_F(\mathbf{x}_0) \neq 0$. Thus at infinity $g(\mathbf{x}(\lambda))$ is determined by only the three terms whose exponents are on the corresponding facet of P. Take F_1 as an example. The limiting form of $g(\mathbf{x}(\lambda))$ is then

$$g_{F_1}(\mathbf{x}(\lambda)) = H^2 + HK - AK \qquad \text{(VI.21)}$$

provided $g_{F_1}(\mathbf{x}_0) \neq 0$. The zeros of this function determine some new critical rays that form a curved surface.

Let us investigate how this surface relates to the planes (VI.19) viewed from within $\mathcal{Q}$. The exponent polytope of the polynomial (VI.21) is a triangle. We may apply all the theory derived so far to this case. Along rays in directions such that the edges are extreme faces, we get two limiting forms that can vanish. These are $H^2 = AK$ and $HK = AK$. Hence the solution curve of (VI.21) in $\mathcal{Q}$ is asymptotic to the planes $h = \frac{1}{2}(a + k)$ and $h = a$, which we found earlier. These planes intersect in the line $h = a = 2k$, and this line is orthogonal to the facet F_1 and parallel to $\hat{F}_1$. The new critical rays have base points on (VI.21) and are in the direction $\hat{F}_1$. The surface formed by these rays is asymptotic to the planes $h = \frac{1}{2}(a + k)$ and $h = a$; however it is a smooth surface that makes a smooth transition from one plane to the other. Therefore it should differ most from the planes along their line of intersection $h = a = 2k$.

The complete picture is summarized in Fig. 65. The three facets of the cone $\mathcal{C}_0(V_4)$ are identified with the "concentrated acid" plane $h = \frac{1}{2}(a + k)$, the "dilute acid" plane $h = a$, and the "buffer" plane $h = k + a - b$. The "concentrated" and "dilute" planes intersect along the line $h = a = 2k$, where a better approximation can be obtained by solving (VI.21). The "concentrated" and "buffer" planes intersect along the line $h = b = \frac{1}{2}(a + k)$, where a better approximation comes from solving $H^2 +$

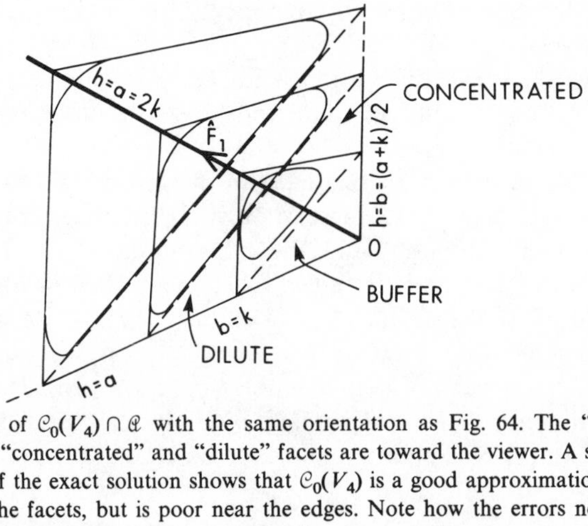

Fig. 65. Sketch of $\mathcal{C}_0(V_4) \cap \mathcal{Q}$ with the same orientation as Fig. 64. The "buffer" facet is behind and the "concentrated" and "dilute" facets are toward the viewer. A sketch of several cross-sections of the exact solution shows that $\mathcal{C}_0(V_4)$ is a good approximation to $g(\mathbf{x}) = 0$ in the middle of the facets, but is poor near the edges. Note how the errors near the edges of $\mathcal{C}_0(V_4)$ remain almost constant going out the edge; however the errors appear less significant relative to the cross-sections beacuse of an increase in scale.

$BH - AK = 0$. The "dilute" and "buffer" planes intersect along the line $h = a$, $b = k$, where a better approximation comes from solving $H(B + K) = AK$. Finally, these three 2-*face approximations* fail near the apex (vertex) of $\mathcal{C}_0(V_4)$, where $h = a = k = b$. An accurate solution in this region requires the solution of the full equation (VI.15).

It can be shown in general that the equation $g_F(\mathbf{x}) = 0$ can be transformed by a change of variables into an equation with precisely the number of variables as F has dimensions. For example, because P was three-dimensional, (VI.15) can be converted from a four-variable equation to a three-variable equation. The equations associated with the facets of P are reducible to two-variable equations, and the equations associated with the edges are reducible to one-variable equations. Hence as the approximation to $g(\mathbf{x}) = 0$ is refined starting with edges and then passing to 2-faces and higher dimensional faces, the number of variables increases steadily. In this pH problem, the general case (VI.15) is quadratic and can be solved as easily as the 2-face (facet) equations, such as (VI.21). In other problems where the polynomials are of higher order, the solutions must be found numerically. It is then an advantage to be able to reduce the equations to the smallest number of true independent variables.

The exponent polytope approach eliminates the need to define dimensionless parameters because for every parameter that can be eliminated using dimensionless parameters, the exponent polytope will have one lower dimension and no actual advantage will result from the use of dimensionless parameters. For example, the polynomial $\alpha_1(\mathbf{h},\mathbf{j})$ is homogeneous first order in $\mathbf{h}$ and $\mathbf{j}$ independently. Conventionally, one would define the dimensionless parameters h_i/h_1 and j_i/j_1, thereby eliminating two parameters. Homogeneity causes the exponent polytope to lie in the intersection of two hyperplanes. This makes P two dimensions less than the $n + g$ dimensions of $\mathbf{h} \in R_+^n$, $\mathbf{j} \in R_+^g$. Defining dimensionless parameters cannot reduce the dimension further. The problem is therefore $(n + g - 2)$-dimensional in both cases. Hence it is preferable to retain symmetry and treat all species (h_i) and all reaction rates (j_i) equally.

The part of the exponent polytope method that is useful for obtaining first approximations to the problems occurring in chemistry can be summarized as follows. To solve the polynomial equation $f(\mathbf{X}) = 0$ for $\mathbf{X} \in R_+^n$:

1. Simplify the polynomial by high school algebra. Gather together all terms with the same exponents and find the coefficient associated with the exponent vector.
2. Find the vertices of the polytope $P \equiv \operatorname{conv} Y$, where Y is the set of points whose Cartesian coordinates in R^n are the exponent vectors.

3. For each vertex V of P whose corresponding polynomial term has a negative coefficient, find all adjacent vertices. That is, find all edges of P that contain V.
4. For each edge found in step 3, select all terms in $f(\mathbf{X})$ whose exponent vectors are points on the edge. Call this subpolynomial $f_E(\mathbf{X})$. Reduce $f_E(\mathbf{X})$ to a polynomial in one variable by a transformation of variables, and calculate the values of this variable that make $f_E(\mathbf{X})$ vanish.
5. For each zero of $f_E(\mathbf{X})$, transform back to the full set of variables and obtain an $\mathbf{X}_0$ such that $f_E(\mathbf{X}_0) = 0$. Define $\mathbf{x}_0 = \log_{10}\mathbf{X}_0$.
6. Let the exponent vector of the "negative" vertex V selected in step 3 be $\mathbf{y}^-$ and let the exponent vector of the adjacent vertex selected in step 4 be $\mathbf{y}^a$. Define the *edge vector* (a column matrix) by $\mathbf{E} = \mathbf{y}^- - \mathbf{y}^a$. Then the hyperplane of critical rays associated with this edge and $\mathbf{X}_0$ has the equation

$$\mathbf{E}^t(\mathbf{x} - \mathbf{x}_0) = 0 \tag{VI.22}$$

7. If $f_E(\mathbf{X})$ in step 4 does not vanish for $\mathbf{X} \in R^n_+$, construct the hyperplane (VI.22) by choosing $\mathbf{x}_0$ so that the hyperplane passes through the midpoint of the edge E.
8. We now have a set of hyperplanes (VI.22) for each edge meeting at V. The *region of dominance* of V, called $\mathcal{C}(V)$, is the convex polyhedral cone, which is defined by replacing the equality sign with $\geqslant$ in (VI.22);

$$\mathcal{C}(V) \equiv \{\mathbf{X} \,|\, \mathbf{X} = 10^{\mathbf{x}}, \mathbf{x} \in R^n, \mathbf{E}^t(\mathbf{x} - \mathbf{x}_0) \geqslant 0$$
$$\text{for all edges meeting at } V \text{ and all } \mathbf{x}_0 \text{ for each edge}\}$$

9. The region $\mathcal{C}(V)$ is an approximation for the set $\mathbf{X} \in R^n_+$, where $f(\mathbf{X}) < 0$. The boundary of $\mathcal{C}(V)$ consists of portions of the hyperplanes (VI.32). This boundary gives an approximation to $f(\mathbf{X}) = 0$.
10. The approximation in step 9 is a generalization of Fig. 62. When there is more than one zero of $f_E(\mathbf{X})$ for an edge, the approximation in step 9 will miss some of the features shown in Fig. 63. The reader can easily discover how to generalize step 9 in those cases.

This method requires computer algorithms to find the vertices of conv Y in step 2 and the edges containing a given vertex in step 3. Von Hohenbalken[43] gives APL computer code for such algorithms. Readers having access to APL are urged to try solving some problems by the exponent polytope method. These algorithms can be typed into the computer in a few minutes.

D. How the Face Structures of the Exponent Polytopes of the Stability Polynomials Are Related to Stoichiometric Network Stability

This section shows that the exponent polytopes of the α_i's and Δ_i's have a very simple structure when the network does not have critical current cycles or other features that make β's vanish. Thus it is possible to use the exponent polytope method to write down explicit approximations for the parameter values that make α_i or Δ_i zero or negative. Recall that the network is exponentially unstable when any of these polynomials is negative. Hence the parameter values for which instability occurs (in these simple cases when all $\beta \neq 0$) can be related to the known face structure of the exponent polytopes.

We begin by reformulating the discussion in the last half of Section III.C into exponent polytope language. Let $f(\mathbf{X})$ be a stability polynomial. Assuming that each term in the polynomial has a unique exponent vector, we replace (III.56) with

$$f(a^{\mathbf{x}}) = g(\mathbf{x}) = \sum_{\mathbf{y} \in Y(f,\mathbf{X})} c(f,\mathbf{X},\mathbf{y}) a^{\mathbf{y}'\mathbf{x}} \tag{VI.23}$$

where $\mathbf{X} = a^{\mathbf{x}}$ and $Y(f,\mathbf{X})$ is the set of exponent vectors that occur in the expansion of f as a polynomial in $\mathbf{X}$. This notation is necessary to distinguish the exponents of $\alpha_i(\mathbf{h},\mathbf{j})$ regarded as a polynomial in $\mathbf{h}$ with coefficients $\beta(\mathbf{h},\mathbf{j})$, from the exponents of $\alpha_i(\mathbf{h},\mathbf{j})$ regarded as a polynomial in $(\mathbf{h},\mathbf{j})$ with constant coefficients. The exponent sets are $Y(\alpha_i,\mathbf{h})$ and $Y(\alpha_i,(\mathbf{h},\mathbf{j}))$, respectively. Similarly, the coefficients are $c(\alpha_i,\mathbf{h},\mathbf{y})$ and $c(\alpha_i,(\mathbf{h},\mathbf{j}),\mathbf{y})$, respectively.

The set of networks whose stability polynomials contain only positive terms is [see (III.57)]

$$\begin{aligned} \mathfrak{N}^+ \equiv \{ N \in \mathfrak{N} \mid c(f,\mathbf{X},\mathbf{y}) > 0 \qquad & \text{for all } \mathbf{y} \in Y(f,\mathbf{X}) \\ Y(f,\mathbf{X}) \neq \emptyset \qquad & \text{for all } f \in F_i(N) \} \end{aligned} \tag{VI.24}$$

The definition of "potentially dominant term" in Section III.C is equivalent to requiring that the exponomial of $f(\mathbf{X})$ have this term as a limiting form along some ray in logarithmic space. Hence this term must be a vertex of $P(f,\mathbf{X}) \equiv \operatorname{conv} Y(f,\mathbf{X})$. Thus the dominant terms have the exponent set $\operatorname{vert} Y(f,\mathbf{X})$, the vertices of $Y(f,\mathbf{X})$. Then

$$\begin{aligned} \mathfrak{N}_i^{D-} = \{ N \in \mathfrak{N} \mid c(f,\mathbf{X},\mathbf{y}) < 0 \qquad & \text{for some } \mathbf{y} \in \operatorname{vert} Y(f,\mathbf{X}), \\ & \text{for some } f \in F_i(N) \} \end{aligned} \tag{VI.25}$$

and these networks are unstable by (III.61). The networks that remain fall into the two principal classes

$$\mathfrak{N}_i^0 = \{ N \in \mathfrak{N} \mid Y(f,\mathbf{X}) = \emptyset \quad \text{for some } f \in F_i(N) \quad c(f,\mathbf{X},\mathbf{y}) > 0 \quad \text{for all } f \in F_i(N)\} \tag{VI.26}$$

$$\mathfrak{N}_i^{I-} \equiv \{ N \in \mathfrak{N} \mid c(f,\mathbf{X},\mathbf{y}) > 0 \ \forall\, \mathbf{y} \in \text{vert}\, Y(f,\mathbf{X}) \ \forall\, f \in F_i \quad \text{and} \quad c(f,\mathbf{X},\mathbf{y}) < 0 \quad \text{for some } \mathbf{y} \in Y(f,\mathbf{X}) \quad \text{for some } f \in F_i(N)\} \tag{VI.27}$$

In Section III.C I conjectured that $\mathfrak{N}_i^{I-}$ contains very few networks. A theoretical reason for this is now given. I think the argument is convincing, but it is a theoretical model, not a proof. Consider polynomials of the form

$$f(X,Y) = \sum_i c_i (X^{\alpha_i} \pm Y^{\beta_i})^{\delta_i} \tag{VI.28}$$

for random choices of $\pm$ and random c_i, α_i, β_i, and $\delta_i \in R$. The polynomials might be representative of the stability polynomials of chemical networks. Since the networks have a tendency toward stability, let us assume that the probability distribution function for $\pm$, c_i, α_i, β_i, and δ_i makes negative signs and negative c_i possible but improbable. When the ith binomial term is multiplied out, it gives a set of exponents that lie along a line perpendicular to the vector (α_i, β_i). Figure 60 is a typical case. Each term in the sum over i contributes a *binomial line* of exponent points to Y. The binomial lines are oriented in random directions. The vertices of Y must all be points that are at the ends of their respective binomial lines. The coefficients of terms of $f(X, Y)$ whose exponents are near the middle of $P = \text{conv}\, Y$ will probably be the sum of contributions from many binomial lines; however the coefficients of points near the surface of P will probably have far fewer contributions. Since relatively few binomial lines will correspond to terms with $c_i < 0$ or with a minus sign in the binomial, it is unlikely that negative coefficients will appear in the middle of P. This is because the positive contributions from many positive binomial lines will probably be much greater than the negative contributions from the few negative binomial lines. Thus the most likely place to find *negative terms* (i.e., exponent points corresponding to terms whose coefficients are negative) is on the surface of P. However the surface consists of facets that are polyhedra of one lower dimension. Negative terms are much more likely to be found on the surface of these facets than in their interior. This argument may be repeated until we conclude that the most likely places to find negative terms are as vertices of P; the next most likely places are as points in the interiors of the edges; the next most likely places are as points in the interiors of 2-faces; and so on.

This description provides an excellent theoretical model of what actually happens. The model suggests that negative terms are found on an edge only when they are also found at the vertices of the edge. It suggests that negative terms are found on a 2-face only when they are also found on the edges and vertices of the 2-face. This is what occurs in practice, with a few exceptions. The model allows exceptions but says they are improbable. The exceptions that do occur are rare; however one should not regard them as random events, but as an aspect of the stability problem that occurs only for certain kinds of network. Exceptions occur (in my experience) only in enzyme networks where a destabilizing feedback cycle involves more than one extreme current. In Selkov's model of glycolysis[26] discussed in Section IV.G, the dominant negative term occurs on an edge of P. In the Goodwin model[34] of a biochemical feedback control system the dominant negative term occurs on a 2-face. In both these cases all other terms on the edge and 2-face are positive. These two networks are the only ones I have found in $\mathfrak{N}^{I-}$, and they are both unstable.

These observations plus the discussion in Section V.I motivate the following conjectures.

1. $\mathfrak{N}^+$ is the set of asymptotically stable networks.
2. $\mathfrak{N}_i^{D-} \cup \mathfrak{N}_i^{I-}$ is the set of unstable networks.
3. $\mathfrak{N}_i^{I-}$ is relatively small and contains only networks where the destabilizing feedback cycles involve promotions and inhibitions coming from more than one extreme subnetwork.
4. $\mathfrak{N}^0$ is the set of stable but not asymptotically stable networks that have the explosion-extinction dynamics discussed in Section V.I.
5. All networks having limit cycle oscillations, chaos (a strange attractor), or multiple steady states are in $\mathfrak{N}_i^{D-} \cup \mathfrak{N}_i^{I-}$.

No counterexamples to any of these conjectures are known.

Every point in the set $Y(\alpha_i, \mathbf{h})$ is a permutation of the column vector $(1, 1, \ldots, 1, 0, 0, \ldots, 0)^t \in R^n$, where i ones are present. Each $\mathbf{y} \in Y(\alpha_i, \mathbf{h})$ satisfies $\mathbf{y}^t\mathbf{e} = i$ and

$$\sum_j y_j^2 = i \tag{VI.29}$$

Hence $\mathbf{y}$ lies in the intersection of a hyperplane with a sphere of radius $\sqrt{i}$ about the origin. Call this intersection sphere $\Sigma(\alpha_i, \mathbf{h})$. The convex hull of a set of points on a sphere is a polytope that is contained in the sphere. The tangent plane to $\Sigma(\alpha_i, \mathbf{h})$ at the point $\mathbf{y}$ is therefore a supporting hyperplane of $P(\alpha_i, \mathbf{h}) \equiv \operatorname{conv} Y(\alpha_i, \mathbf{h})$. Hence $\mathbf{y}$ is a vertex of $P(\alpha_i, \mathbf{h})$. We have proved:

Theorem VI.5. The set of vertices of $P(\alpha_i, \mathbf{h})$ is $Y(\alpha_i, \mathbf{h})$.

Consequently, if there exists $\mathbf{y} \in Y(\alpha_i, \mathbf{h})$ such that $c(\alpha_i, \mathbf{h}, \mathbf{y}) < 0$, the negative vertex $\mathbf{y}$ makes $\alpha_i < 0$ at infinity in the cone $\mathcal{C}_0(\mathbf{y})$; then the network is exponentially unstable. Note that since

$$c(\alpha_i(\mathbf{h}, \mathbf{j}), \mathbf{h}, \mathbf{y}) = \beta_i(\gamma, \mathbf{j}) \tag{VI.30}$$

where γ gives the subscripts of the components of $\mathbf{y}$ that are ones, this result is identical to that obtained in the paragraph following (V.16).

Let $\mathrm{Y}^0(\alpha_i, \mathbf{h})$ be the exponent set that occurs when every $\beta_i(\gamma, \mathbf{j}) \neq 0$. The edge structure of this polytope is very simple. One can prove that $\mathbf{y}$ and $\mathbf{y}'$ lie on an edge if and only if $\mathbf{y} - \mathbf{y}'$ is a permutation of $(1, -1, 0, 0, \ldots, 0)$. If the 1 occurs on the kth position and the -1 occurs in the lth position, the hyperplane dividing the region of dominance of these adjacent vertices is

$$c(\alpha_i, \mathbf{h}, \mathbf{y}) h_k = c(\alpha_i, \mathbf{h}, \mathbf{y}') h_l \tag{VI.31}$$

These equations can be written down without much difficulty for networks of any size.

Example VI.1. Consider the subnetwork E^1 of the Oregonator. Figure 23 shows the diagram of all stabilizers and polygons. The only negative one is the crossed 1-polygon, which represents a positive feedback cycle between X and Y. This term appears in α_2. We now make the supergraph expansion of α_2 as in Fig. 24 and convert the terms to their algebraic form by consulting Fig. 16*b*. The result is

$$\alpha_2 = 2yz - xy$$

The exponents are $(0, 1, 1)'$ and $(1, 1, 0)$, and their difference is $(-1, 0, 1)$. The plane dividing the regions of dominance has the equation

$$2z = x$$

which is an example of (VI.31). Thus E^1 is unstable if $2z < x$.

We now consider $P(\alpha_i, (\mathbf{h}, \mathbf{j})) \equiv \operatorname{conv} Y(\alpha_i, (\mathbf{h}, \mathbf{j}))$. If none of the coefficients of $\alpha_i(\mathbf{h}, \mathbf{j})$ vanish, the resulting polytope and set will be called $P^0(\alpha_i, (\mathbf{h}, \mathbf{j}))$ and $Y^0(\alpha_i, (\mathbf{h}, \mathbf{j}))$, respectively. The set of terms with the same $\mathbf{j}$ exponents will form the polytope $Y^0(\alpha_i, \mathbf{h})$ discussed previously. The set of terms with the same $\mathbf{h}$ exponents are ith order homogeneous polynomials in $\mathbf{j} \in R^g_+$ with no terms missing. The exponents vectors are the $(i + g - 1)!/i!(g - 1)!$ lattice points of a $(g - 1)$-simplex, which we call S. Hence $P^0(\alpha_i, (\mathbf{h}, \mathbf{j})) = P^0(\alpha_i, \mathbf{h}) \times S$. Since S has g vertices (one for each extreme subnetwork), $P^0(\alpha_i, (\mathbf{h}, \mathbf{j}))$ has g times as many vertices as $P^0(\alpha_i, \mathbf{h})$, and each vertex corresponds to a term in $\alpha_i(\mathbf{h}, \mathbf{j})$, which comes entirely from a single extreme subnetwork.

Choose any i species and consider $\beta_i(\boldsymbol{\gamma}, \mathbf{j})$ for those species as $\mathbf{j}$ varies. The exponents of this polynomial form the simplex S. Theorem V.8 showed that if $\{X_{\gamma(1)} \cdots X_{\gamma(i)}\}$ were the species of a simple critical current i-cycle, then $\beta_i(\boldsymbol{\gamma}, \mathbf{j}) = 0$ for some extreme subnetwork. Then we argued (not rigorously) that whenever $\boldsymbol{\gamma}$ is the set of species on an i-cycle of a realistic network, $\beta_i(\boldsymbol{\gamma}, \mathbf{j})$ has its minimum when $\mathbf{Ej}$ is an extreme current. This idea is consistent with the theoretical distribution of negative terms on the surface of S, which we discussed using (VI.28) as a model. According to this model the negative terms with the largest magnitude coefficients would be at the vertices of S, the next largest would be on the edges, then on the 2-faces, and so on. The minimum value of $\beta_i(\boldsymbol{\gamma}, \mathbf{j})$ would then most likely occur at infinity in a vertex cone $\mathcal{C}_0(V)$, where V is a vertex of S. The distant part of this vertex cone in $\log \mathbf{j}$ space represents points near the orthant boundary in $\mathbf{j}$ space. Hence the minimum is approached as $\mathbf{Ej}$ approaches an extreme current.

Now consider the possibility of a destabilizing positive feedback cycle that passes through species labeled by $\boldsymbol{\gamma}$, such that no extreme subnetwork involves all these species. If the cycle contributes a negative term to $\beta_i(\boldsymbol{\gamma}, \mathbf{j})$, the term cannot be a vertex of S, because the term comes from more than one extreme subnetwork. Since positive feedback cycles usually (always to my knowledge) destabilize by making $\beta_i(\boldsymbol{\gamma}, \mathbf{j})$ negative, we assume this term makes $\beta_i < 0$. Since the term's exponent vector is not a vertex of the S, the most likely place the term could appear is on an edge if two extreme subnetworks were involved, on a 2-face if three extreme subnetworks were involved, and so on, up to a k-face if $k + 1$ extreme subnetworks were involved.

In calculations on a biochemical control system whose destabilizing feedback cycle involved three extreme subnetworks, I found that destabilization occurred from a term in the middle of a hexagonal 2-face. (This 2-face was not a triangle (simplex) because the enzyme conservation condition caused the rows of $\underline{\boldsymbol{\nu}}$ to be linearly dependent for each extreme subnetwork. Thus β vanished at the vertices of S by Theorem V.6, thereby causing the triangle to become a hexagon.)

The exponent polytopes of the Hurwitz determinants are only slightly more complicated in the ideal case where no β's vanish. In this case the polytope is called $P^0(\Delta_i, \mathbf{h}) \equiv \operatorname{conv} Y^0(\Delta_i, \mathbf{h})$. The homogeneity of Δ_i makes every $\mathbf{y} \in Y^0(\Delta_i, \mathbf{h})$ satisfy

$$\mathbf{y}^t\mathbf{e} = \frac{i(i+1)}{2} \qquad \text{(VI.32)}$$

Also, every permutation of the components of $\mathbf{y}$ gives a new exponent vector $\mathbf{y}' \in Y^0(\Delta_i, \mathbf{h})$. The sum of squares of the exponents of both $\mathbf{y}$ and $\mathbf{y}'$ are the same [cf. (VI.29)]. Call this number η. Hence $\mathbf{y}$ and all its permutations lie in the intersection of the hyperplane (VI.32) and a hypersphere of radius $\sqrt{\eta}$ about the origin. This intersection is a hypersphere of one lower dimension. Hence the points of $Y^0(\Delta_i, \mathbf{h})$ lie in a finite set of hyperspheres that we call Σ^0, Σ^1, Σ^2, and so on, in order of decreasing radius, that is, in order of decreasing η.

Exponent points on Σ^0 mnst be vertices of $P^0(\Delta_i, \mathbf{h})$. These points have the largest η, and to achieve this η, the corresponding terms must involve the least possible number of variables raised to the highest possible exponents. The form of α_i, given in (V.15) imposes the restriction that i different variables h_i must occur in every term in α_i. We now select the exponent points in vert $P^0(\Delta_i, \mathbf{h})$ by consulting the array A given by (III.45). Take $i = 4$ as an example. The right-hand column of Δ_4 contains α_4, α_5, α_6, and α_7. Since other columns have α's involving fewer parameters, we minimize the number of parameters by selecting α_4 from this column. Without loss of generality we consider only the term proportional to $h_1h_2h_3h_4$ in α_4. The minor in Δ_4 that is associated with this element of A is Δ_3. The same argument applied to Δ_3 forces us to select α_3 on the diagonal of A. We now select the terms in α_3 that produce the highest powers of h_i when combined with the term in $h_1h_2h_3h_4$ from α_4. Clearly, the term must contain three h_i's chosen from h_1, h_2, h_3, and h_4. Without loss of generality we assume that our selection is $h_1h_2h_3$. The product is proportional to $h_1^2h_2^2h_3^2h_4$. Repeating this argument for the next minor yields a choice of α_2 on the diagonal of A. The parameter choice must overlap with the highest power parameters already obtained (which are h_1, h_2, h_3). Suppose we choose h_1h_2. The result is now $h_1^3h_2^3h_3^2h_4$. The final step leads to the choice α_1 with parameter h_1 (say), and the final form of the term in Δ_4 is $h_1^4h_2^3h_3^2h_1^1$. Hence the exponent vector is $(4, 3, 2, 1, 0, \ldots, 0)$. This argument may be generalized to conclude

$$\operatorname{vert} P^0(\Delta_i, \mathbf{h}) = \{\text{set of all permutations of the vector}$$

$$(i, i-1, i-2, \ldots, 2, 1, 0, \ldots, 0)'\}$$

The proof of this result is given in Clarke,[36] where it is shown that no other exponent vectors can be vertices of $P^0(\Delta_i, \mathbf{h})$.

The edges of $P^0(\Delta_i, \mathbf{h})$ are also easily found. To characterize the edges of a vertex V, we give the vertices at the opposite ends of the edges. The edges of $(4, 3, 2, 1, 0)'$ are $(3, 4, 2, 1, 0)'$, $(4, 2, 3, 1, 0)'$, $(4, 3, 1, 2, 0)'$, and $(4, 3, 2, 0, 1)'$. To generate the adjacent vertices, one transposes the components whose values are j and $j-1$ for $j = 1, \ldots, i$. The hyperplanes have a form similar to (VI.31).

The coefficient of the term with exponent vector $(i, i-1, \ldots, 2, 1, \ldots)$ is the product $\beta_i(\gamma^{(0)}, \mathbf{j})\beta_{i-1}(\gamma^{(1)}, \mathbf{j}), \ldots, \beta_1(\gamma^{(i)}, \mathbf{j})$ evaluated on the respective sets of $i, i-1, \ldots, 1$ species. Each β has an SP-diagram expansion as in Fig. 24. The numerical values of the polygons may be read directly from the current matrix diagram. Hence it is easy to calculate by hand the coefficient of any vertex diagrammatically. The enormous numbers of terms shown in Table IV are thus no obstacle because only the coefficients of a few of the (usually) $i!$ vertices are important.

The fact that the vertices of $P^0(\Delta_i, \mathbf{h})$ all come from the product of diagonal elements $\alpha_1\alpha_2 \cdots \alpha_i$ implies that the exponent polytope approach in this case is equivalent to making the approximation $\Delta_i = \alpha_1\alpha_2 \cdots \alpha_i$. Hence the conditions $\Delta_i > 0$ for $i = 1, \ldots, d$ are equivalent to the conditions $\alpha_i > 0$ for $i = 1, \ldots, d$. Thus the conditions for $\alpha_i > 0$ for $i = 1, \ldots, d$ become necessary and *sufficient* conditions for stability.

This conclusion only applies when all $\beta_i(\gamma, \mathbf{j})$ are nonzero, which occurs for example when there are no conservation conditions and all current cycles are weak and not near critical. Conservation conditions cause a very simple kind of truncation of $P^0(\Delta_i, \mathbf{h})$ to occur. This truncation never introduces any new vertices. Hence regardless of whether there are conservation conditions, when all current cycles are weak and $\alpha_i > 0$, for $i = 1, \ldots, d$, the network is stable.

The last statement of the preceding paragraph is not quite rigorous. There are two qualifications to be made. First, a negative term can occur in α_i if a current cycle is "almost" critical. Second, the statement is based on the exponent polytope approach that only determines the signs of $\Delta_i(\mathbf{h}, \mathbf{j})$ near the boundary of the orthant $\mathbf{h} \in R^n_+$. Nevertheless, the only counterexamples to this statement I know of are in the very artifically constructed networks with "almost" critical current cycles.

So far, stability has been proved when there are no nonequilibrium current cycles on $\mathfrak{D}_C$. I believe that stability actually occurs when all current cycles are weak and not too near critical.

E. Hurwitz Nonmixing Networks: How Instability Occurs

In HNM networks, $\alpha_i \geqslant 0$ for $i = 1, \ldots, d$, for all $\mathbf{h} \in R^m_+$. If the network's exponent polytope $P(\Delta_i, \mathbf{h})$ has vertices on the sphere Σ^0, these points must come (proved in Clarke[36]) from the product of diagonal elements $\alpha_1\alpha_2 \cdots \alpha_i$, which is positive for all $\mathbf{h} \in R^n_+$. Hence these vertices of $P(\Delta_i, \mathbf{h})$ cannot make Δ_i negative near the boundary of the orthant $\mathbf{h} \in R^n_+$. There are two other ways Δ_i could be negative. First, the exponent polytope approximation might not be valid. This would mean that the unstable region of $\mathbf{h} \in R_+^n$ does not approach the orthant boundary. I have never seen this occur. The other possibility is that $P(\Delta_i, \mathbf{h})$ has vertices that are not vertices of $P^0(\Delta_i, \mathbf{h})$ and the corresponding terms have negative coefficients. This is what always occurs in my calculations on unstable HNM networks.

The presence of critical current cycles causes β's to vanish, thereby producing the kind of truncation of $P^0(\Delta_i, \mathbf{h})$ that allows terms whose exponents are not permutations of the vector $(i, i-1, \ldots, 2, 1, 0)$ to become vertices of $P(\Delta_i, \mathbf{h})$. The new vertices cannot be on the hypersurface Σ^0.

The expansion of Δ_i in terms of the α's produces a polynomial whose terms are proportional to $\pm\alpha_1^{\delta_1}\alpha_2^{\delta_2}\alpha_3^{\delta_3}, \ldots, \alpha_d^{\delta_d}$. This term is characterized by the exponent vector $\boldsymbol{\delta} \in Y(\Delta_i, \boldsymbol{\alpha})$. We say that $\boldsymbol{\delta}$ is *k steps off-diagonal* if the exponent set of $\alpha_1^{\delta_1}\alpha_2^{\delta_2} \cdots \alpha_d^{\delta_d}$ as a polynomial in $\mathbf{h}$ contains points on Σ^k and no points on Σ^{k-1}. The previous discussion shows that the diagonal product $\alpha_1\alpha_2 \cdots \alpha_i$ of Δ_i is zero steps off-diagonal.

The principal minors of the array A in (III.45) are now discussed using the same concepts used to discuss the determinant $\beta_i(\boldsymbol{\gamma}, \mathbf{j})$ in (V.16). In analogy with the supergraph expansion of β_i in Figure 24, we now make a supergraph expansion of the principal minors of A. These minors are the Hurwitz determinants. The corresponding diagrams must not be confused with the supergraph expansion of Δ_i in Fig. 26. In the latter, a 2-cycle corresponds to a 2-cycle of the matrix M (e.g., $M_{12}M_{21}$), whereas in the case of present interest, a 2-cycle represents a 2-cycle of the matrix A (e.g., $A_{12}A_{21} = \alpha_0\alpha_3$). Consult Fig. 26 and make this reinterpretation. Then the diagram containing only stars represents $\alpha_1\alpha_2 \cdots \alpha_i$ and is the only term zero steps off-diagonal.

The analysis by Clarke[34] proves that the terms that are one step off-diagonal are represented by the supergraph diagrams containing one 2-cycle of A and stars (diagonal elements of A). Diagrams are further off-diagonal when the size of the cycle and the number of cycles increases in Fig. 24. The exponent polytope approach and the binomial model (VI.28) for the distribution of negative terms suggest that most of the points of vert $P(\Delta_i, \mathbf{h})$ will come from terms zero steps off-diagonal, the next most will come from terms one step off-diagonal, the next most will come from terms two steps off-diagonal, and so on. In Hurwitz nonmixing networks the instability cannot come from terms zero steps off-diagonal because $\alpha_1\alpha_2 \cdots \alpha_i > 0$. The next most likely place to find the exponents of destabilizing terms is among the terms which are one step off-diagonal.

We now study instabilities coming from one step off-diagonal terms. From (III.45), the general one step off-diagonal term has the form

$-\alpha_1\alpha_2 \cdots \alpha_{k-4}\alpha_{k-3}^2\alpha_{k+1}^2 \cdots \alpha_i$. The only terms that can dominate over the terms in Σ^1 in this expression are terms in Σ^0 that come from $\alpha_1\alpha_2 \cdots \alpha_i$. We now locate all negative terms in the first expression that are not dominated by terms in the diagonal product. Such terms must be capable of making the polynomial $\alpha_1\alpha_2 \cdots \alpha_i - \alpha_1\alpha_2 \cdots \alpha_{k-4}\alpha_{k-3}^2 \cdot \alpha_k^2\alpha_{k+1} \cdots \alpha_i$ negative near the orthant boundary $\mathbf{h} \in R_+^n$. After dividing by the positive common factors $\alpha_1, \alpha_2, \ldots, \alpha_{k-3}, \alpha_k, \ldots, \alpha_i$, we obtain the expression

$$\Gamma_k(\mathbf{h},\mathbf{j}) = \begin{vmatrix} \alpha_{k-2}(\mathbf{h},\mathbf{j}) & \alpha_k(\mathbf{h},\mathbf{j}) \\ \alpha_{k-3}(\mathbf{h},\mathbf{j}) & \alpha_{k-1}(\mathbf{h},\mathbf{j}) \end{vmatrix} \tag{VI.33}$$

If the off-diagonal term has negative terms in Σ^1 that can dominate all positive terms coming from the diagonal product, then $\Gamma_k(\mathbf{h},\mathbf{j})$ can become negative as $\mathbf{h}$ approaches the orthant boundary along certain curves. Then negative terms coming from $-\alpha_{k-3}\alpha_k$ will produce vertices of $P(\Gamma_k, \mathbf{h})$ on Σ^1. These vertices must also be vertices of $P(\Delta_i, \mathbf{h})$ because all terms with exponents on Σ^0 have been taken into consideration. Thus these negative terms can make $\Delta_i < 0$ for $\mathbf{h}$ near the orthant boundary; hence the network is unstable. We have proved:

Theorem VI.6. A network is exponentially unstable if $P(\Gamma_k, \mathbf{h})$ has a negative vertex on Σ^1.

To see how such instabilities may arise, make a supergraph expansion of $\Gamma_k(\mathbf{h},\mathbf{j})$ in terms of the elements of the matrix M. The simplest case is $k = 3$ ($\Gamma_3 = \Delta_2$), whose expansion is shown in Fig. 26. Note that the 3-cycle occurs with a minus sign, and therefore can destabilize only if it has negative feedback. Note that the 2-cycle terms all have positive coefficients and lie in Σ^0. Hence they all come from $\alpha_1\alpha_2$ and cannot destabilize, since $\alpha_1\alpha_2 > 0$. Note also that the pure stabilizer term in $h_1h_2h_3$ has a combinatorial factor that also makes stability more likely. Hence $P(\Gamma_k, \mathbf{h})$ could have a negative vertex on Σ^1 only if the 3-cycle had negative feedback and if some of the terms in $\alpha_1\alpha_2$ vanished. Note also that the pure stabilizer term in $h_1h_2h_3$ will cancel the triangle unless the former is sufficiently small. To see that an instability usually occurs when a negative feedback 3-cycle passes through a species with a vanishing stabilizer, examine Fig. 66, which proves that $h_1h_2h_3$ becomes a vertex if the parameter h_1 cannot occur in any other term. A vanishing stabilizer at h_1 is not sufficient; there cannot be 2-cycles containing h_1 either. A positive feedback 2-cycle would give $P(\alpha_2, \mathbf{h})$ a negative vertex and cause instability. Hence only a negative feedback 2-cycle through X_1 can prevent an instability when there is a negative feedback 3-cycle and the stabilizer at X_1 vanishes.

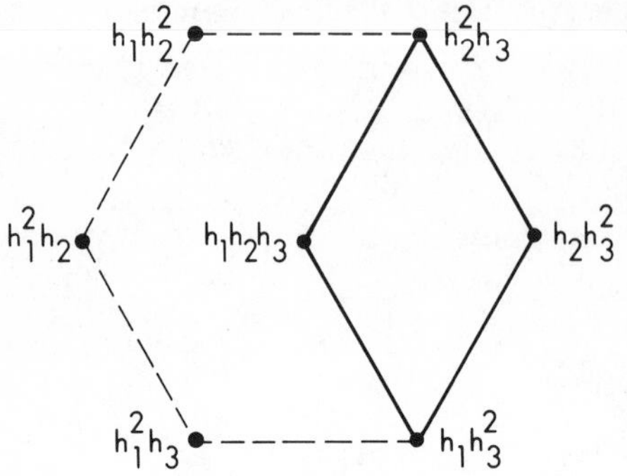

Fig. 66. Truncation of $P^0(\Delta_2, \mathbf{h})$ that occurs when the parameter h_1 cannot come from any source other than the 3-cycle and a 2-cycle between X_1 and X_3.

This picture generalizes to negative feedback cycles of any size. As a general rule, a negative feedback k-cycle can destabilize through $\Gamma_k(\mathbf{h}, \mathbf{j})$. I now prove that there is a pure stabilizer contribution to the same point in $Y(\Gamma_k, \mathbf{h})$ and calculate its coefficient. Then it will follow that destabilization cannot occur in this manner unless a stabilizer on the k-polygon (almost) vanishes. Suppose that the k-polygon passes through the species $X_1 \cdots X_k$. We are interested in the pure stabilizer term in α_k that will be proportional to $h_1h_2 \cdots h_k$. A pure stabilizer term in Γ_k occurs when this term is multiplied by a pure stabilizer term in α_{k-3}. Such terms will only be on Σ^1 if they overlap, so we take the term in α_{k-3} that has the parameters $h_1 \cdots h_{k-3}$ as an example. The product is $h_1^2h_2^2 \cdots h_{k-3}^2h_{k-2}h_{k-1}h_k$. This has its exponent vector in Σ^1. This negative contribution to Γ_k cancels with similar terms coming from $\alpha_{k-2}\alpha_{k-1}$. Let us construct these terms. To obtain the same squared terms, we must choose stabilizers at $X_1 \cdots X_{k-3}$ in the terms coming from both α_{k-2} and α_{k-1}. There is one more stabilizer to choose in α_{k-2} and two more to choose in α_{k-1}. There are three ways to make this choice to achieve the product $h_{k-2}h_{k-1}h_k$. Hence the combinatorial factor 3 must appear. The pure stabilizer term from $-\alpha_k\alpha_{k-3}$ is therefore canceled off in general, and the resulting term in Γ_k always has the combinatorial coefficient 2. Since this term combines with the k-polygon term, destabilization by the k-polygon cannot occur unless this stabilizer term is sufficiently small, that is, almost zero.

The preceding argument is much more easily stated using supergraphs instead of words. Figure 67 shows Γ_4. Destabilization by a negative feedback 3-cycle or 4-cycle can occur via terms in Γ_4 with exponents in Σ^1. In general, destabilization by negative feedback $(k-1)$-cycles and k-cycles occurs through Γ_k. Probably negative feedback j-cycles for $j < k-1$ do not destabilize through Γ_k because the diagonal product contains these cycles in both terms. The reason for the ability of an odd and even

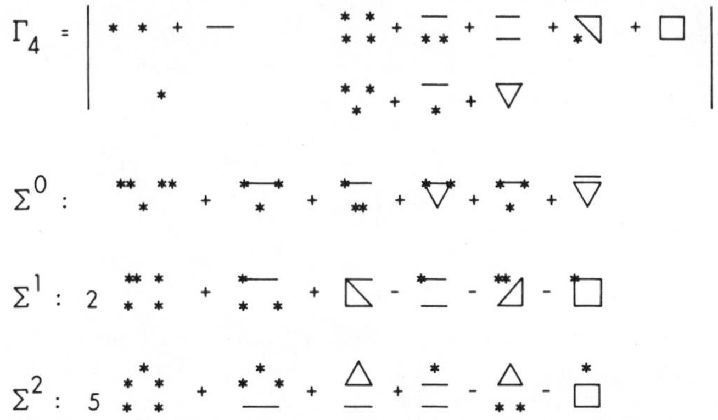

Fig. 67. Supergraph expansion of Γ_4 showing the terms in Σ^0, Σ^1, and Σ^2.

feedback cycle to destabilize through any Γ_k is related to the Liénard-Chipart stability conditions, which state that only Δ_i for i odd or i even need be considered.

Example VI.2. Figure 68 shows $\mathcal{D}_C$ for the network previously discussed in Example III.4. A knot-type cycle has been darkened on the current diagram. This cycle represents a negative feedback 3-cycle that passes through the critical current 1-cycle at X_1. Since the critical 1-cycle causes the stabilizer to vanish at X_1, this negative feedback 3-cycle will probably destabilize via Γ_3 or Γ_4. The reaction $X_3 + X_1 \rightarrow 2X_1$ produces a positive feedback 2-cycle between X_3 and X_1. Since a positive feedback 2-cycle in this location could in principle block destabilization by the negative feedback cycle (NFC), a calculation must be made to be sure that the NFC is a vertex of Γ_3 (Δ_2).

The polynomial $\Delta_2(\mathbf{h}, \mathbf{j})$ was given explicitly in Example III.4. From the exponents we form the matrix

$$\mathsf{Y}(\Delta_2, \mathbf{h}) = \begin{pmatrix} 1 & 1 & 0 & 0 & 0 & 0 & 0 \\ 1 & 0 & 2 & 2 & 1 & 1 & 1 \\ 1 & 2 & 1 & 0 & 2 & 0 & 1 \\ 0 & 0 & 0 & 1 & 0 & 2 & 1 \end{pmatrix}$$

The corresponding coefficients can be expressed as the row matrix

$$\mathbf{c}(\Delta_2, \mathbf{h}) = (-1 \quad 2 \quad 1 \quad 1 \quad 2 \quad 2)$$

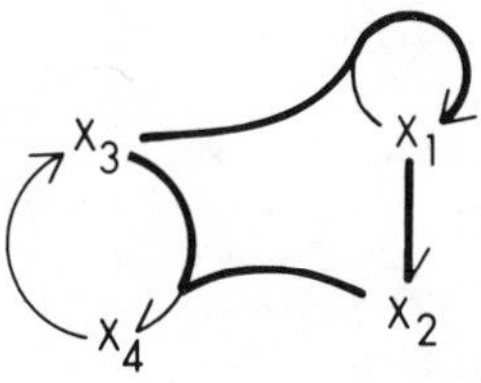

Fig. 68. The unstable Hurwitz nonmixing network discussed in Example VI.2.

The APL algorithm VERTICES of Von Hohenbalken[43] may be used to prove that the first six column vectors of $\mathsf{Y}(\Delta_2, \mathbf{h})$ are the coordinates of vertices of $P(\Delta_2, \mathbf{h})$ and that the seventh is not a vertex. Then the algorithm EDGES[43] gives the *adjacency matrix*, which is defined by the condition that its i, jth element is 1 if and only if the ith and jth column of Y are the coordinates of adjacent vertices of P. The adjacency matrix is

$$\begin{bmatrix} 0 & 1 & 1 & 1 & 0 & 1 \\ 1 & 0 & 0 & 0 & 1 & 1 \\ 1 & 0 & 0 & 1 & 1 & 0 \\ 1 & 0 & 1 & 0 & 0 & 1 \\ 0 & 1 & 1 & 0 & 0 & 1 \\ 1 & 1 & 0 & 1 & 1 & 0 \end{bmatrix}$$

A drawing of P, which was constructed with the help of the algorithm FACES, appears in Fig. 69. The adjacency matrix shows that the negative vertex $\mathbf{y}_1$ is adjacent to $\mathbf{y}_2$, $\mathbf{y}_3$, $\mathbf{y}_4$, and $\mathbf{y}_6$. Thus, in a loose sense, we may say that $-h_1h_2h_3$ is "adjacent" to $2h_1h_3^2$, $h_2^2h_3$, $h_2^2h_4$, and $2h_2h_4^2$. The necessary and sufficient conditions for the negative term to dominate all other terms near the boundary of the orthant $\mathbf{h} \in R_+^4$ are

$$\begin{aligned} h_1h_2h_3 &> 2h_1h_3^2 \\ h_1h_2h_3 &> h_2^2h_3 \\ h_1h_2h_3 &> h_2^2h_4 \\ h_1h_2h_3 &> 2h_2h_4^2 \end{aligned}$$

It is now convenient to define new variables by

$$H_i = \log h_i = -\log X_i^0$$

Thus H_i is the negative logarithm of a concentration. This definition follows the commonly used practice in chemistry (cf. $\text{pH} = -\log[H_3O^+]$). After simplifying and taking logarithms of the inequalities, we get

$$\begin{aligned} H_2 &> H_3 + \log 2 \\ H_1 &> H_2 \\ H_1 + H_3 &> H_2 + H_4 \\ H_1 + H_3 &> 2H_4 + \log 2 \end{aligned} \qquad \text{(VI.34)}$$

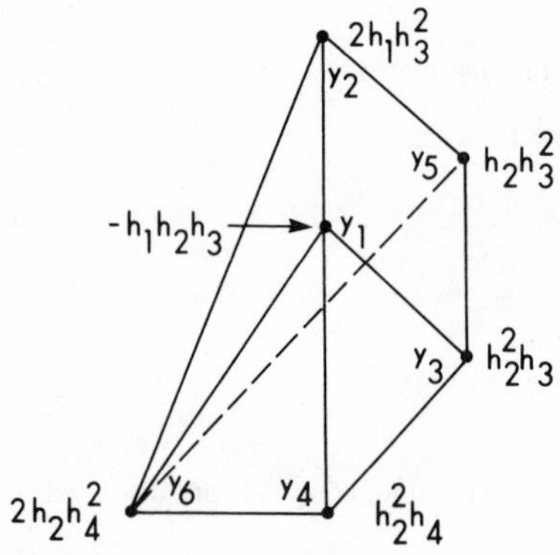

Fig. 69. $P(\Delta_2, \mathbf{h})$ in Example VI.2. Note that all vertices whose exponent vectors are a permutation of $(2, 1, 0, 0)^t$ are in Σ^0. The only vertex in Σ^1 comes from $-h_1h_2h_3$, which corresponds to the negative feedback 3-cycle.

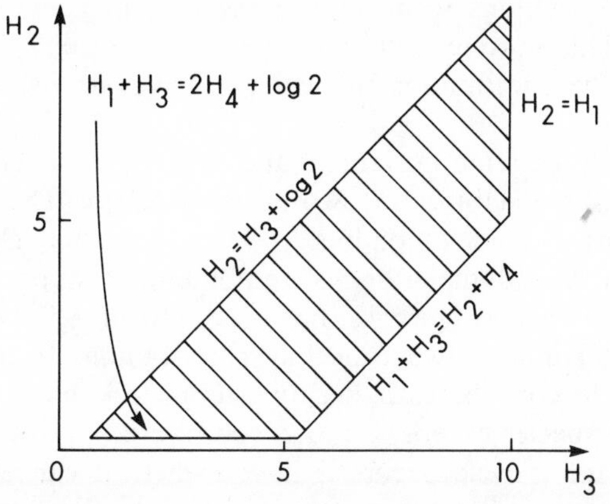

Fig. 70. A cross-section through the cone $\mathcal{C}_1(\mathbf{y}_1)$ defined by the inequalities (VI.34). The plot was made for $H_1 = 10$ and $H_4 = 5$. The shaded region is as good an approximation to the unstable region as the pH equations (VI.19) are for the pH of a weak acid in water. The orientation of this cross-section has been rotated roughly 45° counterclockwise from the orientation of P in Fig. 69.

Each of these equations defines a *half-space* in R^4. The intersection of the four half-spaces is $\mathcal{C}_1^1(\mathbf{y}_1)$. A cross-section through this cone appears in Fig. 70. From Figure 69 one can imagine the intersection of $\mathcal{C}_1^1(\mathbf{y}_1)$ with the affine subspace $\mathcal{A}$ containing P, using the fact that the four plane faces of $\mathcal{C}_1^1(\mathbf{y}_1) \cap \mathcal{A}$ are perpendicular to the edges of P that meet at $\mathbf{y}_1$. Then it is clear that the shaded region in Fig. 70 is a cross-section through this cone.

If Example VI.2 is modified by replacing the reaction $X_1 \rightarrow X_2$ with $X_1 \rightarrow Y_1 \rightarrow Y_2 \rightarrow \cdots \rightarrow Y_{k-2} \rightarrow X_2$, the negative feedback k-cycle destabilizes through Γ_k or Γ_{k+1}.

F. Hurwitz Nonmixing Networks: On the Difficulty of Proving Stability

In the preceding section we saw how instability can occur in HNM networks. Destabilization from positive feedback cycles may occur via terms with exponents in Σ^0 and destabilization from negative feedback cycles may occur via terms with exponents in Σ^1. If this is the only way destabilization may occur, we should be able to prove stability in all other cases. Since constructing Hurwitz polynomials is practical only for small networks, stability can be proved in large networks only by constructing Lyapunov functions. In this section, we see that the method also has major difficulties for HNM networks.

The problem is illustrated by using a set of networks I call *autocatalytic ring networks*. The ring network of order 3 is shown in Fig. 71. The nth order ring network contains n species and is obtained by replacing the reactions $X \rightarrow Y \rightarrow Z$ with the reactions $X \rightarrow Y_1 \rightarrow Y_2 \rightarrow \cdots \rightarrow Y_{n-2} \rightarrow Z$.

The irreversible reactions form an extreme current containing a critical current cycle. This current conserves $X + Y_1 + \cdots + Y_{n-2} + Z$; hence $d = n - 1$, and the contribution this current makes to α_n must cancel out exactly.

The irreversible extreme current of the general autocatalytic ring network has a vanishing stabilizer at X and nowhere else. The diagram $\mathfrak{D}_{\mathrm{ASP}}$ has two polygons, one corresponding to a negative feedback (NF) 2-cycle between X and Z and the other corresponding to a positive feedback n-cycle around the ring. Figure 71 shows $\mathfrak{D}_{\mathrm{ASP}}$ for $n = 3$. The SP-diagram expansion of α_n contains two canceling contributions from this current. The large n-cycle contributes a negative term and the NF-2-cycle with stabilizers at all species except X and Z contributes a positive term. Since the only place the n-cycle occurs is in α_n, where it cancels exactly, the n-cycle will also cancel with certain NF-2-cycle diagrams in the SP-diagram expansions of all the Hurwitz determinants. The remaining diagrams in the expansions of these determinants are built only out of stabilizers and NF-2-cycles. Since an instability in these networks could only be attributed to the NF-2-cycle terms in the SP-diagram expansion of the Hurwitz determinants, we have two reasons for believing that an instability cannot occur. First, NF-2-cycles and stabilizers are both safe with respect to sign stability. Second, an instability would have to be attributed to an NF-2-cycle which destabilizes via terms in Σ^k for $k \neq 0$. If all such networks are stable we cannot hope to prove it by using the Hurwitz determinant expansions. Hence we seek a general Lyapunov function for this set of networks. Let us consider the case $n = 3$ shown in Fig. 71.

This network is not mixing stable. Since $\mathfrak{D}_N$ contains an irreversible cycle, we cannot prove stability using Theorem V.3. Hence none of the simple theorems developed so far can prove stability.

We now try to construct a Lyapunov function of the form

$$L(\zeta) = \zeta^t(\operatorname{diag}\boldsymbol{\rho})(\operatorname{diag}\mathbf{h})\zeta \qquad \text{(VI.35)}$$

for the interior of the simplicial cone spanned by the principal irreversible extreme current and the extreme currents $j_{\mathrm{X}}: \to \mathrm{X} \to$, $j_{\mathrm{Y}}: \to \mathrm{Y} \to$, and

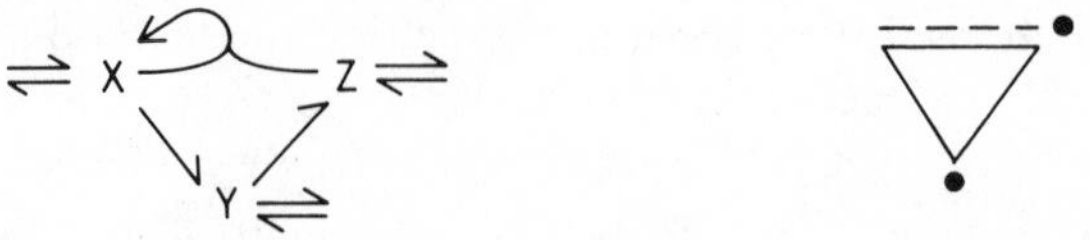

Fig. 71. A network and the $\mathfrak{D}_{ASP}$ of its principal extreme subnetwork.

$j_Z: \to Z \to$. From Fig. 71, one may directly write down

$$V(\mathbf{j}) = \begin{bmatrix} j_X & 0 & -1 \\ -1 & 1+j_Y & 0 \\ 1 & -1 & 1+j_Z \end{bmatrix}$$

From (III.68), we deduce that the symmetric part of $\mathsf{V}(\mathrm{diag}\,\rho)$ must be positive definite for $L(\zeta)$ to be a Lyapunov function. Set $\rho_1 = 1$ and let s_2 be the second principal minor of the symmetrized determinant. Then we can solve this definition to express ρ_2 in terms of j_X, j_Y, j_Z, and s_2. Now consider s_2 to be an arbitrary positive parameter. Only the parameter ρ_3 remains to be fixed. $L(\zeta)$ can be a Lyapunov function only if it is possible to choose ρ_3 so that $s_3 > 0$ (the third principal minor) for all positive values of the parameters j_X, j_Y, j_Z, and s_2. An analysis of this polynomial s_3 using the exponent polytope shows that it is not possible to always have $s_3 > 0$. I have shown that $L(\zeta)$ is a Lyapunov function in the limit $s_2 \to 0$ if

$$j_X(1+j_Z)(1+2j_Y)(1+j_Y) > \tfrac{1}{4}$$

provided $\boldsymbol{\rho}$ is defined by

$$\boldsymbol{\rho} = \left(1, \frac{1}{4j_X(1+j_Y)} + \frac{s_2}{4}, \frac{1+2j_Y}{2(1+j_Y)}\right)^t$$

In most of the region where the inequality is not satisfied, no diagonal Lyapunov function exists.

This example shows that there are asymptotically stable HNM networks whose quadratic Lyapunov functions cannot have diagonal matrices. The eligible Lyapunov functions might be so complex that it may be impractical to find a general method of constructing one for the nth order ring network. This observation raises the question of whether a general proof of stability is always possible when instability has not yet been found in any example. Kurt Gödel has shown that undecidable questions can be stated in the language of arithmetic. Evaluation of the signs of an infinite number of Hurwitz determinants is certainly a problem in arithmetic. Could it be that the positivity of all signs for certain sets of networks is one of Gödel's undecidable questions?

The ring network problem has been resolved by the discovery of a counterexample. In a recent preprint, C. Hyver integrated the $n = 7$ ring network and found oscillations about an unstable steady state. I then constructed the Hurwitz determinants and proved that ring networks are stable if $n \leqslant 4$ and unstable if $n \geqslant 5$.

The instability in the ring networks occurs through the following mechanism. Let the parameter $\mathbf{h}$ have the components $(x, y_1, \ldots, y_{n-2}, z)$. Recall that the expansion of Δ_{n-1} may be expressed in general in terms of the stabilizers and polygons of $\mathfrak{D}_{\text{ASP}}$. The expansion is valid for any algebraic values of the stabilizers and polygons. Three sets of values will be considered. First, let all such objects other than the $n-1$ stabilizers of $\mathfrak{D}_{\text{ASP}}$ take the algebraic value zero and let the stabilizers take their usual values. Then Δ_{n-1} is a polynomial that does not contain x; its exponent set $Y(\Delta_{n-1}, \mathbf{h})$ lies in the coordinate hyperplane $\mathcal{H}$ (say) where the exponent of x vanishes. Second, let the NF-2-cycle take its usual value as well, and recalculate $Y(\Delta_{n-1}, \mathbf{h})$. New points outside of $\mathcal{H}$ must be present; they correspond to terms containing the NF-2-cycle because this cycle is the only object containing the parameter x. Hence there must be new vertices of $P(\Delta_{n-1}, \mathbf{h})$ that do not lie in $\mathcal{H}$. These vertices must be positive for the following reason. If they were negative the parameters in the sign stability problem for a matrix having only stabilizers and NF-2-cycles could be chosen to make the matrix unstable, thereby contradicting the Quirk-Ruppert-Maybee theorem. Third, let all the stabilizers and polygons of $\mathfrak{D}_{\text{ASP}}$ take their usual values. New terms containing the n-cycle are now present; however, they must all cancel with certain 2-cycle terms. The canceled terms are all outside $\mathcal{H}$. To account for the instability, some of the surviving 2-cycle terms must be negative.

Calculations show that such negative terms do occur and that some of them are vertices of $P(\Delta_{n-1}, \mathbf{h})$. The exponents of x in all such terms are relatively small. Hence such terms lie on one of the inner hyperspheres Σ^k, where k is large. These instabilities do not make $\Gamma_k(\mathbf{h}, \mathbf{j})$ negative as discussed in Section VI.E. The coefficients of the negative vertices have contributions coming from many different terms $\alpha_1^{\delta_1}\alpha_2^{\delta_2} \cdots \alpha_d^{\delta_d}$ in the expansion of Δ_{n-1}. Only after the contributions are combined does a negative coefficient finally emerge.

The ring networks provide the first indication that critical current cycles can cause an instability merely because of their *length*. I have investigated this idea in detail by evaluating $Y(\Delta_i, \mathbf{h})$ for networks with up to three critical cycles of various lengths and relative positions. It was not necessary to test every possible network because the theorems on topologically similar networks (Clarke[39]) enabled me to deduce the instability of extensions of an unstable network and the stability of reductions of a stable network. These very important theorems are adequately discussed in the literature and therefore need not be described further here. The general picture that emerged from my calculations is as follows. When a network contains a large and a small critical cycle, the network is unstable if the larger cycle contains at least four more species than the smaller cycle.

Otherwise, the network is stable. When three critical cycles are present, the relative position of the two smaller cycles becomes important. The detailed results are complicated.

In this section we have demonstrated that the stability of a large HNM network cannot be established by proving the stability of a smaller network that "approximates" the larger network. For example, replacing the reactions $Y_2 \rightarrow Y_3 \rightarrow Z$ by the reaction $Y_2 \rightarrow Z$ "approximates" the 5th order ring network by the 4th order ring network. However, the larger network is unstable and the "approximate" network is stable! Since Lyapunov functions for large stable HNM networks are difficult to construct, it appears that the stability of many such networks may never be proved.

G. The Transition from Network Stability to Sign Stability

When the orders of kinetics of the reactions are adjustable positive parameters [in addition to $(\mathbf{h}, \mathbf{j})$], a reaction involving m species on the left has m additional degrees of freedom in the parameters. Usually the reaction contributes more than m arrows to $\mathfrak{D}_{CM}$, so the effect is not exactly the same as making each arrow have its own independent parameter. However the effect is similar. Arrows of $\mathfrak{D}_{CM}$ may be combined arrows coming from more than one reaction, and the $\mathbf{h}, \mathbf{j}, \mathbf{q}$ parameters taken together often provide enough flexibility to be equivalent to each arrow's being independent, as in Example III.5. This section examines how this additional flexibility can produce new instabilities.

The diagrammatic methods need to be reinterpreted in the sign stability case only. If each arrow has its own parameter, the terms in the diagrammatic expansion of $\beta_i(\gamma, \mathbf{j})$ will no longer combine. Hence feedback cycles that would cancel with stabilizers in the network stability problem no longer cancel in the sign stability problem. The question is whether (and when) they can destabilize.

Our analysis gains new insights from the instability proof of Quirk and Ruppert,[46] which goes as follows. Consider any matrix k-cycle, $k \geqslant 3$. Choose the parameters to make all other cycles vanish. This is possible because every other cycle contains an arrow that is not in the k-cycle. Let S_i represent the remaining stabilizer terms in α_i and let K be the k-cycle term in α_k. Let all stabilizers vanish that are not on species through which the k-cycle passes. Then $\alpha_i = 0$ for all $i > k$. For $i < k$ we have $\alpha_i = S_i$, and for $i = k$ we have $\alpha_k = S_k + K$. Now choose the determinant Δ_j that makes α_k appear in the first or second row of the second column from the right (depending on whether k is odd or even). Then α_k will also appear in the right-hand column two rows further down [see the array (III.45)]. Note that K only appears in Δ_j in two places, so the expansion of Δ_j is quadratic in K. The coefficient of the quadratic term can be easily calculated by deleting the right two columns of Δ_j and the rows containing K. The resulting determinant is $-\Delta_{j-4}$! Hence we have

$$\Delta_j = (\cdots) + K(\cdots) - K^2\Delta_{j-4}$$

Clearly, as $K \to \infty$, Δ_j and Δ_{j-4} have opposite signs. Hence the matrix is sign unstable.

Note that Quirk and Ruppert's proof of instability is based on selecting terms from Δ_j that are almost as far off-diagonal as possible. In the chemical network stability problem, these terms would lie far in the interior of P. Such terms could not be made dominant by adjusting **h** or **j**. They could be made dominant only by adjusting the orders of kinetics. The important point is that the orders of kinetics must be made so large that these terms dominate the large number of other terms present in Δ_j. This means the elements of κ_{ij} must differ by very many orders of magnitude, which would be absurd in chemistry and probably even in ecology and economics.

Stoichiometric networks can be classified according to the variability of the elements of $\underline{\kappa}$. At one extreme are chemical networks, farther along are ecosystems and farther along yet are economic systems. One expects a slightly increasing tendency for instability. In all these systems there are important underlying stoichiometric constraints. Systems that are almost free of stoichiometric constraints can be viewed as posing a sign stability problem. The reason is not necessarily that $\underline{\kappa}$ varies over many orders of magnitude, but that the lack of a steady-state restriction (to $\mathcal{C}_v$) enables every interaction to become independent of every other one. Probably human social interaction can be modeled according to sign stability.

The insight that can be gained into economic, political, and social conflicts fits what seems to occur so well that I discuss these topics very briefly even though they do not bear directly on chemical networks. Consider first human relationships, which I think can be modeled by sign stability. The instability caused by all k-cycles for $k \geqslant 3$ may be the underlying reason for the failure of group marriages. The stability of the negative feedback 2-cycle may explain the stability of pairwise marriage. Sign stability implies that groups of more than two people will have highly unstable patterns of social interaction unless the interactions are arranged in the only stable configuration, which is a tree graph whose lines are negative feedback 2-cycles. The stability and frequency of hierarchical forms of social organization from the Catholic Church to dictatorships supports this observation. Dictators can maintain stability only by suppressing all interactions that do not follow the tree structure. Otherwise new k-cycles appear and destabilize.

Economics is highly stoichiometric because the raw materials and parts needed to manufacture a machine (or any other commodity) have flexibility only within very restrictive constraints. Consequently free enterprise economic systems should be fairly stable, even though they contain a great many current cycles as part of the supply and demand feedback system that controls production. It seems to me that government intervention into the market place has the effect of enabling production to take place even when the steady-state restrictions on the inflow and outflow of capital are not met. Thus such intervention must be destabilizing. In countries where production is controlled by a hierarchical social organization rather than by a market mechanism, one expects production to be highly unstable and unpredictable because production rates are controlled by decree (which is equivalent to κ_{ij} being infinite or zero) rather than by the much lower order kinetics of the market.

The early work on the stability of ecological systems[47] was based on sign stability. Since ecological systems are undoubtedly highly stoichiometric, I would expect them to be much more stable than this work suggests. May[47] notes that ecosystems are observed to become more stable as they become more complex. This stability probably results from the elimination of unstable ecosystem structures through oscillations and the extinction of species. May believes that evolution selects "singular strategies" that are rare exceptions to a general tendency toward instability in more complex systems. In contrast, I suspect that the stoichiometric constraint provides nature with a wide choice of possible stable ecosystem structures. The probability of a catastrophic ecosystem collapse is greatly overestimated by the sign stability model.

H. Networks with Diffusion

This section considers the stability of the homogeneous steady states of the reaction-diffusion equations when linearized about steady state. Equation (III.1) generalizes to

$$\frac{d\zeta}{dt} = \mathsf{M}\zeta + (\operatorname{diag}\mathbf{D})\nabla^2\zeta \tag{VI.36}$$

where $\mathbf{D} \in R_+^n$ is the vector of diffusion constants. Decomposition of $\zeta(\mathbf{r}, t)$ into eigenfunctions of the Laplacian with appropriate boundary conditions yields

$$\frac{d\zeta}{dt} = \mathsf{M}^D\zeta$$

where

$$M_{ij}^D = M_{ij} - \delta_{ij} D_i k_i^2 \tag{VI.37}$$

and $\mathbf{k}$ is the wave vector of the eigenfunction of the Laplacian.

The stability problem now has parameters $(\mathbf{h}, \mathbf{j}, \mathbf{D})$. We wish to identify networks whose homogeneous steady states are asymptotically stable for all $\mathbf{h} \in R_+^n$, $\mathbf{j} \in R_+^f$, $\mathbf{D} \in R_+^n$, and $\mathbf{k} \in R^n$. Note that the diffusion term contributes an arbitrary negative element to the diagonal of M. Note also that a pair of external reactions such as $\square \rightleftarrows \mathrm{X}$ also contribute an arbitrary negative diagonal element. Hence asymptotic stability of the homogeneous steady state holds whenever the corresponding problem without diffusion is asymptotically stable when arbitrary amounts of these external reactions are included. Hence we conclude:

Theorem VI.7. If a network is mixing asymptotically stable, the homogeneous state is stable to all inhomogeneous perturbations.

Theorem VI.8. If $\mathcal{D}_N$ has no irreversible cycles, the homogeneous steady state is stable to all inhomogeneous perturbations.

Instability with respect to inhomogeneous perturbations can be treated using the exponent polytope method. A detailed example appears in Clarke.[37]

VII. SUMMARY

This chapter has presented the foundations of a comprehensive theory of chemical reaction networks as a special case of a general theory of stoichiometric networks. Other applications of stoichiometric network theory are in ecology and economics.

Stoichiometric networks are defined by specifying a stoichiometric matrix $\underline{\boldsymbol{\nu}}$ and a kinetic order matrix $\underline{\boldsymbol{\kappa}}$. The latter will be a function of secondary parameters if the "reactions" do not have power law kinetics. The theory is concerned with the dynamics and stability of the systems specified by $\underline{\boldsymbol{\nu}}$ and $\underline{\boldsymbol{\kappa}}$, for all positive values of the proportionality constants in the reaction rate expressions. These constants are called rate constant parameters.

The foundation of this theory is a one-to-one mapping between the set of steady states for all positive rate constant parameters and the set of points $R_+^n \times \mathcal{C}_v$, where R_+^n is the positive orthant of n dimensions and $\mathcal{C}_v$ is the cone of steady-state reaction velocities. This mapping replaces the rate constants and the amounts of the various thermodynamic components by two new parameters $(\mathbf{h}, \mathbf{j})$, where $\mathbf{h} \in R_+^n$, and $\mathbf{j}$ gives the coordinates of a point in $\mathcal{C}_v$ relative to the frame of this cone. The mathematical simplifications that occur in the dynamical and stability problems for stoichiometric networks as a result of this transformation to new parameters are almost revolutionary. This transformation will probably be as important for future theories of reaction networks as the introduction of the concept of entropy was for thermodynamics.

First let us review the consequences for the general dynamical problem. The vector of concentrations **X** evolves within a convex *concentration polyhedron* $\Pi_X(C)$. This new concept is a generalization of the "reaction simplex." We have shown that this polyhedron is no longer a simplex in problems involving the dynamics of the internal species of open systems. The dimension of $\Pi_X(\mathbf{C})$ is d, the rank of $\underline{\boldsymbol{\nu}}$. The possible amounts of the thermodynamic components form a set of dimension $n - d$, and for each choice of the components there is a corresponding concentration polyhedron.

The general equation of motion for fixed secondary parameters is

$$\frac{d\mathbf{x}}{dt} = (\operatorname{diag}\mathbf{h})\,\underline{\boldsymbol{\nu}}(\operatorname{diag}\mathsf{E}\mathbf{j})\exp(\underline{\boldsymbol{\kappa}}^{t}\ln\mathbf{x}) \qquad \text{(VII.1)}$$

where the reduced concentration vector is defined by

$$\mathbf{x} = (\operatorname{diag}\mathbf{h})\mathbf{X}$$

and E is any matrix whose columns are the frame vectors of the convex polyhedral cone $\mathcal{C}_v$. This form of the kinetic equations has many advantages. First, it is in the canonical form studied by singular perturbation theory. The factor diag **h** determines the relative time scales of the motions of all species. Second, we know that it has the steady state $\mathbf{x} = \mathbf{e} \equiv (1, 1, \ldots, 1)^t$. Thus the right-hand side can be factored exactly. Dividing by $\mathbf{x} - \mathbf{e}$ enables one to obtain necessary and sufficient conditions on **j** for the existence of multiple steady states. These conditions do not involve **h**. A comparison with the complicated folds and cusps of catastrophe theory shows that the bifurcation points are far simpler in our parameter system. Sets of steady states that are "equivalent" multiple steady states are represented by sets of equivalent rays in $\mathcal{C}_v$. The *current polytope* Π_v is defined to be $\mathcal{C}_v \cap \{\mathbf{v} \in R^r \,|\, \mathbf{v}^t\mathbf{e} = 1\}$. The equivalent rays become equivalent points in Π_v.

This formalism is completely consistent with both equilibrium and nonequilibrium thermodynamics. Every pattern of topological dynamics that occurs for any $(\mathbf{h}, \mathbf{j})$ can be realized within the upper limits on the concentrations and reaction velocities that result from solubilities and molecular collision rates. Thus morphogenesis has complete access to all that is mathematically possible.

The motivation for our inquiry into steady-state stability is an intuitive conviction that exotic dynamical phenomena such as oscillations, chaos, and evolution cannot occur in networks that have only asymptotically steady states, for all values of $(\mathbf{h}, \mathbf{j}) \in R_+^n \times \mathcal{C}_v$. However we have seen that explosions may occur in certain classes of asymptotically stable networks when a finite perturbation from steady state is made. The theory makes it possible to recognize these networks easily.

The general form of the dynamics linearized about steady state is

$$\frac{d\boldsymbol{\zeta}''}{dt} = (\operatorname{diag}\mathbf{h})\, \underline{\boldsymbol{\nu}}(\operatorname{diag}\mathsf{E}\mathbf{j})\, \underline{\boldsymbol{\kappa}}'\boldsymbol{\zeta}''$$

where $\boldsymbol{\zeta}'' \equiv \mathbf{x} - \mathbf{e}$. The matrix M of this equation is a bilinear form in $\mathbf{h}$ and $\mathbf{j}$; consequently the stability problem is tractable for many general classes of networks. Asymptotic stability has been established for certain classes using the two principal Lyapunov functions:

$$L_M \equiv \boldsymbol{\zeta}^t(\operatorname{diag}\mathbf{h})\boldsymbol{\zeta}$$

$$L_D \equiv \boldsymbol{\zeta}^t(\operatorname{diag}\rho(\mathbf{j}))(\operatorname{diag}\mathbf{h})\boldsymbol{\zeta}$$

With both functions the dependence of the problem on $\mathbf{h}$ drops out immediately. L_M is called the *mixing Lyapunov function* because it has the powerful property that subnetworks with this Lyapunov function may be mixed and still retain this Lyapunov function. In this way we were able to prove asymptotic stability on many cones in a simplicial decomposition of $\mathcal{C}_v$, even for an oscillatory network. A method was devised for using L_M in systems that are not "mixing asymptotically stable" but have L_M as a Lyapunov function when the concentrations are constrained to $\Pi_X(\mathbf{C})$. These systems are called "constrained mixing asymptotically stable." They may also be mixed and retain stability, provided the key constraints are kept. Subnetworks having L_M for a Lyapunov function are easily recognized by calculating the eigenvalues of a few purely numerical matrices. If the eigenvalues are positive, asymptotic stability is proved for all $\mathbf{h} \in R_+^n$ and $\mathsf{E}\mathbf{j} \in \mathcal{C}_v$.

The key concept in determining whether a network is stable or unstable is the *current cycle*. Using the Lyapunov function L_D, we proved that all networks that do not have any current cycles (other than the trivial current cycles found in equilibrium extreme currents) are stable. Thus current cycles are necessary for instability.

Current cycles have been classified into three types—*strong*, *critical*, and *weak*, according to whether a particular determinant called β is negative, zero, or positive. Networks with strong current cycles are unstable. The concentrations of all the species on the cycle grow exponentially in an explosion. Strong current cycles probably do not occur in realistic chemical reaction systems because they are equivalent to higher than first-order autocatalysis.

A method of determining the stability of networks by computer has been used to study many small networks. These studies strongly suggest that networks can be unstable only if they have a current cycle that is strong, critical, or weak but almost critical. An investigation into the problem of proving stability in networks with only weak current cycles led to the discouraging result that the Lyapunov functions cannot have diagonal form (such as L_M and L_D have). These networks also cannot be proved to be stable in general by the computer algorithm because of the enormous size of the required computations.

This chapter has presented a comprehensive theory of how instability results from critical current cycles. All instabilities in realistic systems I have examined can be understood. The recognition of unstable reaction networks is accomplished primarily by means of the current diagrams of the extreme subnetworks. Briefly, the procedure is as follows:

1. Determine E from $\underline{\boldsymbol{\nu}}$.
2. For each column of E, ($\mathbf{E}_i$, say) define an extreme subnetwork stoichiometric matrix by $\underline{\boldsymbol{\nu}}^{(i)} = \underline{\boldsymbol{\nu}}(\operatorname{diag} \mathbf{E}_i)$.
3. Draw a diagram $\mathfrak{D}_C$ for each extreme subnetwork using $\underline{\boldsymbol{\nu}}^{(i)}$ and $\underline{\boldsymbol{\kappa}}$ as discussed in Section V.C.

The rationale for these steps is that the quantities that produce instability are convex functions on $\mathcal{C}_v$. Thus it is only necessary to look at these functions for the extreme subnetworks. The next step is to search for critical current cycles in the subnetwork current diagrams. To do this we take advantage of theorems that prove that critical current cycles cannot pass through species having certain topological features on the diagram.

4. Circle all species that are produced by only one reaction on $\mathfrak{D}_C$. Of these, mark those whose consuming reactions all have the same order of kinetics.
5. Only current cycles through the marked species in step 4 are eligible to be critical current cycles in realistic networks.

Critical current cycles produce instability by interacting with other feedback cycles. If the other feedback cycle has positive feedback, the instability will appear through a negative term in the coefficient of the characteristic equation $\alpha_i(\mathbf{h}, \mathbf{j})$. If the other feedback cycle has negative feedback, the instability will appear through a negative term in the 2×2 determinant $\Gamma_k(\mathbf{h}, \mathbf{j})$, defined by (VI.33). Sometimes more than two feedback cycles are needed to produce instability. Arbitrarily many may be involved in rare cases.

The ring networks are rare exceptions where no other feedback cycle is required, and the destabilization does not occur through α_i or Γ_k. Instead, instability is caused by a larger critical cycle being at least four species *longer* than a smaller critical cycle. One ring network containing the reactions $X \rightarrow Y \rightarrow Z$ is unstable, while a similar network, in which these reactions are replaced by the single reaction $X \rightarrow Z$, is stable. If substituting two consecutive unimolecular reactions by a single one is not valid, then every approximation to a network is suspect. Thus the stability of a complex network cannot in general be established by proving the stability of a simpler "approximate" network.

It is important to know which steady states of an unstable network are unstable. The unstable region in $R_+^n \times \mathcal{C}_v$ can be found by solving polynomials in many variables. A powerful new approach to this problem is called the exponent polytope method. The method is an n-dimensional generalization of a two-dimensional technique called Newton's polygon, and it is related to recent advances in algebraic geometry. The exponent polytope method is particularly valuable in chemistry because when it is applied to solving the law of mass action for a simple equilibrium, it gives the approximations that are familiar to chemists. This method can be used to approximate the unstable region to the accuracy needed in chemistry, even when the boundary of the unstable region must be found by solving for the zeros of polynomials with thousands of terms in many variables.

The exponent polytope method gives a direct connection between network stability and the face structure of the exponent polytope P obtained from the Hurwitz determinants as polynomials in $\mathbf{h}$. Strong critical current cycles and other destabilizing positive feedback cycles produce instability via a vertex of P in a hypersphere called Σ^0. This hypersphere is the outermost of many concentric hyperspheres within which the vertices of P must lie. Negative feedback cycles destabilize via vertices on the next sphere, called Σ^1, however, the ring networks are an exception where destabilization occurs via one of the inner hyperspheres Σ^k.

At the beginning of this summary I stated that the matrix $\underline{\boldsymbol{\kappa}}$ can be a function of "secondary parameters." If the secondary parameters can make the elements of $\underline{\boldsymbol{\kappa}}$ vary over many orders of magnitude, the stoichiometric network stability problem degenerates into the sign stability problem. In the latter problem destabilization occurs from every feedback cycle except two—the negative feedback 1-cycle and 2-cycle. However destabilization occurs through the innermost hypersphere Σ^k of P. Hence the effective orders of kinetics must indeed differ greatly in order of magnitude from each other for instability to occur via the sign stability mechanism.

The generalization of all these results to networks with diffusion is almost trivial. The networks that were proved to be stable using the Lyapunov functions L_M and L_D have also been proved to be stable to all spatially inhomogeneous fluctuations. The exponent polytope method works particularly well for these systems. It can easily find the set in $R_+^n \times \mathcal{C}_v$ where the steady state is stable to homogeneous fluctuations, but unstable to inhomogeneous fluctuations within a range of wave vectors that is easily calculated.

This chapter has discussed many examples that illustrate the important theoretical ideas. At the same time, a number of new results, which are worth mentioning here, have been discovered.

1. Among the reactions of Belousov-Zhabotinski systems we found another source of instability that probably explains some observations by Jwo and Noyes. This instability appears to be capable of operating in tandem with the recognized source of instability.
2. An analysis of the Noyes-Sharma mechanism for the Bray-Liebhafsky system was given. This 12-species system was easily proved to be unstable.
3. We found that stability in enzyme systems is conditional on the conservation condition for enzyme. If an enzyme were to catalyze two different steps along a metabolic pathway, instability would be likely (subject to more detailed other criteria).

Finally, let us discuss the relationship between this theory and other related theoretical developments. Almost all other analyses of chemical networks use rate constants. For theoretical purposes, such approaches are now obsolete. Although the chemist may wish to think in terms of rate constants, it is still more advantageous to calculate using $(\mathbf{h}, \mathbf{j})$ and convert back to the rate constants where necessary. The exponent polytope method holds out the promise of greatly simplifying this interconversion.

Stoichiometry imposes such a rigid constraint on the dynamics that perhaps most of the studies of nonlinear equations in the mathematical literature are not relevant in chemistry. An exception is the theory by Horn et al. which is more powerful than ours because it proves global stability, not just local asymptotic stability. However Horn's results apply to a far narrower class of networks than we have proved to be asymptotically stable. New work along the lines of Horn's approach continues to appear;[79–83] the main thrust is toward extending the proofs of global stability toward a wider class of networks. In a sense, Horn's work merges with this approach because (VII.1) is a generalization of the Horn equation.

A second exception are the papers on knot-tree networks by Hyver, Solimano, Beretta, et al. These theories have been developing very slowly toward the kind of analysis given in this chapter. In fact, the stability investigations by these authors make use of the symmetry properties of the matrix M of the linearized system. As in our theory, they did not solve for the steady states in terms of the rate constants. Our theory has gone very much further by setting up the general mapping to the (**h**, **j**) parameters. Furthermore, the theorems in Section IV include and go far beyond the knot-tree line of development. This line has therefore been completely absorbed by the stoichiometric network approach.

The basic ideas in a theory of molecular evolution have been outlined. Critical current cycles have explosion-extinction dynamics. Instabilities appear when reactions exist that put the cycles into competition with each other. The direction of the fluctuations from this unstable steady state determine which current cycle goes extinct and which explodes.

A recent theory of molecular evolution by M. Eigen[84, 85] focuses attention on what he calls a "hypercycle." In Eigen's theory there is a hierarchy of cycles. At the lowest level is the 2-cycle between the enzyme and its substrate complex in the Michaelis-Menten step. At the next level is autocatalysis. The third level of the hierarchy is the reproduction of a molecule in a series of many steps, each step involving an autocatalytically reproducing species. This cycle, built out of smaller autocatalytic cycles, is called a hypercycle.

From our point of view all these cycles are current cycles that can be critical. Thus they can all lead to instabilities. Networks with autocatalysis or a hypercycle can have explosion-extinction dynamics; however the conservation condition for the enzyme forbids explosions at the enzyme level. The hypercycle differs from the autocatalytic current cycle in that it has many short circuits. These short circuits are the autocatalytic cycles that are systematically arranged around the larger current cycle. We have proved here that a random arrangement of certain kinds of short circuits within the larger cycle also gives it the mathematical property ($\beta = 0$), which leads to explosion-extinction dynamics. Hence the hypercycle is only one of many complicated critical current cycles that potentially could play the role in evolution that Eigen ascribes only to the hypercycle. The important point is that the autocatalytic steps do not help the larger current cycle to become unstable. They do not make it equivalent to a strong current cycle. Thus from the point of view of our theory of explosion-extinction dynamics, there is no advantage to be had from building cycles out of smaller cycles, which are built out of smaller cycles.

Eigen's theory is based primarily on computer simulations of model networks that, he believes, approximate the dynamics of the hypercycle. However, the equations he uses represent the hypercycle by a *strong* current cycle. The hypercycle, being a large critical current cycle with many autocatalytic short circuits, is very different from a strong cycle. Thus Eigen's computer simulations probably give an inaccurate picture of hypercycle dynamics. His calculations show that competition between these strong cycles appears to be the molecular analogy of world conquest by a military power that suppresses all opposition ruthlessly and remains in control forever once it is established. This theory would certainly explain the universality of the genetic code; however, the following theory based on *critical* current cycles leads to a more benign picture than the one described by Eigen.

An explosion in a current cycle causes the concentrations of all species on the cycle to increase until bimolecular reactions between them become dominant and the network has a new steady state elsewhere in $\mathcal{C}_v$. Such jumps from one point in $\mathcal{C}_v$ to another are found in Example II.8. The species on this current cycle that have the lowest steady-state concentration in the new steady state are the most vulnerable point of attack by another current cycle. Contrary to the picture described by Eigen, new species can appear via a slow reaction (a mutation), initiate a competing current cycle, and upset the hegemony of the preestablished current cycle. Some of the short-circuit current cycles of the original cycle may continue to survive and compete. This view of molecular evolution is one of continual revolution from below. It is impossible for any current cycle to protect itself completely from species that might appear and reactions that might occur following a "mutation." Within this viewpoint the universality of the genetic code can also be accounted for. The code is the near optimum code that has been arrived at by many different routes in a very flexible, highly competitive, and benign evolutionary system where no strong current cycle takes over and freezes the evolutionary state of the system at a point far from optimum.

Acknowledgments

I express my deepest appreciation to Prof. B. Von Hohenbalken for inventing the algorithms needed for finding the vertices and edges of polytopes, and the frames of cones. I also thank Prof. I. Prigogine for his hospitality during my stay in Brussels during 1976–1977. The interest and comments of G. Nicolis and J. Stucki were particularly stimulating. In addition, I thank Prof. R. M. Noyes for helping to keep me up to date on the latest experimental mechanisms of oscillating reaction systems.

References

1. *Physical Chemistry of Oscillatory Phenomena*, Symposium of the Faraday Society, Vol. 9, Chemical Society, London, 1974.
2. G. Nicolis and I. Prigogine, *Self-Organization in Nonequilibrium Systems*, Wiley, New York, 1977.
3. J. Z. Hearon, *Bull. Math. Biophys.*, **15**, 121 (1953).
4. J. Z. Hearon, *Ann. N.Y. Acad. Sci.*, **108**, 36 (1963).
5. P. Delattre, *L'Evolution des Systèms Moléculaires*, Maloine, Paris, 1971.
6. C. Hyver, *J. Theor. Biol.*, **42**, 397 (1973).
7. F. Solimano and E. Beretta, *J. Theor. Biol.*, **59**, 159 (1976).
8. F. Solimano, E. Beretta, and E. Piatti, *J. Theor. Biol.*, **64**, 401 (1977).

9. E. Beretta, F. Vetrano, F. Solimano, and C. Lazzari, *Bull. Math. Biol.*, **41**, 641 (1979).
10. D. B. Shear, *J. Chem. Phys.*, **48**, 4144 (1968).
11. D. B. Shear, *J. Theor. Biol.*, **16**, 212 (1967).
12. J. Higgins, *J. Theor. Biol.*, **21**, 293 (1968).
13. F. Horn and R. Jackson, *Arch. Ration. Mech. Anal.*, **47**, 81 (1972).
14. M. Feinberg, *Arch. Ration. Mech. Anal.*, **49**, 187 (1973).
15. F. Horn, *Arch. Ration. Mech. Anal.*, **49**, 172 (1973).
16. M. Feinberg and F. J. M. Horn, *Chem. Eng. Sci.*, **29**, 775 (1974).
17. F. Horn, *Proc. R. Soc. London, Ser. A*, **334**, 299 (1973).
18. F. Horn, *Proc. R. Soc. London, Ser. A*, **334**, 313 (1973).
19. F. Horn, *Proc. R. Soc. London, Ser. A*, **334**, 331 (1973).
20. B. Clarke, *J. Theor. Biol.*, to appear, 1980.
21. G. F. Oster, A. S. Perelson, and A. Katchalsky, *Q. Rev. Biophys.*, **6**, 1 (1973).
22. J. Schnakenberg, *Thermodynamic Network Analysis of Biological Systems*, Springer-Verlag, New York, 1977.
23. A. Lotka, *J. Phys. Chem.*, **14**, 271 (1910).
24. A. Lotka, *Proc. Nat. Acad. Sci. (U.S.)*, **6**, 410 (1920).
25. A. M. Turing, *Phil. Trans. R. Soc. London, Ser. B*, **237**, 37 (1952).
26. E. E. Selkov, *Eur. J. Biochem.*, **4**, 79 (1968).
27. F. Schlögl, *Z. Phys.*, **248**, 446 (1971).
28. F. Schlögl, *Z. Phys.*, **253**, 147 (1972).
29. R. Field, *J. Chem. Phys.*, **63**, 2289 (1975).
30. R. J. Field and R. M. Noyes, *J. Chem. Phys.*, **60**, 1877 (1974).
31. B. Edelstein, *J. Theor. Biol.*, **29**, 57 (1970).
32. B. Edelstein, *J. Theor. Biol.*, **37**, 221 (1972).
33. B. C. Goodwin, *Temporal Organization in Cells*, Academic Press, New York, 1963.
34. H. G. Othmer, *J. Math. Biol.*, **3**, 53 (1976).
35. B. L. Clarke, *J. Chem. Phys.*, **58**, 5605 (1973).
36. B. L. Clarke, *J. Chem. Phys.*, **60**, 1481 (1974).
37. B. L. Clarke, *J. Chem. Phys.*, **60**, 1493 (1974).
38. B. L. Clarke, *J. Chem. Phys.*, **62**, 773 (1975).
39. B. L. Clarke, *J. Chem. Phys.*, **62**, 3726 (1975).
40. B. L. Clarke, *J. Chem. Phys.*, **64**, 4165 (1976).
41. B. L. Clarke, *J. Chem. Phys.*, **64**, 4179 (1976).
42. B. L. Clarke, *SIAM J. Appl. Math.*, **35**, 755 (1978).
43. B. Von Hohenbalken, *Math. Program.*, **15**, 1 (1978).
44. C. Jefferies, V. Klee, and P. Van den Driessche, *Can. J. Math.*, **29**, 315 (1977).
45. C. Jefferies, *Ecology*, **55**, 1415 (1974).
46. J. Quirk and R. Ruppert, *Rev. Econ. Stud.*, **32**, 311 (1965).
47. R. M. May, *Stability and Complexity in Model Ecosystems*, Princeton University Press, Princeton, NJ, 1973.
48. J. J. Tyson, *J. Chem. Phys.*, **62**, 1010 (1975).
49. G. Hadley, *Linear Algebra*, Addison-Wesley, Reading, MA, 1961.
50. G. Hadley, *Linear Programming*, Addison-Wesley, Reading, MA, 1962.
51. R. T. Rockafellar, *Convex Analysis*, Princeton University Press, Princeton, NJ, 1970.
52. G. Kirchhoff, *Ann. Phys.*, **72**, 497 (1847).
53. E. L. King and C. Altman, *J. Phys. Chem.*, **60**, 1375 (1956).
54. T. L. Hill, *Free Energy Transduction in Biology, The Steady-State Kinetic and Thermodynamic Formalism*, Academic Press, New York, 1977.
55. R. E. O'Malley, *Introduction to Singular Perturbations*, Academic Press, New York, 1974.

56. R. M. Noyes, *J. Chem. Phys.*, **64**, 1266 (1976).
57. M. W. Hirsch and S. Smale, *Differential Equations, Dynamical Systems and Linear Algebra*, Academic Press, New York, 1974.
58. F. R. Gantmacher, *Applications of the Theory of Matrices*, Interscience, New York, 1959.
59. S. Barnett and C. Storey, *Matrix Methods in Stability Theory*, Nelson, London, 1970.
60. J. Maybee and J. Quirk, *SIAM Rev.*, **11**, 30 (1969).
61. F. Harary, *Graph Theory*, Addison-Wesley, Reading, MA, 1969.
62. O. Sinanoglu, *J. Am. Chem. Soc.*, **97**, 2309 (1975).
63. D. Gibbs, *Instability Domains of the Reversible Oregonator*, M.Sc. Thesis, University of Alberta, Edmonton, 1978.
64. P. Glansdorff and I. Prigogine, *Thermodynamic Theory of Structure Stability and Fluctuations*, Wiley-Interscience, New York, 1971.
65. D. R. Chillingworth, *Differential Geometry with a View to Applications*, Pitman, London, 1976.
66. G. W. Beadle, *Sci. Am.*, **179**, 30 (1948).
67. K. J. Laidler, *Reaction Kinetics*, Vol. 1, Pergamon, Oxford, 1963.
68. M. Eigen, *Naturwissenschaften*, **58**, 465 (1971).
69. R. J. Field, E. Körös, and R. M. Noyes, *J. Am. Chem. Soc.*, **94**, 8649 (1972).
70. K. R. Sharma and R. M. Noyes, *J. Am. Chem. Soc.*, **97**, 202 (1975).
71. K. R. Sharma and R. M. Noyes, *J. Am. Chem. Soc.*, **98**, 4345 (1976).
72. J. J. Jwo and R. M. Noyes, *J. Am. Chem. Soc.*, **97**, 5422 (1975).
73. C. N. Hinshelwood, *Proc. R. Soc. London, Ser. A*, **188**, 1 (1946).
74. S. S. Abhyankar, *Am. Math. Mon.*, 409 (1976).
75. J. Dieudonné, *Infinitesimal Calculus*, Houghton Mifflin, Boston, 1971.
76. H. Hironaka, *J. Math. Kyoto Univ.*, **7**, 251 (1967).
77. J. P. de Gua de Malves, *Usages de l'Analyse de Descartes pour Découvrir sans le Secours du Calcul Différential, les Propriétés ou Affections Principales des Lignes Géométriques de Tous des Ordres*, Paris, 1740.
78. I. Newton, in *The Mathematical Papers of Isaac Newton*, Vol. 3, 1670–1673, D. T. Whiteside, Ed., Cambridge University Press, Cambridge, 1969, p. 50.
79. U. Müller-Herold, *Chem. Phys. Lett.*, **33**, 467 (1975).
80. K. D. Williamowski and O. E. Rösseler, *Z. Naturforsch.*, **31**, 408 (1976).
81. K. D. Williamowski and O. E. Rösseler, *Z. Naturforsch.*, **33**, 827 (1978).
82. K. D. Williamowski and O. E. Rösseler, *Z. Naturforsch.*, **33**, 983 (1978).
83. K. D. Williamowski and O. E. Rösseler, *Z. Naturforsch.*, **33**, 989 (1978).
84. M. Eigen and P. Schuster, *Naturwissenschaften*, **64**, 541 (1977).
85. M. Eigen and P. Schuster, *Naturwissenschaften*, **65**, 7 (1978).

CHEMICAL INSTABILITIES

PETER H. RICHTER, ITAMAR PROCACCIA,* and JOHN ROSS

Department of Chemistry
Massachusetts Institute of Technology
Cambridge, Massachusetts 02139

CONTENTS

We review some aspects of the subject of chemical instabilities, that is, the properties of certain chemical reaction systems, possibly coupled with other processes (such as diffusion), under far-from-equilibrium conditions. The topic includes the interesting phenomena of multiple stationary states, oscillatory reactions, and formation of macroscopic spatial structures. A number of books and reviews have been published on this subject.[1–18] This chapter emphasizes theory, both deterministic and stochastic; we cite experiments to indicate progress and challenges.

I. INTRODUCTION

The fundamental difference between equilibrium systems and those that are subject to persistent fluxes of matter or energy is their behavior under time reversal. In equilibrium, by definition, every flow in one direction is

* Permanent address: Department of Chemical Physics, Weizmann Institute of Science, Rehovot, Israel.

compensated by a reverse flow; any production is canceled by a corresponding loss, and the system is in a "detailed balance" of active tendencies; that is, it is invariant under time reversal. This symmetry is broken as soon as fluxes are imposed that penetrate the system and thereby drive it away from the equilibrium situation. Close to equilibrium reaction systems are stable in that any imposed perturbation decays in time. This remains true for many cases no matter how far the system is driven from equilibrium. However for certain auto- (or cross-) catalytic systems instability occurs at some point sufficiently far from equilibrium: the system becomes unstable to small perturbations and a transition, say, to oscillatory variation of concentration, occurs. As a result, spontaneous ordering grows out of internal fluctuations. The combination of such reaction mechanisms with dissipative processes may lead to the formation of macroscopic spatial structures. As the system is driven farther away from equilibrium, a series of instabilities may appear until finally chaotic behavior[19] may set in.

The term "chemical instabilities" is used whenever chemical reactions play an essential part in such transitions. In particular, a purely chemical instability is brought about by chemical rate processes alone, without any contribution from hydrodynamic transport processes. The most well-known example of oscillatory reactions, the Belousov-Zhabotinsky reaction,[2] has been interpreted with that simplification. For a review of oscillatory reactions systems, see Refs. 1, 5, and 18.

Under suitable conditions the same systems often develop spatial structures[2,18] and even wave propagation.[21,22] Obviously, some account of transport processes is required for an understanding of such phenomena, and the simplest conceivable addition to the purely chemical kinetic equations consists of diffusional currents. Therefore most of the theoretical models have been formulated in terms of reaction-diffusion equations, and the present review reflects this state of affairs.

A common feature of the example for structure formation just given is that the broken symmetries are continuous; for instance, in the transition to a limit cycle there is a choice of an infinite number of phases. There are many cases, however, where the choice is among a finite number of possibilities only. These are usually treated under the heading of "multiple stationary states". Typically, when a system is driven away from equilibrium a bifurcation takes place in the plot of a characteristic observable versus the "pump parameter" which forces the system far from equilibrium. This offers the system an alternative to being on the "equilibrium branch", and at some point a transition occurs. If the bifurcation emerges from the old state as in Fig. 1*a* the transition is smooth, whereas a situation as in Fig. 1*b* entails abrupt changeovers and hysteresis effects. The inorganic subsystem of the Belousov-Zhabotinsky[23] reaction is of the

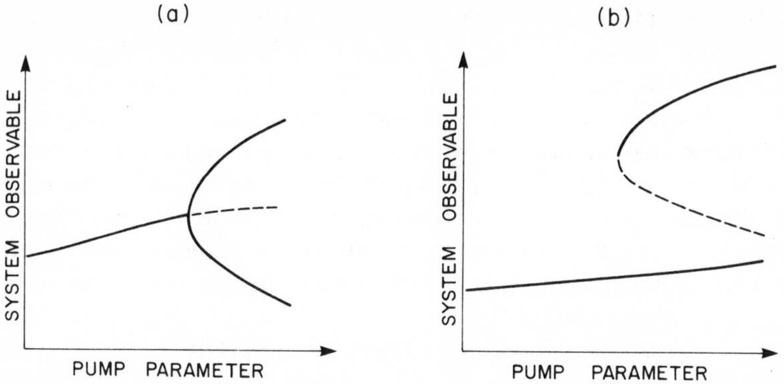

Fig. 1. Two types of bifurcation process. Solid lines represent stable branches, broken lines unstable branches. Equilibrium holds for vanishing pumping, and the thermodynamic branch extends from there to the first bifurcation.

type as in Fig. 1*b*. Another example that under appropriate conditions can show oscillations or bistability is the oxidation of NADH by oxygen in the presence of horseradish peroxidase,[24] and even chaotic behavior may be observed in this case if the reaction is driven by a sufficient amount of enzyme.[25]

These phenomena are reminiscent of phase transitions in physical equilibrium systems.[26–28] For instance, Fig. 1*a* resembles the behavior of an Ising ferromagnet near its Curie point (i.e., a phase transition of second order), whereas Fig. 1*b* reminds one of the first-order transition in a liquid-gas system. Increasing the fluxes in a nonequilibrium situation is analogous to lowering the temperature in those equilibrium systems. The analogy draws attention to the role of fluctuations in the vicinity of a transition point, where they tend to increase in strength, to slow down in their decay, and to become coherent over a long range. The universality of critical phenomena[29,30] in diverse systems has its origins in these characteristics of the fluctuations; therefore their detailed investigation may be crucial for an understanding of chemical instabilities as well.

Theoretical progress so far has been made along different lines. Our review cannot attempt to be exhaustive but focuses on what we feel are pertinent ideas, methods, and models when dealing with the phenomena mentioned. This seems all the more justified insofar as there are other reviews with different emphases. Some stress the variety of instability phenomena,[2,4,12–14,31] others make a strong point of their similarity to ordinary phase transitions.[9,32] There are classifications according to the topology in state variable phase space;[33,34] there are computer simulations[35,36] and the use of highly abstract mathematics[37] to discuss the possibility of "chaotic" noise[38] generated by deterministic equations.

This review is organized as follows. The major division concerns the levels of description. Thus in the first part we deal with deterministic (macroscopic) model equations and in the second we go into some detail on the stochastic (mesoscopic) behavior. Within each section the systems to be discussed are arranged according to the type of order that they exhibit; that is, we start with homogeneous steady states, go on to oscillating but still homogeneous systems, and finally include spatial structures.

II. MACROSCOPIC BEHAVIOR

A. General Comments

We are concerned with multicomponent fluids in which chemical reactions take place and transform the components $X_j (j = 1, \ldots, n)$ into each other according to mass action laws:

$$\sum_{j=1}^{n} \nu_{\lambda j} X_j \underset{\bar{k}_\lambda}{\overset{k_\lambda}{\rightleftharpoons}} \sum_{j=1}^{n} \bar{\nu}_{\lambda j} X_j, \qquad \lambda = 1, \ldots, l \tag{II.1}$$

With the forward and backward rates for the l reactions (the symbol X_j is used both for the jth component and the corresponding particle number)

$$R_\lambda = k_\lambda X_1^{\nu_{21}} \cdots X_n^{\nu_{\lambda n}}, \qquad \bar{R}_\lambda = \bar{k}_\lambda X_1^{\bar{\nu}_{\lambda 1}} \cdots X_n^{\bar{\nu}_{\lambda n}} \tag{II.2}$$

we write the standard equations of reaction kinetics as

$$\frac{dX_j}{dt} = \sum_{\lambda=1}^{l} (\bar{\nu}_{\lambda j} - \nu_{\lambda j})(R_\lambda - \bar{R}_\lambda) \tag{II.3}$$

These deterministic equations are fundamental, and the next two sections of this chapter do not go beyond them. However a word about their limitations is in order.

Chemical equations of the type (II.1) account for total mass conservation during the reaction events, but there are other conservation laws to be satisfied. Individual masses are conserved in the flow of the components relative to each other (i.e., in diffusion), and the total mass must also be conserved in the convective flow of the fluid. Such flow is subject to momentum conservation as expressed in the Navier-Stokes equation, and finally there is the requirement of conservation of energy. In addition, chemical rates are usually strongly temperature dependent and also vary with pressure. In general, therefore, a complete set of kinetic equations contains appreciable coupling of chemistry to hydrodynamics. In a non-equilibrium situation, where there is a steady turnover of chemicals, it is then quite likely that there are hydrodynamic fluxes as well, and therefore inhomogeneities. Indeed, experimental examples like the Zhabotinsky system[39–41] or the glycolytic oscillations[3] do show periodic heat production.

Under what conditions is a separate treatment of the chemical subset of equations (to which most of the work in the field has been confined) justified? The hydrodynamic fluxes must be sufficiently small not to affect the chemistry appreciably. For instance, if the reactions have small enthalpy changes, no large temperature changes occur, and thermal conduction need not be considered. Likewise, volume effects should be small enough not to alter the rates via their pressure dependence. It is furthermore desirable to avoid the onset of convection, which might confuse the study of chemical effects.

It is important to realize that all this depends very much on the spatial and temporal scales of the observed phenomena. To be specific, let us discuss three hydrodynamic modes and their relation to chemistry: sound propagation, which is responsible for mechanical equilibration, heat conduction, which homogenizes temperature, and diffusion, which tends to remove gradients in composition. The time scales of these processes depend on the spatial scale chosen for observation. Sound waves behave in time like $\exp\{\pm ikct - \Gamma k^2 t\}$, where c is the sound velocity ($\sim 10^5$ cm/sec in water) and Γ the attenuation constant ($\sim 10^{-2}$ cm^2/sec). For wavelengths much larger than 10 Å the oscillation period is much shorter than the characteristic damping time, $c/\Gamma k \gg 1$. Therefore the typical frequency at which coupling between sound and chemistry might take place is $|\omega| \sim ck$.[42] Heat modes decay like $\exp\{-D_T k^2 t\}$, where the thermal diffusivity D_{T} is $\sim 10^{-3}$ cm^2/sec for water. A similar k^2-dependence of the relaxation frequency holds for diffusion, with $D \sim 10^{-5}$ cm^2/sec for small molecules.

Diffusion and chemical reactions appear together in the same equation: $\dot{X}_i = D_i \nabla^2 X_i + f(X_1, \ldots, X_n)$, for a species X_i, with f taking account of the chemical reactions. Thus if λ is the chemical rate as derived from (II.3), then an expression of the form $|\lambda| + Dk^2$ describes the total rate of composition relaxation. Heat conduction couples to composition via the reaction enthalpy, and the temperature dependence of the rate coefficients; sound couples via the reaction volume and the pressure dependence of the rate coefficients. Assuming that these couplings are not negligible let us discuss an example where $\lambda = 10$ sec^{-1} (Fig. 2). On sufficiently small spatial scales ($k \gtrsim 10^3$ cm^{-1}) this chemical relaxation is slower than all the hydrodynamic modes. Equilibrium is always established by means of the usual transport processes, within times λ^{-1}, and fluctuations behave in the ordinary way. In this sense there appears to be "local equilibrium" even if there are persistent chemical fluxes—just because chemistry does not

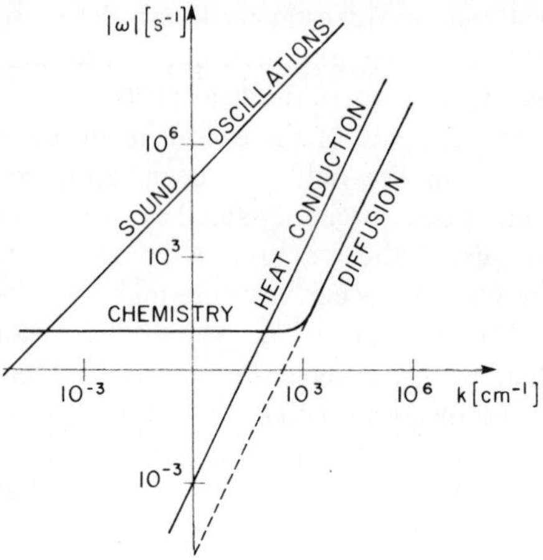

Fig. 2. Typical frequencies $|\omega|$ for hydrodynamic modes as a function of the wave number k for dilute aqueous solutions of small molecules. Sound velocity $\sim 10^5$ cm/sec, thermal diffusivity $\sim 10^{-3}$ cm^2/sec, diffusion constant $\sim 10^{-5}$ cm^2/sec.

matter. On scales such that chemistry is faster than diffusion ($k \lesssim 10^3$ cm^{-1}) the system may develop spatial gradients in composition, since Dk^2 is unimportant compared to λ. Thus diffusion must in principle be included to test for such a possibility. If it turns out that the steady state is still homogeneous, the macroscopic behavior may indeed be described by the purely chemical equations (provided $k \gtrsim 10^2$ cm^{-1} to ensure thermal equilibrium). If not, stirring may help. Next, on scales $k \lesssim 10^2$ cm^{-1} thermal gradients may build up because the processes are effectively adiabatic, and considerable temperature changes may occur in the system and affect the chemistry. Thus heat conduction must be considered. Finally, at very much larger wavelengths the coupling of chemistry to sound must be taken into account because a mechanical nonequilibrium may now be sustained. (With reaction times in the microsecond range, this coupling occurs at millimeter wavelengths.) The Zhabotinsky systems typically work in the range $\omega \sim 10^{-2}$ sec^{-1}; thus in a 1 cm test tube neither homogeneity in composition nor thermal equilibrium is necessarily established within a relaxation time.

In gases the dissipative processes tend to occur all on the same time scale ($D, D_T, \Gamma \sim 10^{-1}$ cm^2/sec). A separation of time scales is then clearly not possible, and greater care is needed.

B. Multiple Stationary States

1. Experiments

Multiple stationary states are known in various driven systems, be they chemical or otherwise.[6,43] Some of the chemical examples involve heterogeneous reactions. Here we consider only examples that are predominantly chemical and occur in the bulk. Schmitz[8] summarized the known experi-

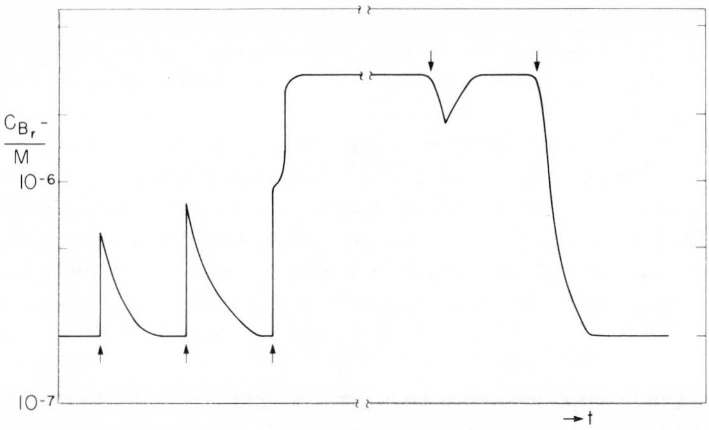

Fig. 3. Experiments on transitions between two stable stationary states: bistable behavior in the inorganic subsystem of the Belousov-Zhabotinsky reaction carried out in a continuously stirred reactor tank. The arrows indicate the imposition of external perturbations. Small perturbations decay to the original stationary state; a sufficiently large perturbation brings about a transition. The experiment measures bromide ion concentration versus time. Redrawn from Geiseler and Föllner.[23]

ments up to 1975. We cite briefly three experiments as recent examples. Geiseler and Föllner[23] have reported on experiments in a continuously stirred tank reactor (CSTR) with the inorganic subsystem of the Belousov-Zhabotinsky reaction, which consists of sulfuric acid solutions of bromate, bromide, and cerium (III). The bistability involved is shown in Fig. 3, which gives the concentration of Br^- as a function of time. At the various points indicated by arrows the system is perturbed by changing the Br concentration abruptly. One sees that the two stationary states are stable to small perturbations, but a sufficiently large perturbation triggers a transition from one stationary state to the other.

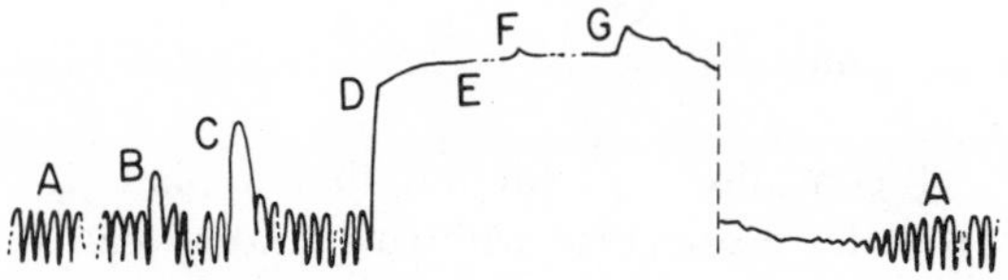

Fig. 4. Example of bistability involving a stationary state (E) and a limit cycle (A). Electrochemical potential appears as a function of time. Systems: KIO_3, 0.024 M; malonic acid, 0.056 M; H_2O_2, 1.2 M; $HClO_4$, 0.058 M; $MnSO_4$, 0.004 M; 25°C, in a CSTR. At points B and C the limit cycle was perturbed by fast injection of I^- ions; the system returned to the limit cycle. At point D the perturbation was sufficient for the system to reach a new stationary state E. At points F and G malonic acid is injected. The last perturbation is sufficient to make the system go back to limit cycle A. After Pacault, de Kepper, and Hanusse.[45]

Another example was reported by Creel and Ross.[44] In this case a hysteresis loop was observed in a system composed of an $2NO_2 \rightleftharpoons N_2O_4$ mixture that is pumped by green laser light. The hysteresis is obtained in the NO_2 stationary state concentration as a function of the laser intensity.

An interesting observation was reported by Pacault, deKepper, and Hanusse.[45] They studied a version of the Belousov-Zhabotinsky reaction in which a stationary state and an oscillatory state are involved in the bistability phenomena (Fig. 4). Starting with the oscillatory state, increasing perturbations result in a transition to a stationary state and vice versa.

Other experimental observations are reported in Refs. 24 and 46–48.

2. *Theory*

Among the components of a chemical reaction system let us distinguish between control variables $A_1, \ldots, A_m$, which are determined externally, and free variables $X_1, \ldots, X_n$, which adjust themselves according to the reaction mechanism. Writing (II.1) and (II.2) in the slightly more elaborate form (summation over repeated index implied as usual)

$$\mu_{\lambda i} A_i + \nu_{\lambda j} X_j \underset{\bar{k}_\lambda}{\overset{k_\lambda}{\rightleftharpoons}} \bar{\mu}_{\lambda i} A_i + \bar{\nu}_{\lambda j} X_j, \qquad \lambda = 1, \ldots, l$$

$$R_\lambda = k_\lambda \prod_{i,j} A_i^{\mu_{\lambda i}} X_j^{\nu_{\lambda j}}, \qquad \bar{R}_\lambda = \bar{k}_\lambda \prod_{i,j} A_i^{\bar{\mu}_{\lambda i}} X_j^{\bar{\nu}_{\lambda j}} \tag{II.4}$$

we obtain for the control variables

$$\frac{dA_i}{dt} = -\sum_\lambda (\bar{\mu}_{\lambda i} - \mu_{\lambda i})(\bar{R}_\lambda - R_\lambda) \tag{II.5}$$

and for the free variables

$$\frac{dX_j}{dt} = -\sum_\lambda (\bar{\nu}_{\lambda j} - \nu_{\lambda j})(\bar{R}_\lambda - R_\lambda) \tag{II.6}$$

A natural way of controlling a reaction that applies, for example, to biological cells, is to have exchange with surrounding baths in proportion to differences of the concentrations $a_i \equiv A_i / V$, and to add terms $k_i(a_i - a_i^{\text{bath}})V$ to (II.5); see Ref. 49. Alternatively, one may think of constant fluxes dA_i/dt. From the point of view of computational convenience, however, it is by far more desirable to have the A_i themselves kept constant so that (II.5) need not be considered at all. Since most of the literature has used that simplification, we shall do so as well, in all the following. In any case, when we talk about stationary states X_j^0 of the system of free components, given by

$$\frac{dX_j}{dt} = 0, \qquad j = 1, \ldots, n \tag{II.7}$$

the distinction between equilibrium and nonequilibrium resides in the absence or presence, respectively, of the fluxes dA_i/dt.

The analysis is simplest when there is only one free variable. In this case, the dynamics of the system is described by an equation like

$$\dot{X} = f(X, A_1, \ldots, A_m) \tag{II.8}$$

where f is arbitrarily nonlinear in X and the control parameters A_i. We therefore choose two simple examples that will be of interest later on because they allow for an exact stochastic treatment of their static properties. They were introduced by Schlögl[50] to simulate phase transitions of the second and first order.

The chemical reactions for the two cases are

$$A + nX \underset{k_2}{\overset{k_1}{\rightleftharpoons}} (n+1)X, \qquad B \overset{k_3}{\rightarrow} X, \qquad X \overset{k_4}{\rightarrow} C \tag{II.9}$$

with $n = 1$ and 2, respectively ($n > 2$ does not give qualitatively new behavior). The first part is autocatalytic and represents the system's reactivity; taken alone it would lead to an equilibrium state $X_0^{(1)} = k_1 A/k_2$. The rest are ordinary first-order reactions that by themselves tend to establish the state $X_0^{(2)} = k_3 B/k_4$. The competition between these two aspects is represented by a parameter κ,

$$\kappa = \frac{k_3 B/k_4}{k_1 A/k_2} \tag{II.10}$$

which is zero in the limit of purely autocatalytic behavior, infinite for dominating first-order turnover, and 1 in case of equilibrium between the two. If we measure time in units of k_4^{-1} and particle numbers in units of $N \equiv (k_4/k_2)^{1/n}$, which is an appropriate extensivity parameter,

$$t = k_4^{-1}\tau, \qquad X = N\tilde{x} \tag{II.11}$$

we get the kinetic equations in terms of dimensionless quantities

$$\frac{d\tilde{x}}{d\tau} = (q - \tilde{x})\tilde{x}^n + \kappa q - \tilde{x} \tag{II.12}$$

where q represents the driving force

$$q = \frac{k_1 A}{k_2 N} \tag{II.13}$$

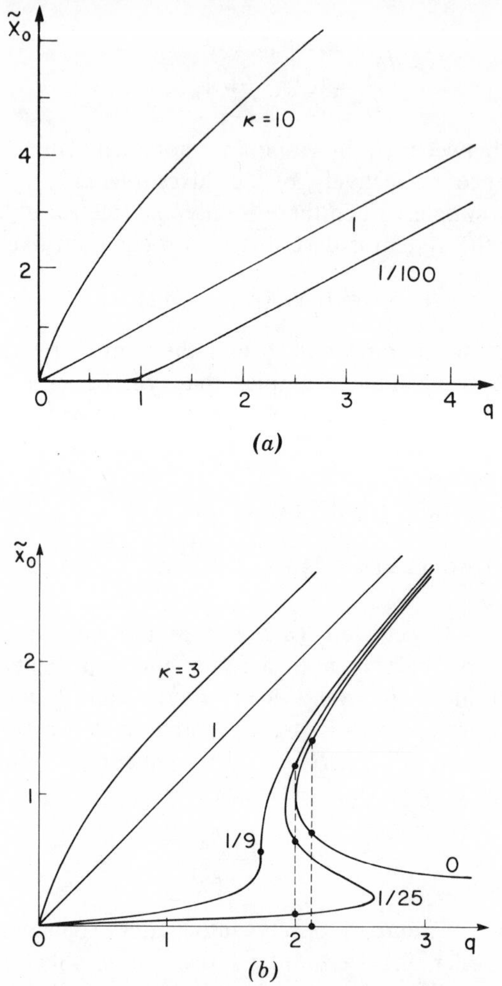

Fig. 5. Stationary state concentration, $\tilde{x}$, of chemical species X in the Schlögl model (II.9) as a function of the pump parameter q (II.13): (*a*) $n = 1$; (*b*) $n = 2$. The curves are labeled by values of the parameter κ (II.10). As κ decreases, the autocatalytic reaction in the mechanism leads to instability. The dots in Fig. 5*b* indicate coexistence of two stable stationary states according to (II.62).

and may be called "pump parameter" in analogy to the terminology of laser physics.[51, 52] Evaluation of (II.12) for the steady states is shown in Figs. 5*a*, *b* for $n = 1$, 2, respectively. Figure 5*a* ($n = 1$) shows a smooth transition from $\tilde{x} = 0$ to $\tilde{x}_0 = q - 1$, at $q = 1$, in the limit $\kappa \to 0$. Figure 5*b*, on the other hand, shows that the $n = 2$ case gives a first-order transition for $\kappa < 1/9$; the critical point is $\kappa = 1/9$, $q = \sqrt{3}$, $\tilde{x}_0 = \sqrt{1/3}$.

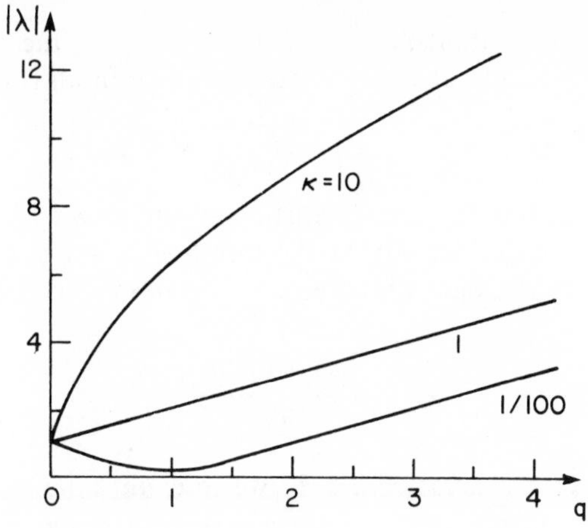

(a)

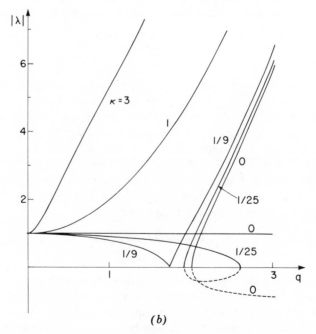

(b)

Fig. 6. Temporal eigenvalues of the Schlögl model (II.9) as a function of the pump parameter q (II.13): (*a*) $n = 1$; (*b*) $n = 2$. The curves are labeled by the values of the parameter κ (II.10). Notice the critical slowing down ($\lambda \to 0$) at marginal stability.

There is only one stationary state, for a given q, in the case $n = 1$. A small perturbation would decay at a rate $|\lambda|$, which in units of k_4 is

$$|\lambda| = 2\tilde{x}_0 - q + 1 = \sqrt{(q-1)^2 + 4\kappa q} \tag{II.14}$$

For $\kappa \to 0$ this exhibits the characteristic slowing down of second-order phase transitions, in the vicinity of the threshold $q = 1$ (Fig. 6*a*). In the case $n = 2$, for $\kappa < 1/9$, there are three steady states in a certain range of q. From

$$\left.\frac{\partial \tilde{x}}{\partial q}\right|_{\dot{\tilde{x}}} = -\frac{\partial \dot{\tilde{x}}/\partial q|_{\tilde{x}}}{\partial \dot{\tilde{x}}/\partial \tilde{x}|_q} \tag{II.15}$$

and the positivity of $\partial \dot{\tilde{x}}/\partial q|_{\tilde{x}} = \kappa + \tilde{x}^n$ we infer that the stable branches are those where $\tilde{x}_0$ increases with q, since that is where the relaxation rates

$$\lambda = \left.\frac{\partial \dot{\tilde{x}}}{\partial \tilde{x}}\right|_q = 3\tilde{x}^2 + 1 - 2q\tilde{x} \tag{II.16}$$

are negative. Figure 6*b* displays these rates against q, for various κ. There is critical slowing down at the points of marginal stability, $\partial \tilde{x}_0/\partial q = \infty$:

$$\tilde{x} = \frac{q}{3} \pm \frac{\sqrt{q^2 - 3}}{3} \tag{II.17}$$

As q is pushed beyond these points, the system jumps from one stable branch to the other one. Thus there is hysteresis in the response to alternately increasing and decreasing the pump parameter. Note, however, that we are severely restricted in discussing this hysteresis for two reasons: (1) we discarded all transport processes that would allow us to treat inhomogeneities, and in particular nucleation phenomena; (2) if we were to include spontaneous fluctuations, the transition between the two stable branches could occur before the marginal stability points were reached. This problem is taken up again subsequently.

C. Temporal Oscillations

1. Experiments

Examples of oscillatory reactions are known both in well-stirred conditions, where diffusion is believed to be unimportant, and in unstirred systems where diffusion may play an important role.

A considerable amount of work has been directed toward the understanding of the Belousov-Zhabotinsky reaction and its variants. The experimental results have been reviewed extensively[1, 5, 10, 18, 53] and are not repeated here. Other examples include systems that are driven by means

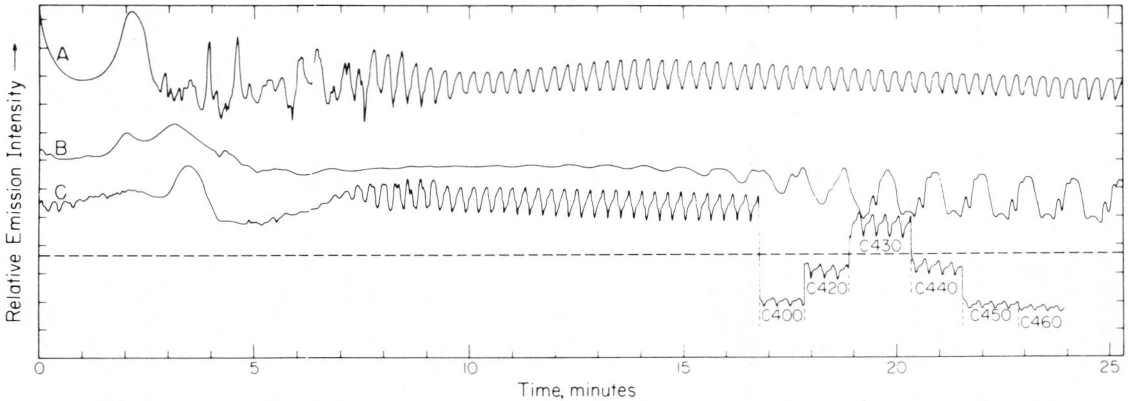

Fig. 7. Plots of 9, 10-dimethylanthracene (DMA) fluorescence intensity against time in ~3 cc of $CHCl_3$ solution at 298°K. All plots were obtained using a Perkin-Elmer MPF3 emission spectrophotometer, and the sample was in a square (1.0 × 1.0 cm i.d.) fluorescence cell. All curves are for an excitation wavelength of 260 nm (slit 16 nm) and a monitoring wavelength of 410 nm (slit 6 nm) unless specified otherwise. Curve *A*: the baseline is the dashed horizontal line; $[DMA] = 2.0 \times 10^{-5}$ *M*. Curve *B*: the baseline is at the solid horizontal line; $[DMA] = 4.0 \times 10^{-5}$ *M*. Curve *C*: the baseline is as in curve *B*, and the sample is just a fresh solution of that used in curve *A*; the excitation and emission slits are the same, but the monitoring wavelength is 410 nm only up to T 16.7 min (during the intervals marked C400, C420, C430, C440, C450, and C460 the monitoring wavelength was 400, 420, 430, 440, 450 and 460 nm, repectively). From Bose et al.[54]

other than chemical fluxes. An oscillatory reaction and diffusion mechanism induced by radiation has recently been reported by Bose et al.[54] Figure 7 reproduces their fluorescence signal obtained from a solution of 9,10-dimethylanthracene in chloroform that is excited with light at 260 nm. Initially the fluorescence is aperiodic, but it transforms eventually to a periodic oscillation. Note the fine structure in the fluorescence peaks. The details of the mechanism are not known.

Other interesting recent experiments involve coupling of two oscillatory reactions.[55–57] Phenomena of synchronization of oscillations of a common frequency, synchronization of multiples of common frequency, rhythm splitting, and amplitude amplification were observed.

For experimental results on entrainment of oscillations in biological systems (yeast cells and slime molds) see the review of Hess, Goldbeter, and Lefever in Ref. 16.

2. *Theory*

Let us consider a reaction mechanism that may become oscillatory. Close to equilibrium, the dynamics of a perturbation can be described in terms of equilibrium correlation functions. Their symmetry with respect to time reversal rules out the possibility of oscillations.[58,59] As the reaction is driven farther from equilibrium, there is usually a range of pump parameter (the fluxes of reactants in and products out of the system) for which the temporal eigenvalues of the linearized dynamics that govern the decay of a perturbation are complex. The system is still stable in that the real part of the eigenvalues is negative. The existence of nonzero imaginary parts, as obtained by a linear stability analysis, is an indication, but no proof, that the system may make a transition to a stable oscillatory state when the real part of the eigenvalue becomes positive. The system is then said to exhibit a "hard instability" as opposed to the "soft instabilities," where the imaginary part of the critical eigenfrequency is zero. (The adjectives "hard" and "soft" have sometimes been used as substitutes for "first" and "second" order in designating transitions.) Thus there are two aspects to be discussed: (1) the origin of an oscillatory tendency per se, and (2) the peculiarities, if any, of a hard transition.

Let us first focus on the nature of the "spring" of a chemical oscillator.[60] Linearizing (II.6) with respect to deviations from the steady state, $\delta X_i \equiv X_i - X_i^0$, we obtain

$$\delta \dot{X}_j = -\frac{1}{X_k}\sum_{\lambda}(\bar{\nu}_{\lambda j} - \nu_{\lambda j})(\bar{\nu}_{\lambda k}\bar{R}_{\lambda} - \nu_{\lambda k}R_{\lambda})\delta X_k := \Lambda_{jk}\delta X_k \qquad \text{(II.18)}$$

or

$$\delta \dot{X}_i = -L_{ij}\Gamma_{jk}^{-1}\delta X_k \qquad \text{(II.19)}$$

where the matrix of static correlations $\Gamma_{ij} = \langle \delta X_i \delta X_j \rangle$ is by definition symmetric, and L_{ij} is the analogue of Onsager's transport matrix. Linear algebra tells that the spectrum of (II.19) is real for symmetric L_{ij} and imaginary for antisymmetric L_{ij}. This suggests that we call

$$D_{ij} \equiv \frac{(L_{ij} + L_{ji})}{2} \tag{II.20}$$

the dissipative and

$$A_{ij} \equiv \frac{(L_{ij} - L_{ji})}{2} \tag{II.21}$$

the "elastic" part of L_{ij}. Oscillations can only occur if A_{ij} is not zero.

In general, there are two distinctly different reasons for nonzero A_{ij}. One is that among the set of variables $(X_1, \ldots, X_n)$ there are some that are even and others that are odd under time reversal. Then Onsager's principle requires that in equilibrium the corresponding coefficients be antisymmetric. The sound waves of hydrodynamics, for example, are of this type, since they arise from a coupling between density (even) and velocity (odd). In a purely chemical system, however, the X_i are all particle numbers and therefore even. The other possibility for nonzero A_{ij} must then be invoked, which is symmetry breaking from the outside by imposing a flux. In a system with electric charges, this can be achieved by a magnetic field **H** as well: at fixed **H** the transport coefficients need no longer be symmetric, $L_{ij}(\mathbf{H}) \neq L_{ji}(\mathbf{H})$. As a consequence, there are cyclotron oscillations that do not exist without **H**. The persistent fluxes sufficiently far from equilibrium in chemical reaction systems can have the same effect: they violate time reversal symmetry, thereby invalidating the arguments that lead to Onsager's reciprocity.

For the sake of the discussion above it was necessary to evoke a decomposition of the kinetic relaxation matrix Λ_{ij} into two parts L_{ij} and Γ_{ij}^{-1} that reflect the system's stochastic behavior. This is the subject of the following section; here we briefly outline the strictly kinetic properties of a hard mode instability and the region beyond it. For that purpose we use the two prototype reaction models to which most of the theoretical work has hitherto been directed, the Brusselator and the Lotka-Volterra model.

Brusselator Model. Following the pioneering work of Turing,[61] the Brusselator was developed by Prigogine and Lefever[62] and got its name from Tyson.[63] The system is defined by the following set of "chemical reactions":

$$A \xrightarrow{k_1} X, \qquad X + B \xrightarrow{k_2} Y + D, \qquad 2X + Y \xrightarrow{k_3} 3X, \qquad X \xrightarrow{k_4} E \tag{II.22}$$

The third step is autocatalytic and has a tendency to produce X in an explosive manner. The second step can be viewed as an antagonist that after a burst of X formation, uses that X to build up Y. Thereby, however, enough Y may eventually be accumulated to ignite the explosion again.

As always, it is convenient to formulate the kinetic equations in terms of dimensionless variables. Following Ref. 12, we use the extensivity parameter $N \equiv \sqrt{k_4/k_3}$ and write

$$t = k_4^{-1}\tau, \qquad X = N\tilde{x}, \qquad Y = N\tilde{y}, \qquad A = N\frac{k_4}{k_1}a, \qquad B = \frac{k_4}{k_2}b \tag{II.23}$$

The kinetic equations then read

$$\begin{aligned}\frac{d\tilde{x}}{d\tau} &= a - \tilde{x} + \tilde{x}^2\tilde{y} - b\tilde{x}\\ \frac{d\tilde{y}}{d\tau} &= -\tilde{x}^2\tilde{y} + b\tilde{x}\end{aligned} \tag{II.24}$$

with steady-state solutions

$$\tilde{x}_0 = a, \qquad \tilde{y}_0 = \frac{b}{a} \tag{II.25}$$

Linearization leads to the relaxation matrix

$$\Lambda = -\begin{pmatrix} 1-b & -a^2 \\ b & a^2 \end{pmatrix} \tag{II.26}$$

which has eigenvalues

$$\lambda_{1,2} = -\tfrac{1}{2}(1 + a^2 - b) \pm \tfrac{1}{2}\sqrt{(1 + a^2 - b)^2 - 4a^2} \tag{II.27}$$

Figure 8 shows how these relaxation frequencies move through the complex plane as b is increased at fixed a. There are two real roots $-1, -a^2$ for $b = 0$ that coalesce at a as $b \to (1-a)^2$. They then depart into the complex plane, where they follow a circle around the origin of radius a. At $b = 1 + a^2$ they cross the imaginary axis, which means that the steady state becomes unstable.Thus we may draw the phase diagram of Fig. 9, where the lower parabola indicates the onset of oscillatory relaxation and the upper one connects the points of marginal stability.

The question then is: what happens at $b > 1 + a^2$ where there is no single stable steady state? Obviously, the linearized equations cannot tell, and we have to go back to the original nonlinear (II.24). Numerical

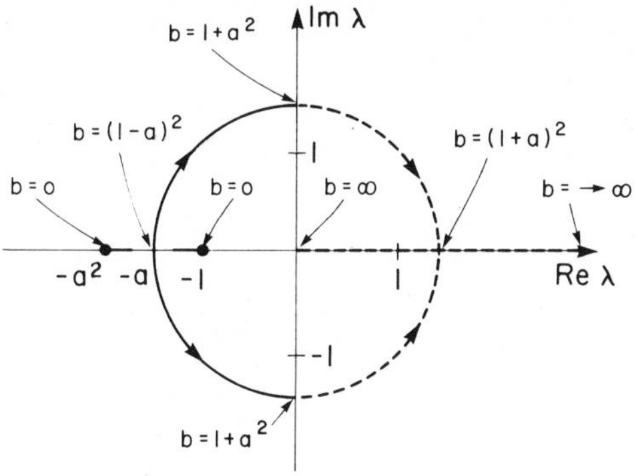

Fig. 8. Complex eigenfrequencies (II.27) of the Brusselator reaction model for $a = \sqrt{2}$ and indicated values of b. As b increases from zero, the two real roots, $-a^2$ and -1, coalesce. At $b = (1 - a)^2$, the eigenfrequencies become complex and their locus is a circle of radius a. At $b = 1 + a^2$ there occurs a hard instability.

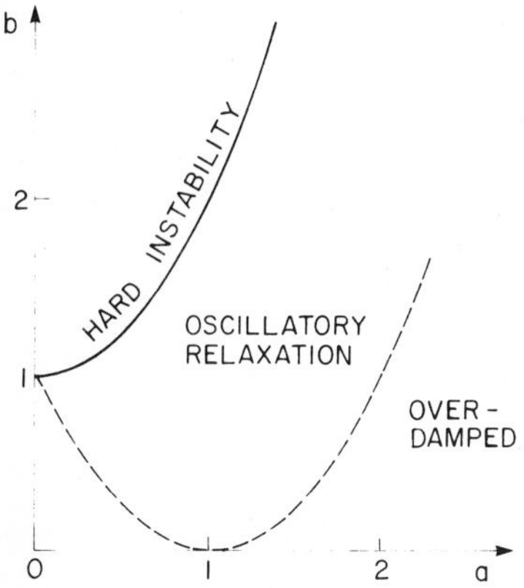

Fig. 9. Phase diagram for the Brusselator reaction model in the parameter space (a, b). Chemical equilibrium occurs at $a = 0$, $b = 0$.

integration is one straightforward way to see that limit cycle oscillations take place; that is, for given a, b there is a well-defined closed trajectory in the $\tilde{x}$, $\tilde{y}$-plane that is traced in a periodic manner, and from every initial point that trajectory will eventually be reached after the transient process. The periodic variation of concentrations breaks time translational invariance. All that is left from this symmetry is the equivalence in time of points that are separated by integral multiples of the limit cycle period. Correspondingly, there is a degeneracy in the phase along the trajectory that is not fixed by the deterministic equations (II.24): any phase is as good as any other; the important point is that one choice has to be made, either deliberately by specified initial conditions or randomly as a result of internal fluctuations.

In the immediate neighborhood of the transition, $|\varepsilon_H| \ll 1$ where

$$\varepsilon_H \equiv \frac{b-(1+a^2)}{1+a^2} \tag{II.28}$$

the nonlinear equations can be treated analytically using the Hopf bifurcation theory.[64] The idea is that the limit cycle grows continuously out of the critical eigenmodes. The task then consists of evaluating the nonlinearities for their role in limiting the amplitude. For the Brusselator, this was first done by Kuramoto and Tsuzuki,[65] see also Refs. 66 and 67. The procedure is as follows.

Write (II.24) in terms of coordinates ψ, ψ^*, which are expected to be slowly varying because the dominant time dependence of the eigensolutions, for $\varepsilon_H = 0$, is explicitly taken care of:

$$\begin{aligned} \tilde{x} &= a + \psi a e^{-ia\tau} + \psi^* a e^{ia\tau} \\ \tilde{y} &= \frac{b}{a} - \psi(a+i)e^{-ia\tau} - \psi^*(a-i)e^{ia\tau} \end{aligned} \tag{II.29}$$

Inserting this into (II.24), we get

$$\begin{aligned} \dot{\psi} = {} & \frac{1+a^2}{2}\varepsilon_H\left(\psi + e^{2ia\tau}\psi^*\right) \\ & + e^{ia\tau}\Big\{\big(1 - a^2 + (1+a^2)\varepsilon_H\big)|\psi|^2 \\ & - \frac{a}{2}\left((3a+i)e^{-ia\tau}|\psi|^2\psi + \text{c.c.}\right) \\ & + \tfrac{1}{2}\Big(\big((1-ia)^2 + (1+a^2)\varepsilon_H\big)e^{-2ia\tau}\psi^2 + \text{c.c.}\Big) \\ & - \frac{a}{2}\left((a+i)e^{-3ia\tau}\psi^3 + \text{c.c.}\right)\Big\} \end{aligned} \tag{II.30}$$

which is still exact. Notice that there are two time scales involved, one relatively fast, which is associated with the oscillations $e^{ia\tau}$, and a slower one where the rates are of order $\varepsilon_H (\ll a)$, as suggested by the first term of (II.30). We now extract the slowly varying part from (II.30), noting a combination of certain exponentials $e^{nia\tau}$ may be slowly varying. Using the ansatz

$$\psi = \varepsilon_H^{1/2}\psi_1 + \varepsilon_H\psi_2 + \varepsilon_H^{3/2}\psi_3 + \cdots \tag{II.31}$$

and analyzing successive orders of ε_H, we find to order $\varepsilon_H^{1/2}$ that $\dot{\psi}_1 = 0$ or

$\psi_1 = \text{const}$ on the fast scale, $\dot{\psi}_1 = O(\varepsilon_H)$. To order ε_H we get

$$\dot{\psi}_2 = e^{ia\tau}\left\{(1-a^2)|\psi_1|^2 + \tfrac{1}{2}\left((1-ia)^2 e^{-2ia\tau}\psi_1^2 + \text{c.c.}\right)\right\} \qquad \text{(II.32)}$$

which is immediately integratable. The order $\varepsilon_H^{3/2}$ gives

$$\begin{aligned}\dot{\psi}_3 + \frac{1}{\varepsilon_H}\dot{\psi}_1 = \frac{1+a^2}{2}&(\psi_1 + e^{2ia\tau}\psi_1^*)\\ &+ e^{ia\tau}\Big\{(1-a^2)(\psi_1\psi_2^* + \text{c.c.})\\ &\qquad - \frac{a}{2}\left((3a+i)e^{-ia\tau}|\psi_1|^2\psi_1 + \text{c.c.}\right)\\ &\qquad + \left((1-ia)^2 e^{-2ia\tau}\psi_1\psi_2 + \text{c.c.}\right)\\ &\qquad - \frac{a}{2}\left((a+i)e^{-3ia\tau}\psi_1^3 + \text{c.c.}\right)\Big\}\end{aligned} \qquad \text{(II.33)}$$

The slowly varying part of this equation determines ψ_1, therefore, to leading order of ε_H, the complex amplitude ψ:

$$\frac{d\psi}{d\tau} = \frac{1+a^2}{2}\varepsilon_H\psi - \tfrac{1}{2}\left(2 + a^2 - \frac{i}{3a}(4 - 7a^2 + 4a^4)\right)|\psi|^2\psi \qquad \text{(II.34)}$$

This equation together with (II.29) describes the hard mode instability as a transition to an ellipse-shaped limit cycle. Below threshold, for $\varepsilon_H < 0$, a perturbation $\psi \neq 0$ decays at a rate $(1+a^2)\varepsilon_H/2$ in agreement with the linear stability analysis (II.27). For $\varepsilon_H > 0$, however, we get a stable amplitude

$$\psi_0 = \left(\frac{1+a^2}{2+a^2}\varepsilon_H\right)^{1/2} \qquad \text{(II.35)}$$

which is consistent with the ansatz (II.31). In addition, there is a frequency shift as the oscillations build up,

$$\psi = \psi_0 e^{-i\nu\tau}, \qquad \nu = -\frac{4 - 7a^2 + 4a^4}{6a}\psi_0^2 \qquad \text{(II.36)}$$

The general interest in (II.34) is that it represents a Landau-type description[68] for the transition under consideration. The complex amplitude ψ plays the role of an order parameter, which in the mean is zero below, and of order $\varepsilon_H^{1/2}$ above the transition, so that an equation

$$\frac{d\psi}{d\tau} = \alpha\varepsilon_H\psi - \beta|\psi|^2\psi \qquad \text{(II.37)}$$

may be postulated a priori. Incidentally, the very same equation, with a complex β, has been used to describe the threshold of a detuned laser.[69] The applicability of Landau's concepts shows that from a phenomenological point of view there is no difference between phase transitions under equilibrium or far-from-equilibrium conditions. This has been advocated by Haken[9] and recently also in Refs. 26 and 27. In subsequent sections we shall see that inclusion of spatial inhomogeneities or internal fluctuations tends to make the analogies even closer.

Lotka-Volterra Model. We now turn to another model that has been widely used in the discussion of oscillatory phenomena and is much older than the Brusselator: the predator-prey system of Lotka and Volterra.[70,71] The reactions are set up in a way that macroscopic oscillations occur for all values of the parameters, which means that the transition from an equilibrium branch to the oscillatory regime has been suppressed and cannot be discussed.[72] It is, however, interesting to analyze the nature of the "ordering".

The "reaction" scheme is

$$\mathrm{A} + \mathrm{X} \xrightarrow{k_1} 2\mathrm{X}, \qquad \mathrm{X} + \mathrm{Y} \xrightarrow{k_2} 2\mathrm{Y}, \qquad \mathrm{Y} \xrightarrow{k_3} \mathrm{E} \tag{II.38}$$

Using $\mathrm{N} = k_3/k_2$ as an extensivity parameter and dimensionless variables defined by

$$t = k_3^{-1}\tau, \qquad \mathrm{X} = N\tilde{x}, \qquad \mathrm{Y} = N\tilde{y}, \qquad \mathrm{A} = \frac{k_3}{k_1} a \tag{II.39}$$

we get the kinetic equations

$$\begin{aligned} \dot{\tilde{x}} &= (a - \tilde{y})\tilde{x} \\ \dot{\tilde{y}} &= (\tilde{x} - 1)\tilde{y} \end{aligned} \tag{II.40}$$

Their nontrivial steady-state solution is

$$\tilde{x}_0 = 1 \qquad \tilde{y}_0 = a \tag{II.41}$$

and the relaxation matrix for small perturbations reads

$$\Lambda = -\begin{pmatrix} 0 & 1 \\ -a & 0 \end{pmatrix} \tag{II.42}$$

which has purely imaginary eigenvalues $\lambda_{1,2} = \pm i\sqrt{a}$: this system has the peculiar property that for all values of the pump parameter a, the steady state is marginally stable. In view of our general statements at the beginning of this section, the spectrum of $\mathbf{\Lambda}$ seems to indicate that the transport matrix L is antisymmetric, or purely reactive. In fact, however, as section III.C.2 demonstrates, L does have a dissipative part D, but the antisymmetric part A diverges, as do the corresponding Γ; the conservative aspect is therefore overwhelmingly dominant. This happens to be true even outside the linear neighborhood of the steady state, since it can be shown that in addition to the phase, the system remembers its initial amplitude: Writing (II.38) in the form

$$\frac{\tilde{x}-1}{\tilde{x}}\,d\tilde{x} = \frac{a-\tilde{y}}{\tilde{y}}\,d\tilde{y} \tag{II.43}$$

we find that

$$\frac{1}{a}(\tilde{x}-1-\ln\tilde{x}) + \frac{\tilde{y}}{a} - 1 - \ln\frac{\tilde{y}}{a} \equiv \frac{1}{2}c^2 \tag{II.44}$$

is a constant of the motion.[12,73] Thus there is an infinity of closed trajectories in the $\tilde{x}$, $\tilde{y}$-plane that can be labeled by c, and with oscillation frequencies that depend on c. An approximate formula derived by Frame[74] is

$$\lambda = \frac{\pm i\sqrt{a}}{I_0\left(c/6\sqrt{1+a}\right)} \tag{II.45}$$

where I_0 is the modified Bessel function. Compared to the Brusselator, the Lotka-Volterra system exhibits more ordering, i.e., more information is preserved during the motion. However the accidental character of the conservation law (II.44) implies that the system is structurally unstable.[75]

Other Models. Helpful as they doubtless are in understanding general features of oscillating nonequilibrium systems, neither the Brusselator nor the Lotka-Volterra model represents known chemical reaction systems. Thus it becomes desirable to think about more realistic models that allow for quantitative comparison of experiments and theoretical predictions. Considerable effort has been made to elucidate the detailed chemistry involved in the Zhabotinsky reaction.[76] This work has led to the formulation of the "Oregonator,"[77] which is believed to contain the essence of the real system while at the same time being mathematically tractable. It has three free components and five reactions that are at most of second order. A comprehensive review of theoretical studies on that model has been given by Tyson.[78] For the peroxidase-oxidase reaction, Olsen and Degn[79] have tentatively proposed a model that compares well to their data. Finally, we mention the glycolytic oscillations in which the allosteric enzyme phosphofructokinase embodies the reactive aspects. An elaborate investigation of those oscillations, reviewed in Ref. 12, has been performed by Goldbeter and Lefever.[80]

D. Spatial Structures

Chemical reactions coupled with transport processes or mechanical modes (sound) may become unstable to inhomogeneous (nonuniform) perturbations and develop macroscopic spatial structures.[1, 2, 9, 12, 14, 61] These structures may be stationary patterns or they may be time dependent; in the latter case we distinguish between chemical waves that are traveling concentration profiles, and oscillatory but nontraveling structures ("standing waves"). A number of reviews have appeared on these topics;[6, 12–14, 81, 82] hence we restrict the presentation to the citation of some representative experiments and to some introductory material on theoretical ideas and methods.

A large variety of chemical wave types have been predicted and analyzed theoretically:[83–95] single fronts of concentration profiles, which turn a given stable stationary state into another such state; pulses, which leave the initial state unaltered; infinite wave trains; planar, circular, spiral, and complex scroll waves; kinematic waves, which involve no mass transport; diffusion-type waves, which attenuate in time; relaxation-oscillation waves, which do not attenuate; threshold excitation waves; waves that originate from a stable stationary reference state, and others that grow out of a homogeneous oscillatory state; and bulk and surface waves. The techniques used in these studies are likewise varied: stability analysis, bifurcation or singular perturbation theory, catastrophe theory, and perturbation expansions based on multiple time and length scales or small wave amplitudes.

1. Experiments

Systems that tend toward temporal oscillations are good candidates for spatial pattern formation because their autocatalytic properties may counteract the homogenizing action of diffusion and thermal conduction and help maintain characteristic spatial gradients. From the analysis of many examples there seems to emerge a rule that this structure formation requires a certain localization of the catalytic activity, and a somewhat more extended inhibitory influence. This may be accomplished, for example, by differences in the diffusion constants for the respective components.[96]

An interesting and readily explainable case of the formation of time-independent macroscopic structures occurs in systems undergoing a phase transition, as in the aging of a colloid. Consider for example the precipitation of PbI_2 from a water solution with the initial condition that the nucleation process has occurred, small colloidal particles have been formed (of size such that the solution is not yet cloudy), and the process of colloidal growth is about to begin. The colloidal particles at this stage are homogeneously distributed, and gel may be added to help avoid convection. No macroscopic gradients exist in the system, and no gradients (of temperature, concentration, etc.) are imposed. Within 10 to 30 hr the colloidal particles have grown and the system looks yellow. In addition, the colloidal particles are found to be arranged in somewhat irregular patterns that do, however, exhibit spatial periodicity of wavelength of the order of 1 mm.[11, 97, 98] The experiment is clearly related to the periodic precipitation process of Liesegang ring formation, in which regularly spaced rings of precipitate are formed, however, in the presence of oppositely directed gradients of the positive and negative ion forming the precipitate.

The driving force for the formation of the macroscopic structure is in the Gibbs free energy difference between the colloid in a finely divided form and a macroscopic crystal of the salt. There is a simple explanation for the existence of an instability with respect to inhomogeneous perturbations.[99] The essential point is that the small colloidal particle is more soluble than a larger particle.[100] Now consider a uniform distribution of colloidal particles and perturb the system (or think of a fluctuation) at one point by addition of PbI_2 molecules. This enhances the supersaturation, or reduces the solubility at that point; hence the colloidal particles grow, with the result that now the solubility is further lowered compared to the neighboring region. A spatial gradient in monomer concentration is set up, and diffusion brings in more particles from the surroundings. Thus there is in effect an autocatalytic mechanism that allows for the development of patterned precipitation. When these ideas are formulated in terms of rate equations, coupled to diffusion, linear stability analysis confirms the existence of this type of instability and also gives a wavelength of the order of that observed.

For examples of spatial pattern formation in biological systems, see Refs. 101 and 102.

Chemical waves have been observed and characterized[82] by measurements of velocity, spacing of wave crests, and the contribution of mass transport (see, e.g., Refs. 103 and 104). Experimental progress is substantially behind theoretical advances. The reader is warned of a pervading lack of agreement, hence confusion, on nomenclature of various types of wave (see the four articles dealing with chemical waves in Ref. 17).

2. *Theory*

Let us turn to some theoretical aspects of the coupling of chemistry to transport processes. We shall abstain from including the full hydrodynamic complexity to which we alluded in the introduction (see, e.g., Ref. 105). Almost all work on spatial structures has been confined to reaction-diffusion equations of the type

$$\frac{\partial X_j}{\partial t} = f_j(X_1, \ldots, X_n; A_1, \ldots, A_m) + D_j \nabla^2 X_j \tag{II.46}$$

It is assumed that each component diffuses independently. To get a feeling for the influence of diffusion on a chemical reaction system, let us start with the Schlögl models (II.12). Taking $D \equiv k_4 d$, we have

$$\frac{\partial \tilde{x}}{\partial \tau} = (q - \tilde{x})\tilde{x}^n + \kappa q - \tilde{x} + d\nabla^2 \tilde{x} \tag{II.47}$$

For homogeneous boundary conditions at infinity, this equation has the same steady states as (II.12). How does diffusion affect their stability? Linearizing (II.47) in $\delta\tilde{x}(r) \equiv \tilde{x}(r) - \tilde{x}_0$ and using the Fourier transformation

$$\delta\tilde{x}(\mathbf{k}) = \int \delta\tilde{x}(\mathbf{r}) e^{-i\mathbf{k}\cdot\mathbf{r}} d^3\mathbf{r} \tag{II.48}$$

we find that the relaxation frequency for a perturbation of wave vector **k** is

$$\lambda(\mathbf{k}) = \lambda - dk^2 \tag{II.49}$$

where λ is the same as before; see, for example, (II.14) or (II.16). Thus the effect of diffusion is to speed up relaxation rates, and on sufficiently small spatial scales, $k \gtrsim \sqrt{|\lambda|/d}$, homogenization becomes the dominant feature of relaxation. There is no tendency thus far toward spatial ordering; on the contrary, the larger dk^2, the more a given steady state $\tilde{x}_0$ will appear to be stable. Therefore as a one-component system is driven away from equilibrium by continuously increased pumping, the first mode that becomes unstable is the homogeneous one ($k = 0$). The same is obviously true for a multicomponent system (II.46) with equal diffusion constants, $D_j \equiv D$. Of course this does not say anything about the system's behavior beyond that first instability, where linear stability analysis would have to be done with respect to the new state of the system. However, as we shall see next, with appropriately chosen unequal diffusion constants, even the first instability may occur at a well-defined, finite wave number.

The Brusselator: Soft Versus Hard Instability. We now return to the Brusselator (II.22), which indeed shows spatial ordering if the reactive component X is less mobile than its antagonist Y (i.e., for $D_Y > D_X$). To see this in detail, we add diffusive terms to (II.24), and with $D_{X,Y} \equiv k_4 d_{X,Y}$ we determine eigenfrequencies for perturbations around the steady state

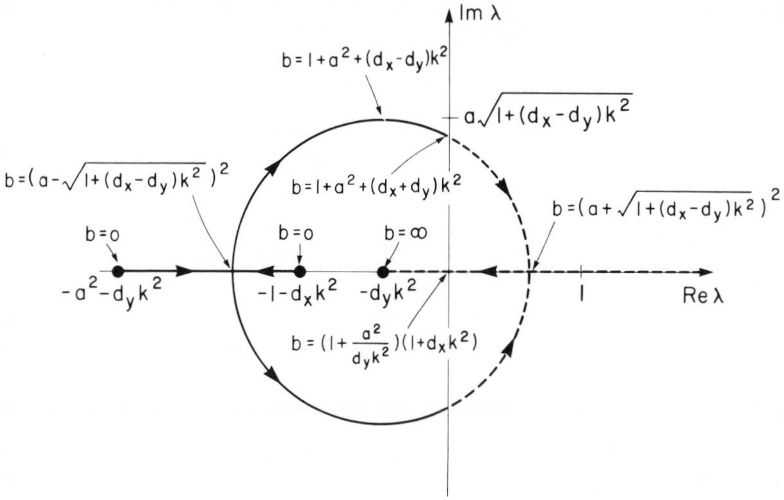

Fig. 10. Complex eigenfrequencies (II.50) of the Brusselator reaction model including diffusion, for $a = \sqrt{2}$ and $d_x k^2 = 1/8$, $d_y k^2 = 1/2$ [see definition preceding (II.50)]. For $k = 0$ this figure reduces to Fig. 8. Note the displacement of the center of the circle to $-d_y k^2$, and the dependence of the radius on the diffusion coefficients d of the species X and Y.

(II.25):

$$\lambda(\mathbf{k}) = -\tfrac{1}{2}\left(1 + a^2 - b + d_X k^2 + d_Y k^2\right)$$
$$\pm \tfrac{1}{2}\sqrt{\left(1 - a^2 - b + d_X k^2 - d_Y k^2\right)^2 - 4a^2 b} \qquad \text{(II.50)}$$

Figure 10 shows their location in the complex plane for fixed a, k^2, and varying b. As before (cf. Fig. 8), we find a circle that this time, however, is centered around $-d_Y k^2$ rather than the origin—a reflection of the increased dissipation. For $d_X = d_Y \equiv d$ the circle is merely shifted by an amount dk^2, in agreement with (II.49). In case $d_X > d_Y$ it is furthermore enlarged, which can be interpreted as an enhancement in elasticity. The hard instability at $b = 1 + a^2 + (d_X + d_Y)k^2$ is delayed as k^2 increases. Therefore homogeneous temporal oscillations set in before any of the wavelike perturbations becomes critical.

The most interesting situation, as stated above, is $d_X < d_Y$, whereupon the circle shrinks and for $a < d_Y k^2/\sqrt{1 + (d_X - d_Y)k^2}$ lies entirely to the left of the imaginary axis. In that case, there is a soft instability at

$$b = \left(1 + \frac{a^2}{d_Y k^2}\right)\left(1 + d_X k^2\right) \qquad \text{(II.51)}$$

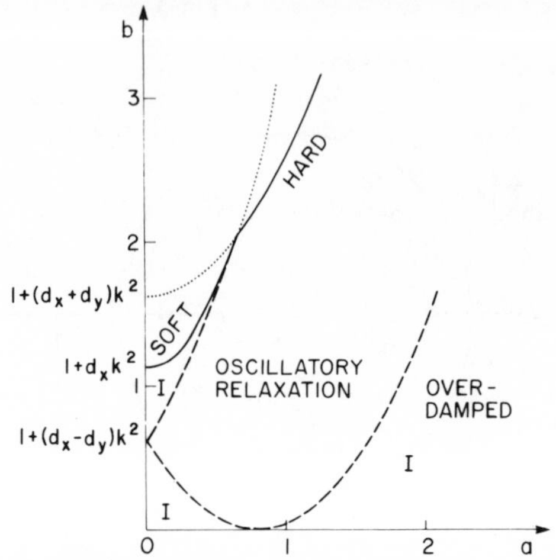

Fig. 11. Phase diagram for the Brusselator reaction model, including diffusion, in the parameter space (a,b) for one value of the wave vector **k**, see Fig. 10. Roman numeral I specifies overdamped relaxation.

and for fixed k^2 the phase diagram analogous to Fig. 9 is shown in Fig. 11. For given a, d_X, d_Y, expression (II.51) has a minimum at $k^2 = k_s^2$,

$$k_s^2 = \frac{a}{\sqrt{d_X d_Y}}, \qquad b = b_s \equiv (1 + \eta a)^2 \tag{II.52}$$

where we used the abbreviation $\eta \equiv \sqrt{d_X / d_Y}$. The important point is that for

$$a > \frac{2\eta}{(1 - \eta^2)} \tag{II.53}$$

this soft transition occurs before the hard one, since $b_s < 1 + a^2$. Thus if we put to test all perturbations of the form $\exp\{\lambda t + i\mathbf{k} \cdot \mathbf{r}\}$ the steady state $\tilde{x}_0 = a$, $\tilde{y} = b/a$ is stable below the line drawn in Fig. 12. For small a, the instability is at $b = 1 + a^2$ and leads to a homogeneous limit cycle ($k = 0$, $\lambda = \pm ia$), whereas for large values of a, a soft transition at $b = b_s$ brings about spatial patterns ($k = k_s$, $\lambda = 0$). Our intuitive picture of this pattern formation is that it exploits the same autocatalytic features that tend to generate limit cycle oscillations; the difference is that here the outbursts of X production are spatially confined and frozen in. It is plausible that this can only happen for $d_Y > d_X$, which means that the antagonist Y would

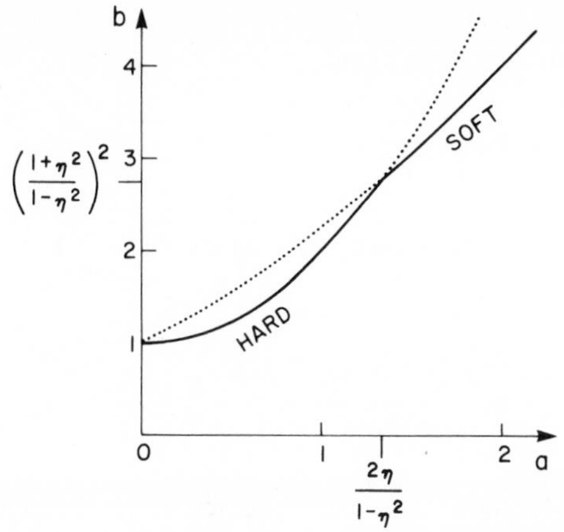

Fig. 12. Phase diagram for the Brusselator reaction model, including diffusion, in the parameter space (a,b) and the value of k that leads to the first instability at each a and increasing b. The symbol η denotes $(D_X/D_Y)^{1/2}$.

leave a center of activity faster than X and would surround it with a wall of inhibitory influence on further growth. This mechanism is very much like the principles underlying primary pattern formation in morphogenesis,[102] where it is essential that the inhibitor have a longer spatial range than the activator.

The immediate neighborhood of the transition line in Fig. 12 can again be described in a Landau framework, using the technique of reductive perturbation.[65–67] For the hard instability, inclusion of diffusion modifies (II.34)

$$\frac{\partial\psi}{\partial\tau} = \alpha\varepsilon_H\psi - \beta|\psi|^2\psi + \frac{1}{2}\left(d_X + d_Y - ia(d_Y - d_X)\right)\nabla^2\psi \qquad \text{(II.54)}$$

with α, β as before. The soft instability with respect to a mode $\sim\exp\{i\mathbf{k}_s\cdot\mathbf{r}\}$ is given by

$$\tilde{x} = a + a(\psi e^{i\mathbf{k}_s\cdot\mathbf{r}} + \text{c.c.})$$

$$\tilde{y} = \frac{b}{a} - \frac{b\eta}{1 + a\eta}(\psi e^{i\mathbf{k}_s\cdot\mathbf{r}} + \text{c.c.}) \qquad \text{(II.55)}$$

and a Landau equation for the amplitude ψ,

$$\frac{\partial\psi}{\partial\tau} = \alpha\varepsilon_S\psi - \beta\psi^3 + \frac{\delta}{k_s^2}\left(\mathbf{k}_s\cdot\nabla - \frac{i}{2}\nabla_\perp^2\right)^2\psi \qquad \text{(II.56)}$$

where $\varepsilon_S = (b - b_s)/b_s$ is the relative distance to the instability, $\nabla_\perp$ is the part of ∇ perpendicular to $\mathbf{k}_s$, and

$$\alpha = \frac{1 + a\eta}{1 - \eta^2},$$

$$\beta = \frac{(2 + a\eta)(-4 + 21a\eta - 8(a\eta)^2)}{9a\eta(1 - \eta^2)},$$

$$\delta = \frac{4dx}{(1 + a\eta)(1 - \eta^2)} \tag{II.57}$$

Note that the instabilities displayed in Fig. 12 first lead either to space-independent temporal structures or to time-independent spatial structures. Wave propagation appears beyond the hard instability only where (II.54) predicts the existence of stable modes of the form $\psi = \psi_0(k) \exp\{i\mathbf{k}\cdot\mathbf{r} - i\nu\tau\}$ even for $d_X \geqslant d_Y$:

$$\psi_0^2(k) = \left(\alpha\varepsilon_H - \frac{d_X + d_Y}{2} k^2\right)/\text{Re}\ \beta,$$

$$\nu = \psi_0^2(k)\text{Im}\ \beta - \frac{a}{2}(d_Y - d_X)k^2 \tag{II.58}$$

For given $\varepsilon_H > 0$, there is a band of stable waves $0 < k^2 < 2\alpha\varepsilon_H/(d_X + d_Y)$ with amplitudes A decreasing as k grows, and with a frequency shift ν that in comparison to (II.36) has an additional contribution $\propto k^2$. Some caution is in order because at $\varepsilon_H > 0$ stability should really be checked with reference to the $k = 0$ limit cycle. But more general and detailed studies on wave formation confirm the results of our discussion.[83, 84] When the system is driven sufficiently strongly, the coupling among waves of different k that is mediated by the nonlinearities may even generate chaotic behavior.[85–87] Beyond the soft instability, we get a band of stable standing waves $\psi = \psi_0(\mathbf{k})\exp\{i(\mathbf{k}_\parallel + \mathbf{k}_\perp)\cdot\mathbf{r}\}$ with

$$\psi_0^2(\mathbf{k}) = \frac{\left(\alpha\varepsilon_S - \delta\left(k_\parallel + \frac{1}{2}k_\perp^2/k_s\right)^2\right)}{\beta} \tag{II.59}$$

The bandwidth is $\propto \varepsilon_S^{1/2}$ for $\mathbf{k}$ parallel to $\mathbf{k}_s$ and $\propto \varepsilon_S^{1/4}$ in the perpendicular direction. There are no temporal oscillations, consequently no propagating waves, in the immediate neighborhood of the soft transition.

Front Propagation. So far we have discussed diffusion from the point of view of its role in determining the nature of the system's stable state. Another interesting effect occurs when a system has the choice between two or more stable stationary states, for a given set of parameters. When two such systems, under identical conditions but in different states, are brought into contact, one phase usually grows and annihilates the other—unless the conditions happen to allow for coexistence. Thus diffusion represents a mechanism whereby one may test for the relative stability of metastable states.[50] Let us illustrate this by using Schlögl's model (II.47) again, with $n = 2$. Here (and only here) we assume nonhomogeneous boundary conditions. We restrict ourselves to a one-dimensional problem and take r as the spatial coordinate. Writing (II.47) as

$$\frac{\partial \tilde{x}}{\partial \tau} = d\frac{\partial^2 \tilde{x}}{\partial r^2} - (\tilde{x} - \tilde{x}_1)(\tilde{x} - \tilde{x}_2)(\tilde{x} - \tilde{x}_3) \qquad \text{(II.60)}$$

where $\tilde{x}_1 < \tilde{x}_2 < \tilde{x}_3$ are the three steady states ($\kappa < \frac{1}{9}$, q in the appropriate range) we require that there is pure phase 1 to the right and phase 3 to the left,

$$\tilde{x} \to \tilde{x}_{1,3} \qquad \text{as} \qquad r \to \pm\infty \qquad \text{(II.61)}$$

The solution of (II.60) that will be approached from any initial situation is a soliton that sweeps out whichever phase is closer to the unstable state $\tilde{x}_2$. In detail,[88,89] we get

$$\tilde{x}(r, \tau) \to \tilde{x}_1 + \frac{\tilde{x}_3 - \tilde{x}_1}{1 + e^{(r - v\tau)/\xi}} \qquad \text{(II.62)}$$

where $\xi = \sqrt{2d}\,/(\tilde{x}_3 - \tilde{x}_1)$, $v = \sqrt{2d}\,[\tilde{x}_2 - (\tilde{x}_1 + \tilde{x}_3/2]$. Coexistence occurs for $v = 0$, in which case the unstable state $\tilde{x}_2$ is midway between $\tilde{x}_1$ and $\tilde{x}_3$. [For $\kappa = 0$ where $\tilde{x}_1 = 0, \tilde{x}_{2,3} = \frac{1}{2}(q \pm \sqrt{q^2 - 4}\,)$, we find coexistence for $q = 3/\sqrt{2}\,$, $x_3 = \sqrt{2}\,$, see Fig. 5*b*.] This condition can be formulated in analogy to a Maxwell construction if (II.60) is written as $\dot{\tilde{x}} = d\tilde{x}_{rr} - \partial F(\tilde{x})/\partial \tilde{x}$: the coexistence line is then determined by $F(\tilde{x}_1) = F(\tilde{x}_3)$. This has led Schlögl[50] to consider $F(\tilde{x})$ as a generalization of the thermodynamic potential. However we shall see in Section III that this $F(\tilde{x})$ does not fulfill one essential role of thermodynamic potentials, which is to measure the strength of fluctuations. Thus it can be used only as a kinetic potential. In systems with more than one component, further complications may arise: (1) the rate equations cannot generally be derived from a kinetic potential, and even more intriguingly, (2) the form and direction of the soliton solution may also depend on the initial form of the boundary layer between the two phases.[90, 91]

III. FLUCTUATIONS

A. General Comments

The macroscopic kinetic equations represent a first level in our understanding of a given system, and in fact, for most practical purposes there is no need to go into further detail. However as our discussion of the possible existence of oscillations indicated (Section II.C), there are good reasons also to study the behavior of spontaneous fluctuations. First, thermodynamics has always intimately been concerned with those fluctuations in the sense that derivatives such as specific heats and compressibilities can be interpreted in terms of fluctuations of entropy, volume, and so on, respectively.[106] Second, statistical physics tells us that fluctuations are of major importance in the neighborhood of a phase transition;[29] a comprehensive analysis of chemical instabilities cannot ignore them. Third, there has been much concern about small systems[107] where fluctuations are expected to be relatively large; this is not a very relevant point in macroscopic chemical reaction systems, but it gains in significance as we approach biological applications where typical particle numbers (say of enzyme molecules in a cell) are indeed far from the thermodynamic limit. For discussions of the effects of externally imposed fluctuations (noise) see Refs. 12 and 44.

Fluctuations around thermodynamic equilibrium states are amenable to standard methods that need not be reviewed here. Canonical statistics give their static correlations in terms of thermodynamic derivatives (via the fluctuation-dissipation theorem), and together with the kinetic equation these determine the correlations in time.[108] Things are more involved in nonequilibrium chemical systems for which canonical statistics does not provide a recipe to calculate the static correlations.

The usual treatment invokes a stochastic description that explicitly accounts not only for the microstates but also for the elementary processes that characterize the system (i.e., for the chemical reactions). The standard approach is based on Pauli's master equation,[109] which expresses the rate of change of the probability $P(i,t)$ of finding a system in state i at time t, in terms of transition rates W_{ij}:

$$\dot{P}(i,t) = \sum_j \left(W_{ij} P(j,t) - W_{ji} P(i,t) \right) \qquad \text{(III.1)}$$

The physics is contained in the choice of states i and processes W_{ij}. For purely chemical systems it is customary to specify a state by means of the particle numbers, $i \equiv (X_i, \ldots, X_n)$, and to assume mass action kinetics for the elementary processes.[110] For the general reaction scheme (II.4) we write

$$W_{\mathbf{XX}'} = \sum_{\lambda=1}^{l} \left(W_{\mathbf{XX}'}^{\lambda} + \overline{W}_{\mathbf{XX}'}^{\lambda} \right) \tag{III.2}$$

where the forward and backward transition rates are

$$W_{\mathbf{XX}'}^{\lambda} = R_{\lambda} \prod_{j=1}^{n} \delta(X_j, X_j' + \bar{\nu}_{\lambda j} - \nu_{\lambda j})$$

$$\overline{W}_{\mathbf{XX}'}^{\lambda} = \overline{R}_{\lambda} \prod_{j=1}^{n} \delta(X_j, X_j' - \bar{\nu}_{\lambda j} + \nu_{\lambda j}) \tag{III.3}$$

Equations III.1 to III.3 completely determine the stochastic behavior of the system (II.4) both in the stationary case, when $P(X, t)$ is constant in time, and during the transient phase, when an initially given distribution $P(X, 0)$ approaches the stationary solution. There are a few interesting cases where the master equation can be solved rigorously, especially for the steady state. The Schlögl models (II.9) belong to that class, as we shall see in the next section (for a general discussion, see, e.g., Ref. 111). As a rule, however, exact solutions are not available. Thus it is fortunate that several techniques have been developed recently to construct approximate solutions.[112–114] Yet in practice there is no need for the complete probability distribution; the interesting quantities are the correlations $\langle \delta X_i(t) \delta X_j(0) \rangle$ around steady states, and standard methods are available for their computation. There are two steps involved: first, one determines the equal time (static) correlations $\langle \delta X_i(t) \delta X_j(t) \rangle$; then the dynamic correlations are derived using the equations of motion.

For obtaining the static correlations, it is extremely useful to start out with the so called $1/N$-expansion, where N is an appropriate extensivity parameter of the system. Van Kampen[115, 116] and others[117, 118] pointed out that for macroscopic systems the transition rate $W_{\mathbf{XX}'}$ should be an extensive quantity $W_{\mathbf{XX}'} \propto N$. Since N is proportional to the system's volume V, we may specifically write

$$W_{\mathbf{XX}'} = V w_{\mathbf{xx}'} \tag{III.4}$$

where $x_j = X_j / V$ are the particle densities. If we formally expand the master equation (III.1) up to order $1/V$, we get the Fokker-Planck equation

$$\frac{\partial P(\mathbf{x}, t)}{\partial t} = - \frac{\partial}{\partial x_i} (j_i(\mathbf{x}) P(\mathbf{x}, t)) + \frac{1}{V} \frac{\partial}{\partial x_i} \frac{\partial}{\partial x_k} (d_{ik}(\mathbf{x}) P(\mathbf{x}, t)) \tag{III.5}$$

where the drift term is the first moment of the transition rates,

$$j_i(\mathbf{x}) = \sum_{\mathbf{X}'} (X_i' - X_i) w_{\mathbf{x}'\mathbf{x}} = \sum_{\lambda} (\bar{\nu}_{\lambda i} - \nu_{\lambda i}) \frac{R_\lambda - \bar{R}_\lambda}{V} \tag{III.6}$$

and the "diffusion" matrix is the second moment

$$\begin{aligned} d_{ik}(\mathbf{x}) &= \frac{1}{2} \sum_{\mathbf{X}'} (X_i' - X_i)(X_k' - X_k) w_{\mathbf{x}'\mathbf{x}} \\ &= \frac{1}{2} \sum_{\lambda} (\bar{\nu}_{\lambda i} - \nu_{\lambda i})(\bar{\nu}_{\lambda k} - \nu_{\lambda k}) \frac{R_\lambda + \bar{R}_\lambda}{V} \end{aligned} \tag{III.7}$$

From (III.5) it is straightforward to derive a stationary distribution, for one-component systems, and to leading order in V; we have

$$P_0(x) \propto \exp\left\{ V \int \frac{j(x)}{d(x)}\, dx \right\} \tag{III.8}$$

Some caution is, however, appropriate when using (III.8) or (III.5). As van Kampen strongly emphasizes, only the Gaussian part around the maximum of (III.8) or the linear part of $j(x)$ around its zeros, is consistently given by (III.8). The flanks of the distribution are influenced by higher moments of $w_{\mathbf{xx}'}$ to the same order as they are by the nonlinearities of $j(x)$.[115–117] This point has been carefully analyzed by Blomberg,[119] who shows that some relief may nonetheless be drawn from the fact that the higher moments are usually fairly small. The fully nonlinear Fokker-Planck equation (III.5) should therefore be a reasonable approximation.

To zeroth order in $1/V$, (III.5) leads to the kinetic equation

$$\dot{x}_k(t) = j_k(\mathbf{x}) \tag{III.9}$$

which is identical to (II.3). To first order, we linearize around this secular motion[117] and obtain an equation for the variance $\sigma_{ik}(t) \equiv \langle \delta x_i(t) \delta x_k(t) \rangle$:

$$\dot{\sigma}_{ik}(t) = \frac{2}{V} d_{ik}(\mathbf{x}) + \frac{\partial j_i}{\partial x_l} \sigma_{lk} + \sigma_{il} \frac{\partial j_k}{\partial x_l} \tag{III.10}$$

In case the system is in a stable stationary state, $\mathbf{j}(\mathbf{x}_0) = 0$, the static correlations are given by

$$\boldsymbol{\Lambda\sigma} + \boldsymbol{\sigma\Lambda}^T = -\frac{2}{V} \mathsf{d}(\mathbf{x}_0) \tag{III.11}$$

where $\boldsymbol{\Lambda} \equiv \partial \mathbf{j}/\partial \mathbf{x}$ is the relaxation matrix. Equation III.11 is a set of

$n(n+1)/2$ linear equations, for an n-component system, due to the symmetry of $\boldsymbol{\sigma}$ and $\mathbf{d}$.

Equations III.10 and III.11 are easily generalized to include the effects of diffusion. From the work of Gardiner[120] and Grossman[121] it is clear that in addition to the local chemical fluctuating forces d_{ik} there is a contribution associated with the random walk of the particles,

$$d_{ik}(\mathbf{r},\mathbf{r}') = (d_{ik} + D_i\delta_{ik}\nabla_{\mathbf{r}}\cdot\nabla_{\mathbf{r}'}x_i(\mathbf{r}))\delta(\mathbf{r},\mathbf{r}') \qquad \text{(III.12)}$$

The equation of motion for the correlation function $\sigma_{ik}(\mathbf{r}t,\mathbf{r}'t) \equiv \langle \delta x_i(\mathbf{r},t)\, \delta x_k(\mathbf{r}',t)\rangle$ then reads

$$\dot{\sigma}_{ik}(\mathbf{r}t,\mathbf{r}'t) = 2d_{ik}(\mathbf{r},\mathbf{r}') + \Lambda_{il}(\mathbf{r})\sigma_{lj}(\mathbf{r}t,\mathbf{r}'t) + \Lambda_{kl}(\mathbf{r}')\sigma_{il}(\mathbf{r}t,\mathbf{r}'t) \qquad \text{(III.13)}$$

where $\boldsymbol{\Lambda}(\mathbf{r})$ contains diffusion in the usual way,

$$\Lambda_{ik}(\mathbf{r}) = \Lambda_{ik} + D_i\delta_{ik}\nabla_{\mathbf{r}}^2 \qquad \text{(III.14)}$$

The procedure above is expected to yield good results whenever V is sufficiently large and the system is not too near an instability. We shall not discuss small V where fluctuations may sometimes dominate over the mean values;[122] this may be of great interest in population dynamics but not for usual chemical reaction systems. Some pertinent comments on the neighborhood of phase transitions are made in connection with specific examples.

Equivalent treatments of static correlations have appeared under various names. Lax[123] called his the quasilinear approximation.[124] With d_{ik} interpreted as the strength of fluctuating forces, a Langevin formulation can be set up. Keizer arrives at (III.10) on the basis of a fluctuation-dissipation postulate.[125] When Görtz applies his method to the calculation of $\sigma_{ik}(t)$ he gets the same results. When the equations are generalized, as in (III.13), to account for the presence of transport processes such as diffusion, they lead to Ornstein-Zernike type of correlations in space, as discussed later. And needless to say, the relations above definitely hold for equilibrium systems, where they merely represent a form of the fluctuation-dissipation theorem. They also hold in systems that are locally in equilibrium in the sense that the fluctuating forces are locally the same as in true equilibrium; that is, they are connected to the local thermodynamic derivatives via the fluctuation-dissipation theorem. This property is usually assumed to hold in hydrodynamic nonequilibrium situations where the pumping is represented by an externally imposed gradient (of temperature, concentration, electric potential, etc.).[2, 121, 126, 127] The correlation functions are then obtained by integration over the local fluctuating forces, using the response function associated with the kinetic equations.

With a knowledge of the static correlations at hand, one may proceed to calculating the dynamic correlations

$$g_{ik}(t,t') \equiv \langle \delta x_i(t)\delta x_k(t')\rangle \tag{III.15}$$

Within the quasilinear regime it is consistent to insert the solution of the linearized (III.9), and to average over the initial conditions only.[123] For a steady state (which is homogeneous in time) this leads to

$$g_{ik}(t-t') = \begin{cases} e^{-\Lambda(t-t')}\boldsymbol{\sigma} & (t \geqslant t') \\ \boldsymbol{\sigma} e^{\Lambda^T(t-t')} & (t \leqslant t') \end{cases} \tag{III.16}$$

In other words, the spectrum of the dynamical correlation function is given by the relaxation frequencies of macroscopic perturbations. This corresponds to the regression hypothesis, which holds in the context of linear response theory. A more systematic treatment, which is based on the Zwanzig-Mori projection operator technique, has recently been introduced by Grossmann and Schranner.[128] Their method can be applied very easily to the original master equation, and it even allows one to treat small V and the vicinity of a phase transition. It relies, however, on the availability of higher order static correlations.

B. Homogeneous Stationary States

The purely chemical part of the fluctuations in the Schlögl models (II.9) is well understood because it is possible to obtain the steady-state probability distributions exactly.[129, 130] With the notation of (II.9) to (II.13), the master equation is defined by the transition probabilities

$$\frac{V}{N} w_{xx'} = \begin{cases} q\left[x'\left(x'-\frac{1}{V}\right)\cdots\left(x'-\frac{n-1}{V}\right)+\kappa\right] & x = x'+\frac{1}{V} \\ x'\left[\left(x'-\frac{1}{V}\right)\cdots\left(x'-\frac{n}{V}\right)+1\right] & x = x'-\frac{1}{V} \end{cases} \tag{III.17}$$

A standard method of solving it is to introduce a generating function $f(z)$,

$$f(z) = \sum_{X=0}^{\infty} P_0(X) z^X \tag{III.18}$$

and to convert the master equation into an ordinary differential equation for $f(z)$. For the two cases $n = 1, 2$, the solutions can be written as

$f(z) = g(z)/g(1)$, where[128, 131, 132]

$$g(z) = \begin{cases} {}_1F_1(\kappa N, N; qNz), & (n=1) \\ {}_2F_2\Big(-\tfrac{1}{2} + \tfrac{1}{2}\sqrt{1-4\kappa N^2}\,, -\tfrac{1}{2} - \tfrac{1}{2}\sqrt{1-4\kappa N^2}\,; & \\ \qquad -\tfrac{1}{2} + \tfrac{1}{2}\sqrt{1-4N^2}\,, -\tfrac{1}{2} - \tfrac{1}{2}\sqrt{1-4N^2}\,; qNz\Big), & (n=2) \end{cases} \tag{III.19}$$

The distribution $P_0(X)$ is then given by the well-known Taylor expansion of the hypergeometric functions ${}_1F_1$ and ${}_2F_2$, respectively. Its moments can be expressed directly in terms of derivatives of $f(z)$, for example,

$$\langle X\rangle = f'(1), \langle(\delta X)^2\rangle = f''(1) + f'(1)(1 - f'(1)) \tag{III.20}$$

The two cases $n = 1, 2$ are very different, as we know already from Section II.B.2. For $n = 1$, $P_0(X)$ is a unimodal distribution with its peak near the mean values as shown in Fig. 5*a* (this is true for sufficiently large N). The width of that distribution is given in Fig. 13, for $N = 1000$ and different κ. The critical enhancement near the transition point, as $\kappa \to 0$, is well borne out. For $n = 2$, the steady-state distribution $P_0(X)$ is bimodal in the range where Fig. 5*b* shows bistability. This means that a given system will be in either of the two stable states with a certain probability. The mean value $\langle X\rangle$ computed from (III.20) must be interpreted as the average over an ensemble of such systems; its dependence on q for $\kappa = \frac{1}{25}$ is given in Fig. 14. For $N \to \infty$ a jump develops at the point where the two peaks have equal height.[130] This is not the same point at which the soliton velocity discussed in Section II.D.2.b vanishes.[133] Whether this jump has any meaning for a single given system depends on the time it takes to cross over from one stable state to the other, the so-called mean first passage

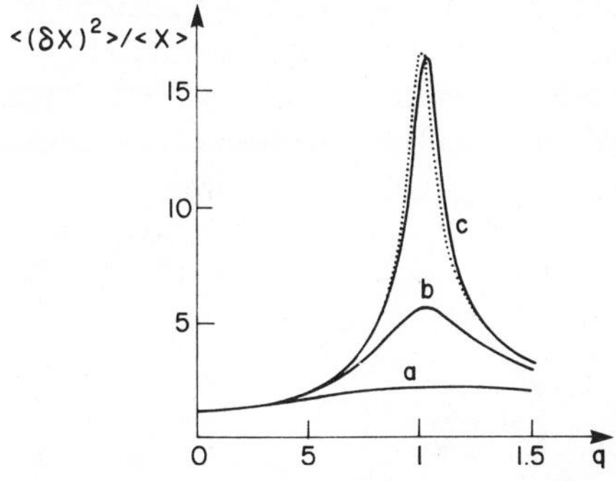

Fig. 13. Mean square fluctuation of the concentration of the species X in the Schlögl model, (II.9), $n = 1$, as a function of the pump parameter q, (II.13), and for three values of the parameter κ (II.10): a, $\kappa = 10^{-1}$; b, $\kappa = 10^{-2}$; c, $\kappa = 10^{-3}$. The solid lines are an exact evaluation of (III. 19), for $N = 1000$, and the dotted line represents the quasilinear approximation (III. 22), which does not depend on N. Redrawn from Grossmann and Schranner[128] and Jähnig and Richter.[131]

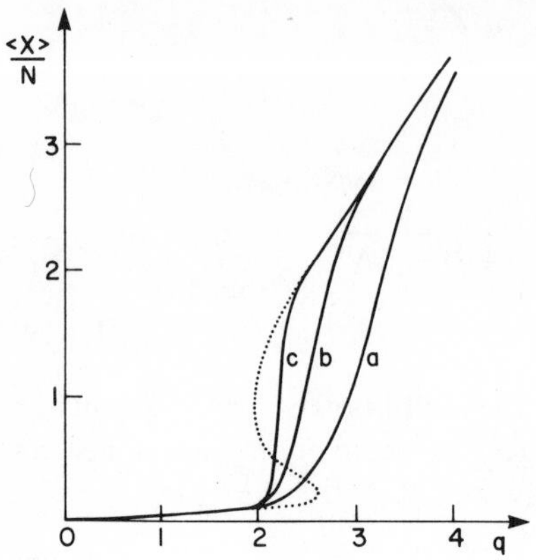

Fig. 14. Mean value of the concentration of species X in the Schlögl model, (II.9), $n = 2$, as a function of the pump parameter q (II.13) for $\kappa = \frac{1}{25}$ (II.10) and the values of the extensivity parameter N: a, $N = 4.5$; b, $N = 11.2$; c, $N = 44.7$. The dotted line is from Fig. 5b and corresponds to the mean-field result in the thermodynamic limit, $N \to \infty$. Redrawn from Grossmann and Schranner.[128]

time.[134, 135] As has been shown by several authors,[130, 136] this time diverges exponentially with N. This means that the steady state becomes degenerate in the thermodynamic limit, as $\kappa < \frac{1}{9}$. Any particular system has then to make a choice among the two possibilities, thereby inevitably breaking a symmetry. Thus it cannot be considered as representative for an ensemble defined by $P_0(X)$—just as a single magnet, below the Curie temperature, is not adequately described by a partition function that gives equal weight to all possible directions. It is of course another matter whether transport processes such as diffusion can make use of internal fluctuations to restore ergodicity. A large amount of work is being devoted to this question, which is intimately related to the nucleation problem.[88, 99, 137–140] Our model, however, assumes homogeneity throughout the system and therefore cannot possibly deal with that matter.

Similar remarks of caution have to be made with respect to the variance $\langle(\delta X)^2\rangle$ as derived via (III.20). For the ensemble we obviously get a peak $\langle(\delta X)^2\rangle \propto (X_3 - X_1)^2 \propto N^2$ at the "coexistence point", X_1 and X_3 being the two stable steady states,[141] see Fig. 15. This does not imply, however, that a given macroscopic system ever exhibits fluctuations of such size. The crucial point is again that the time scale on which this happens is virtually infinite, very much in analogy to the recurrence time of Poincaré cycles.

Let us now briefly discuss the quasilinear approximation, for comparison, and let us include diffusion as outlined in (III.12) to (III.14). In a homogeneous steady state we have $\sigma(\mathbf{r}, \mathbf{r}') = \sigma(\mathbf{r} - \mathbf{r}')$, which suggests the introduction of the Fourier transform

$$\sigma(\mathbf{k}) = \int \sigma(\mathbf{r}) e^{-i\mathbf{k}\cdot\mathbf{r}} d^3\mathbf{r} \qquad \text{(III.21)}$$

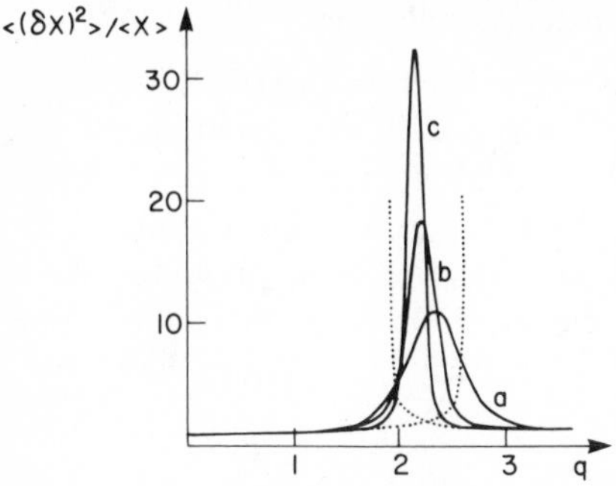

Fig. 15. Mean square fluctuation of the concentration of the species X in the Schlögl model (II.9), $n = 2$, as a function of the pump parameter q, and for the parameters as listed in the legend of Fig. 14. In the thermodynamic limit the central peak narrows to zero and grows to infinity. The dotted lines represent the quasilinear approximation (II. 22) for the fluctuations around each stationary state. These approach infinity near marginal stability. (Redrawn from Grossmann and Schranner.[128]

With the relaxation frequency given by (II.12) and (II.49), and using (III.7) and (III.12) to determine the strength of the fluctuating forces, we get the variance

$$\sigma(\mathbf{k}) = x_0 + n\,\frac{x_0 - \kappa q}{|\lambda| + dk^2} \qquad \text{(III.22)}$$

For the equilibrium case, $\kappa = 1$, $x_0 = q$, the fluctuations are Poissonian as they should be, $\sigma(\mathbf{k}) = x_0$, $\sigma(\mathbf{r}) = x_0\delta(\mathbf{r})$. The same is true for fluctuations on a sufficiently small spatial scale for diffusion to dominate, $dk^2 \gg |\lambda|$. In case of the second-order transition ($n = 1$) we find critical enhancement of the fluctuations for $k = 0$ and $\kappa \to 0$; at the transition point $q = 1$ we have $\sigma/x_0 = (1 + \sqrt{\kappa}\,)/2N\sqrt{\kappa}$, which diverges as $\kappa^{-1/2}$. This has been discussed in Ref. 142, where real space rather than wave number space is used. Fourier transformation of (III.22) shows that $\xi = \sqrt{d/|\lambda|}$ is the spatial coherence length of the fluctuations. In the limiting case $\kappa = 0$, where the critical slowing down is $|\lambda| = |q - 1|$, the correlation length diverges according to the classical law $\xi \propto |q - 1|^{-1/2}$.

For the first-order transition ($n = 2$), the result (III.22) describes the fluctuations around a given stable or metastable steady state x_0, without taking into account possible transitions between the branches. Thus a peak $\sim N^2$ does not appear. Instead there are divergencies near the marginal stability points $q = q_M(\lambda = 0)$. For $\kappa < \frac{1}{9}$ we find $N\sigma/x_0 \propto |q_M - q|^{-1/2}$, and on the critical curve $\kappa = \frac{1}{9}$ the behavior near $q = \sqrt{3}$ is $N\sigma/x_0 \approx (\frac{3}{2}|q - \sqrt{3}\,|)^{-2/3}$. In terms of correlation lengths this means $\xi \propto |q_M - q|^{-1/4}$ near a point of marginal stability, and $\xi \propto |q - \sqrt{3}\,|^{-1/3}$ along the critical curve.

C. Temporal and Spatial Oscillations

Correlations of chemical fluctuations in the Brusselator (II.22) have been calculated by various authors,[66, 143–145] using different techniques. The results obtained so far are easily rederived along the lines described in Section III.A,[146] both in the steady-state regime and in the vicinity of the phase transitions. No analytical treatment has yet been presented, however, for the correlations in a well-developed limit cycle, or in the regime where wave propagation takes place; numerical simulations have been made.[85–87, 147, 148]

1. *Brusselator Model*

The main feature of the hard mode transition can be summarized as follows. Below the instability, $b < 1 + a^2$, the correlations of the fluctuations $\delta x, \delta y$ around the steady state (II.25) are of order N^{-1} provided $|\varepsilon_H| \gtrsim N^{-1/2}$. Above the transition they become quantities of order N^0, which reflects the establishment of a macroscopic limit cycle. The fluctuations of the cycle's amplitude are again of order N^{-1}, and they decrease as the amplitude builds up. There is no restoring force acting on the phase of the cycle, which therefore exhibits diffusive behavior.[92, 94] This phase diffusion can be identified as the Goldstone mode[108] associated with the symmetry breaking instability: it has a tendency to restore homogeneity in time by gradually dephasing the clocks of an initially synchronous ensemble.[149] In a soft transition the behavior of the fluctuations is qualitatively the same.

In detail, let us first discuss the correlations below the transition. We take the relaxation matrix $\mathbf{\Lambda}$ from (II.26) and add diffusion:

$$\mathbf{\Lambda}(\mathbf{k}) = -\begin{pmatrix} 1 - b + d_X k^2 & -a^2 \\ b & a^2 + d_Y k^2 \end{pmatrix} \tag{III.23}$$

The fluctuating forces are derived from (III.7) and the appropriate procedure to include diffusion[121, 141]

$$\mathsf{d}(\mathbf{k}) = \frac{N}{V} a \begin{bmatrix} 1 + b + d_X k^2 & -b \\ -b & b + \dfrac{b}{a^2} d_Y k^2 \end{bmatrix} \tag{III.24}$$

It is then straightforward to determine the static correlations from (III.11). They can be written as a sum of three contributions,

$$\boldsymbol{\sigma}(\mathbf{k}) = \boldsymbol{\sigma}^0 + \boldsymbol{\sigma}^H(\mathbf{k}) + \boldsymbol{\sigma}^S(\mathbf{k}) \tag{III.25}$$

where $\boldsymbol{\sigma}^0$ is the Poissonian part characteristic of equilibrium

$$\boldsymbol{\sigma}^0 = \begin{pmatrix} x_0 & 0 \\ 0 & y_0 \end{pmatrix} \tag{III.26}$$

and $\boldsymbol{\sigma}^H$, $\boldsymbol{\sigma}^S$ diverge at the hard and soft instabilities, respectively. The matrix part of both $\boldsymbol{\sigma}^H, \boldsymbol{\sigma}^S$ is

$$\mathsf{M} = \frac{1}{a^2}\begin{bmatrix} \dfrac{\left[a^2(1 - k^2/k_s^2)^2 + (a^2 + d_Y k^2 + b_s - b)d_Y k^2\right]}{(1 + d_X k^2)} & -(a^2 + d_Y k^2) \\ -(a^2 + d_Y k^2) & b \end{bmatrix} \tag{III.27}$$

which reduces to

$$\mathsf{M}_H = \begin{bmatrix} 1 & -1 \\ -1 & \dfrac{b_H}{a^2} \end{bmatrix}, \qquad \det \mathsf{M}_H = \frac{1}{a^2} \tag{III.28}$$

at the hard instability $b = b_H \equiv 1 + a^2$, $k^2 = 0$, and to

$$\mathsf{M}_S = \begin{bmatrix} \dfrac{1}{\eta^2} & \dfrac{-(1 + \eta a)}{\eta a} \\ \dfrac{-(1 + \eta a)}{\eta a} & \dfrac{b_S}{a^2} \end{bmatrix}, \qquad \det \mathsf{M}_S = 0 \tag{III.29}$$

at the soft instability, $b = b_S \equiv (1 + \eta a)^2$, $k^2 = k_s^2 \equiv a/\sqrt{d_X d_Y}$. Introducing two lengths ξ_H, ξ_S by means of the definitions

$$\xi_H = \sqrt{\frac{d_X + d_Y}{|b - b_H|}} \qquad \xi_S = \sqrt{\frac{4d_X}{|b - b_S|}} \tag{III.30}$$

we can write the results for the correlations as

$$\boldsymbol{\sigma}^H(\mathbf{k}) = \frac{N}{V}\,\frac{2a}{|\varepsilon_H|}\,\frac{1}{1 + \xi_H^2 k^2}\, g_H \mathsf{M} \tag{III.31}$$

$$\boldsymbol{\sigma}^s(\mathbf{k}) = \frac{N}{V}\,\frac{2a}{|\varepsilon_S|}\,\frac{\eta^2}{1 - \eta^2}\,\frac{k_s^2}{k^2 + \frac{1}{4}\xi_S^2(k^2 - k_s^2)^2}\, g_S \mathsf{M} \tag{III.32}$$

where g_H and g_S are not interesting factors that approach 1 at the respective instabilities,

$$g_H \equiv (1+d_X k^2)\left[\left(1+\frac{1}{\xi_H^2 k_s^2}\right)^2 - \frac{4}{\xi_H^2 \xi_S^2 k_s^4}\right]^{-1} \qquad \text{(III.33)}$$

$$g_S \equiv \frac{1+\eta a k^2/k_s^2}{1+\eta a}\;\frac{1+\xi_H^2(2k_s^2-k^2-4\xi_S^{-2})}{1+\xi_H^2 k_s^2 - 4\xi_H^2\xi_S^{-2}/(1+\xi_H^2 k_s^2)} \qquad \text{(III.34)}$$

Equation III.31 shows that the quasilinear method leads to a classical Ornstein-Zernike result[29] for the fluctuation enhancement near the hard instability. According to (III.30) the spatial correlation length ξ_H diverges as $|\varepsilon_H|^{-1/2}$. The fluctuations near the soft instability (III.32) behave similarly to those in the Bénard problem;[127] they are correlated over a distance $\xi_S \propto |\varepsilon_S|^{-1/2}$ that also diverges, but here the correlation involves spatial oscillations of wave number k_s. Note that (III.31) and (III.32) exhibit the spatial symmetries of the steady state, namely, homogeneity and isotropy. The very fact that there are stationary solutions to (III.10) also reflects temporal homogeneity.

Below the transitions the dynamic fluctuations are obtained—within the quasilinear approximation—by means of (III.16), which in frequency space reads

$$\boldsymbol{\sigma}(k,\omega) = (-i\omega - \boldsymbol{\Lambda}(k))^{-1}\boldsymbol{\sigma}(k) + \boldsymbol{\sigma}(k)(i\omega - \boldsymbol{\Lambda}^T(k))^{-1} \qquad \text{(III.35)}$$

For explicit expressions we refer to Chaturvedi et al.[145]

The immediate vicinity of the instability points, both below and above the transition, can be handled with the Landau equations (II.54) and (II.56). To identify the proper fluctuating forces associated with these equations, we simply express the d-matrix of (III.24) in the representation that makes $\boldsymbol{\Lambda}(k)$ diagonal at the critical point. For the hard instability, the corresponding transformation is given in (II.29):

$$\begin{pmatrix}\psi\\ \psi^*\end{pmatrix} = \frac{1}{2a}\begin{pmatrix}(1+ia)e^{ia\tau} & iae^{ia\tau}\\ (1-ia)e^{-ia\tau} & -iae^{-ia\tau}\end{pmatrix}\begin{pmatrix}\delta\tilde{x}\\ \delta\tilde{y}\end{pmatrix} \equiv \mathsf{T}_H\begin{pmatrix}\delta\tilde{x}\\ \delta\tilde{y}\end{pmatrix} \qquad \text{(III.36)}$$

so that the fluctuating forces are characterized by

$$\mathsf{d}_H = \mathsf{T}_H \mathsf{d}(b=b_H, \mathbf{k}=0)\mathsf{T}_H^T = \frac{V}{N}\,\frac{1+a^2}{2a}\begin{pmatrix}\dfrac{e^{2ia\tau}}{1-ia} & 1\\ 1 & \dfrac{e^{-2ia\tau}}{1+ia}\end{pmatrix} \qquad \text{(III.37)}$$

Similarly, the critical mode at the soft instability is subject to fluctuating forces of strength

$$d_S = \frac{V}{N}\,\frac{2(1 + a\eta)}{a(1 - \eta^2)^2} \tag{III.38}$$

In principle, this may be used in connection with the nonlinear Landau equations to calculate the correlations all through the transition region. So far, however, only the linear regimes below and above the instability have been treated where the quasilinear method is again applicable. Below the transition, the critical part of the results (III.25) to (III.32) is recovered. Let us therefore turn to the behavior above the transition and start with the hard instability. Linearizing (II.54) about the homogeneous limit cycle (II.57),

$$\psi = (\psi_0 + \delta\psi)e^{-i\nu\tau - i\varphi} \tag{III.39}$$

we obtain the relaxation equations

$$\begin{pmatrix} \delta\dot{\psi} \\ \psi_0\dot{\varphi} \end{pmatrix} = \tilde{\Lambda}\begin{pmatrix} \delta\psi \\ \psi_0\varphi \end{pmatrix} \qquad \tilde{\Lambda} = -\begin{bmatrix} 2\alpha\varepsilon_H + d_1k^2 & d_2k^2 \\ -2\alpha\varepsilon_H\dfrac{\beta_2}{\beta_1} - d_2k^2 & d_1k^2 \end{bmatrix} \tag{III.40}$$

where $d_1 \equiv \frac{1}{2}(d_X + d_Y)$, $d_2 \equiv (a/2)(d_X - d_Y)$, and $\beta_1 \equiv \operatorname{Re}\beta$, $\beta_2 \equiv \operatorname{Im}\beta$. In terms of the new fluctuating variables $\delta\psi, \psi_0\varphi$ the d-matrix (III.37) reads

$$\tilde{\mathsf{d}} = \frac{1}{4a}\,\frac{V}{N}\begin{bmatrix} 1 + a^2 + \cos 2a\tau - a\sin 2a\tau & -(\sin 2a\tau + a\cos 2a\tau) \\ -(\sin 2a\tau + a\cos 2a\tau) & 1 + a^2 - \cos 2a\tau + a\sin 2a\tau \end{bmatrix} \tag{III.41}$$

Thus the fluctuating forces oscillate with twice the limit cycle frequency. As a result, (III.10) predicts an oscillatory relaxation of the equal time correlations, the relaxation times being given by $\tilde{\Lambda}$ as usual. In the limit $d_{X,Y}k^2 \ll \alpha\varepsilon_H$ there are two times involved in $\tilde{\Lambda}$: one for the relaxation of the amplitudes $\delta\psi$, which occurs at a rate $2\alpha\varepsilon_H$, thus exhibits slowing down as the instability is approached from above; the other characterizes the phase motion as being diffusive, $\lambda \approx -d_1(1 + \beta_2 d_2/\beta_1 d_1)k^2$. On these time scales the oscillations contained in (III.41) are fast; therefore they do not contribute to the critical part of the correlations for which (III.10) or (III.11) give the following results:[66, 143, 144]

$$\langle(\delta\psi)^2\rangle = \frac{V}{4a\varepsilon_H N}\,\frac{1}{1 + \frac{1}{2}\xi_H^2k^2(1 + c_1)} \tag{III.42}$$

$$\langle\psi_0\varphi\delta\psi\rangle = \frac{V}{4a\varepsilon_H N}\,c_2 \tag{III.43}$$

$$\langle\psi_0^2\varphi^2\rangle = \frac{(1 + a^2)V}{4aN}\,\frac{1 + \beta_2^2/\beta_1^2}{d_1k^2}\,c_3 \tag{III.44}$$

where the constants c_i are

$$c_3 = \frac{1}{1 + d_2\beta_2/d_1\beta_1},$$

$$c_2 = -\frac{d_2}{d_1}\left(1 - \frac{d_1\beta_2}{d_2\beta_1}\right)c_3, \qquad \text{(III.45)}$$

$$c_1 = \left(\frac{d_2}{d_1}\right)^2\left(1 + \frac{d_1\beta_2}{d_2\beta_1}\right)c_3$$

Thus we see that the amplitude $\delta\psi$ behaves similarly to that below the transition; its fluctuations show critical enhancement as $k \to 0$, $\varepsilon_H \to 0$. The phase correlations, on the other hand, diverge regardless of ε_H, as $k \to 0$. This is an expression of long-range order in the system; (III.44) tells us that although the phase itself is not subject to restoring forces, its gradients are $\langle|\nabla\varphi|^2\rangle \propto 1/\varepsilon_H$. The more the pump parameter is increased, the stronger the tendency toward phase coherence. The analogy to a laser threshold is striking.[51, 52] The phase diffusion plays the role of a Goldstone mode[108] for the present symmetry breaking transition because (1) φ is the symmetry restoring variable in the Brusselator, (2) it exhibits long range order as borne out by (III.44), and (3) it relaxes infinitely slowly as $k \to 0$ ($\omega \propto k^2$).

Concerning the dynamic correlations, we refer to Ref. 66, where a procedure equivalent to (III.16) has been employed. It should be noted, however, that on time scales of the order of the limit cycle period the expressions (III.42) to (III.44) are insufficient because they neglect the time dependence in the fluctuating forces. For a more systematic treatment, see Ref. 148.

Let us now turn to a soft instability where spatial ordering sets in. Assuming that a planar standing wave develops that is characterized by a vector $\mathbf{k}_s$ and an amplitude $\psi_0^2 = 2\alpha\varepsilon_S/\beta$, see (II.59), we linearize (II.56) in $\delta\psi = \psi - \psi_0$ and get the relaxation behavior

$$\delta\dot{\psi} = -2\alpha\varepsilon_S\left(1 + \tfrac{1}{2}\xi_S^2\left(k_\parallel + \tfrac{1}{2}k_\perp^2/k_s\right)^2\right)\delta\psi \qquad \text{(III.46)}$$

(By superposition of eigenmodes with $\mathbf{k}_s$ pointing in different directions, we can discuss more involved structures like periodic lattices in three dimensions.) Using (III.38), we immediately get the static correlations of the amplitude

$$\langle(\delta\psi)^2\rangle = \frac{V/N}{\alpha\varepsilon_S(1-\eta^2)}\,\frac{1}{1 + \frac{1}{2}\xi_S^2\left(k_\parallel + \frac{1}{2}k_\perp^2/k_s\right)^2} \qquad \text{(III.47)}$$

This is similar to the result for the amplitude fluctuations above the hard instability except that (III.47) exhibits an anisotropy that is inevitably associated with the existence of a vector $\mathbf{k}_s$. We could of course allow for an additional phase factor $e^{i\varphi}$ in the fluctuating order parameter ψ. Again we would find that φ behaves essentially diffusively, its correlations reflecting the long-range order.

2. *Lotka-Volterra Model*

We conclude this section with a brief analysis of the fluctuations in the Lotka-Volterra model. All we need is the relaxation matrix (II.42), to which we add diffusion,

$$\Lambda(\mathbf{k}) = -\begin{pmatrix} d_X k^2 & 1 \\ -a & d_Y k^2 \end{pmatrix} \tag{III.48}$$

and the strength of the fluctuating forces, which we get by means of (III.7),

$$\mathsf{d}(\mathbf{k}) = \frac{Na}{V}\begin{bmatrix} 1+\dfrac{d_X k^2}{a} & -\frac{1}{2} \\ -\frac{1}{2} & 1+d_Y k^2 \end{bmatrix} \tag{III.49}$$

Using (III.11) we then compute the static correlations in the limit $d_{X,Y}k^2 \ll 1, \sqrt{a}$:

$$\boldsymbol{\sigma}(\mathbf{k}) = \frac{N/V}{(d_X - d_Y)k^2}\begin{bmatrix} 1+a+(d_X+2d_Y)k^2 & (ad_Y - d_X)k^2 \\ (ad_Y - d_X)k^2 & a(1+a+d_Y k^2) \end{bmatrix} \tag{III.50}$$

The main difference to the Brusselator is that here both autocorrelations diverge as $k \to 0$. This is of course an expected consequence of the existence of two conserved quantities rather than only one. As $k \to 0$, we may still use (III.10) to study the time development of an initially given finite correlation matrix[12, 150, 151]

$$\begin{bmatrix} \dot\sigma_{xx} \\ \dot\sigma_{xy} \\ \dot\sigma_{yy} \end{bmatrix} = \frac{N}{V}\begin{bmatrix} 2 \\ -1 \\ 2 \end{bmatrix} - \begin{bmatrix} 0 & 2 & 0 \\ a & 0 & -1 \\ 0 & -2a & 0 \end{bmatrix}\begin{bmatrix} \sigma_{xx} \\ \sigma_{xy} \\ \sigma_{yy} \end{bmatrix} \tag{III.51}$$

Since the relaxation matrix here has eigenvalues $0, \pm 2i\sqrt{a}$, there is an oscillatory component with twice the frequency of the individual x, y-relaxation, and a linear increase due to the action of the fluctuating forces.

Note, however, that this broadening is relatively less significant as the system size increases, since $\langle(\delta X)^2\rangle = V\sigma_{xx}\ (k=0)$, and therefore $\sqrt{\langle(\delta X)^2\rangle}\ /X_0 \propto V^{-1/2}$.

So much for the equal time correlations. If we combine (III.50) with the spectrum of $\boldsymbol{\Lambda}(\mathbf{k})$, which is

$$\lambda_{1,2} = -\tfrac{1}{2}(d_X + d_Y)k^2 \pm i\sqrt{a - \tfrac{1}{4}(d_X - d_Y)^2 k^4} \qquad \text{(III.52)}$$

we get the dynamic correlations from (III.16):

$$\boldsymbol{\sigma}(\mathbf{k},\omega) = \frac{2aN/V}{(\omega^2 - a)^2 + \omega^2(d_X + d_Y)^2 k^4} \begin{bmatrix} 1+\omega^2 & \tfrac{1}{2}(a-\omega^2) - i\omega(1+a) \\ \tfrac{1}{2}(a-\omega^2) + i\omega(1+a) & a^2 + \omega^2 \end{bmatrix} \qquad \text{(III.53)}$$

Note that as in (III.50) we have taken only the leading terms, given that $d_{X,Y}k^2 \ll a$. This correlation function exhibits a typical feature of chemical nonequilibrium steady states, namely, an antisymmetric and imaginary part. It can be shown[124] that this is an immediate consequence of the violation of time reversal invariance, and in fact there is a sum rule that relates this part to the antisymmetric component of the generalized Onsager matrix (II.19):

$$\mathsf{A} \equiv \frac{(\mathsf{L} - \mathsf{L}^T)}{2} = iV\int \frac{d\omega}{2\pi}\,\omega\boldsymbol{\sigma}(\mathbf{k},\omega) = N\frac{a(1+a)}{(d_X + d_Y)k^2}\begin{pmatrix} 0 & 1 \\ -1 & 0 \end{pmatrix} \qquad \text{(III.54)}$$

As we discussed in Section II.C.2, a nonvanishing A can be interpreted as the "elasticity" of a chemical nonequilibrium system. Comparing (III.54) to the dissipative part $\mathsf{D} \equiv \frac{1}{2}(\mathsf{L} + \mathsf{L}^T) = V\mathsf{d}$ as given by (III.49), we can now see why the Lotka-Volterra model appears to be conservative although the dissipation does not vanish. The point is that in the limit $k \to 0$ D stays finite while A diverges. Thus in the relaxation spectrum the influence of A becomes overwhelming. Equation III.51 however tells us that dissipation is not completely eliminated; it does show up in the correlations, for example, in the form of Kubo relations

$$\mathsf{D} = V \lim_{\omega \to \infty} \frac{\omega^2}{2}\boldsymbol{\sigma}(\mathbf{k},\omega) \qquad \text{(III.55)}$$

At high wave numbers, of course, the situation must be reversed and $\mathsf{D} \propto k^2$ will dominate the system's behavior.

D. Chemical Fluctuations and Experiments

As the preceding sections demonstrate, considerable effort has been made in recent years to develop the theory of fluctuations in chemical reaction systems. In marked contrast, there are as yet no experiments that put the theory to test. A major reason for this discrepancy seems to be the fact, mentioned in the introduction, that chemical relaxation processes are completely obscured by diffusion unless the wave vector is smaller than the inverse correlation length $\xi = \sqrt{D/\lambda}$, where D is the diffusion constant and λ the chemical relaxation rate. Thus in a light scattering experiment[152] that measures the structure factor $\sigma(\mathbf{k},\omega)$, one has to go to extremely low k-values to see chemical fluctuations. These small k vectors are associated with small scattering angles that render the separation of scattered light from incoming light very difficult. A second difficulty associated with light scattering is that the changes in the dielectric constant that accompany chemical fluctuations are usually very small.

In the context of correlation spectroscopy the small k-values imply small relative signals. There are $N = \xi^3 cN_A$ particles in a volume of diameter ξ, c being their molarity and N_A Avogadro's number; the relative noise is then $\sqrt{\langle(\delta N)^2\rangle}\,/N = (cN_A\xi^3)^{-1/2}$. To take an example, in an equilibrium system with $D = 10^{-5}$ cm^2/sec, $\lambda = 10^3$/sec, we have $N \approx 10^{12}c$ which, even for a micromolar solution would still be 10^6, with a relative noise of 10^{-3}. This is just on the brink of detectability of experiments using fluorescence correlation spectroscopy[153, 154] and conductivity fluctuation analysis.[155, 156]

On the other hand, conventional relaxation methods that observe the response to external perturbations of temperature, pressure, electric field, and so on,[157] are comparatively easy to perform. Since the fluctuation-dissipation theorem guarantees the equivalence of the two approaches for equilibrium systems, there has not been a great need for direct fluctuation analysis. In far-from-equilibrium situations the usual fluctuation-dissipation theorem does not hold. Here it becomes a real challenge to determine correlations and response behavior independently. Relaxation methods may yield interesting information on critical slowing down in nonequilibrium systems near marginal stability.[158] However to measure fluctuations directly, consider the possibility of the use of "reduced dimensionality" (i.e., thin films, capillaries, or beads) to make the effects of fluctuations more pronounced.

Theory predicts the following features of correlations near an instability: (1) enhancement of fluctuations near the critical k-value, (2) slowing down of the relaxation frequency (i.e., sharpening of the spectrum), and (3) asymmetry with respect to time or frequency reversal in cross-correlation functions. Let us discuss these points in some detail.

1. For simplicity, we take a system with only one free component, and with critical k-value $k = 0$. From (III.12) to (III.14) we have, in general,

$$\sigma(\mathbf{k}) = -\frac{d(k)}{\lambda(k)} = \frac{d + Dk^2x_0}{|\lambda| + Dk^2} \tag{III.56}$$

If ε measures the relative distance to the marginal stability point, we have $\lambda = \lambda_0|\varepsilon|$. Let us assume that far away from the transition (i.e., for $|\varepsilon| = 1$), the system behaves essentially like an equilibrium system, $d = |\lambda_0|x_0$. Then there are two spatial scales to be considered:

$$\xi_0 = \sqrt{\frac{D}{|\lambda_0|}} \qquad \xi = \sqrt{\frac{D}{|\lambda|}} = \xi_0\varepsilon^{-1/2} \tag{III.57}$$

The correlations are Poissonian for $k > \xi_0^{-1}$, they grow as $x_0/\xi_0^2k^2$ in an intermediate range $\xi^{-1} < k < \xi_0^{-1}$, and for the smallest $k < \xi^{-1}$ they are again constant, x_0/ε (Fig. 16). Would this improve the feasibility of a correlation spectroscopy experiment as the integrated spectrum $\sigma(\mathbf{k}) = \int \sigma(\mathbf{k}, \omega)\, d\omega/2\pi$ is now enhanced for $k < \xi_0^{-1}$? Unfortunately the answer is negative because with k smaller than ξ_0^{-1} by a factor $\alpha < 1$, the particle numbers involved increase by a factor α^{-3}, while their root-mean-square

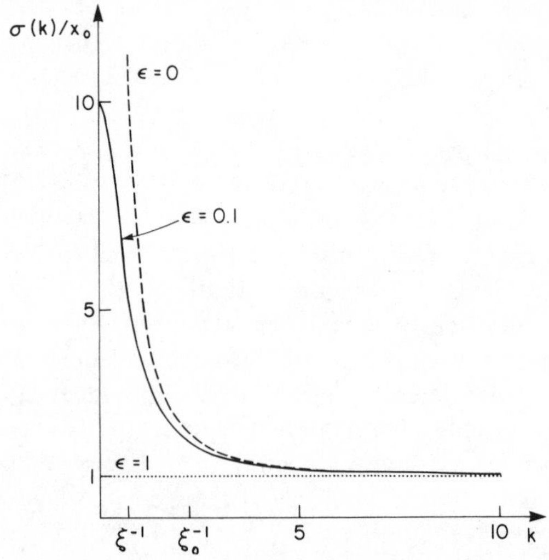

Fig. 16. Static correlation function versus the wave vector $\mathbf{k}$ near a marginal stability. The dotted line represents Poissonian behavior. The quantity ϵ is a measure of the deviation from marginal stability; ξ is a correlation length defined in (III.57). The broken line holds for the limit of marginal stability.

fluctuations $(V\sigma(\mathbf{k}))^{1/2}$ grow only by $\alpha^{-5/2}$. Consequently, if the fluctuations cannot be seen at $k \sim \xi_0^{-1}$, they will remain hidden at smaller k.

The outlook is much better, however, for correlation spectroscopy with systems of reduced dimensionality. For example, take a film of thickness $l < \xi_0$ and let the boundary conditions be such that there is no flux across the boundary $\mathbf{n} \cdot \nabla x = 0$. Then in the direction perpendicular to the film there are only discrete k-values, $k = 0, \pm\pi/l, \pm 2\pi/l, \ldots$, to be considered, but (III.56) is still applicable and gives $\sigma(k) = x_0 + \delta_{k,0}(1-\varepsilon)x_0/\varepsilon$. For capillaries the k-values are discrete in two directions, and in case of a small cube (size $l < \xi_0$) this holds for the whole k-space. A convenient measure for the role of the fluctuations in the respective cases is given by the leading ε-dependence of the local correlations[159]

$$\sigma(\mathbf{r}t, \mathbf{r}t) = \frac{1}{L^{3-d}} \sum_k \int \frac{d^d\mathbf{k}}{(2\pi)^d} \sigma(\mathbf{k}) = c_1(d) + c_2(d)|\varepsilon|^{(d-2)/2} \qquad \text{(III.58)}$$

where $c_1(d)$ and $c_2(d)$ are constants. We see that in the bulk $(d=3)$ the ε-dependence is $|\varepsilon|^{1/2} \to 0$, which corroborates our earlier reasoning. In a film $(d=2)$ matters are at least not worse as the transition is approached, and in a capillary $(d=1)$ a fluctuation enhancement is at last strongly borne out, $\sigma(\mathbf{r},\mathbf{r}) \propto |\varepsilon|^{-1/2}$. The largest effects are obviously expected in "zero-dimensional" systems where $\sigma(\mathbf{r},\mathbf{r}) \propto |\varepsilon|^{-1}$. All this is highly reminiscent of the fluctuation behavior in superconductors,[160] where the analogue to ξ_0 is the coherence length of a Cooper pair, which is also very large compared to molecular distances, necessitating the use of films and filaments for studies of fluctuations.

The analogy to superconductors can be further explored for a discussion of the width of the critical region where the quasilinear (or mean field) theory is expected to break down because of the essentially nonlinear character of the system. When a transition can be described by a Landau equation $\dot{\psi} = \alpha\varepsilon\psi - \beta|\psi|^2\psi + D\nabla^2\psi$, see (II.54) or (II.56), with corresponding fluctuating forces d_ψ, see (III.37) or (III.38), then in addition to $\xi = \sqrt{D/\alpha|\varepsilon|}$ there is a length[159]

$$\xi_G \equiv \frac{D^2}{\beta d_\psi} \qquad \text{(III.59)}$$

that emerges from a comparison of diffusion and the nonlinearity proportional to β. The quasilinear approximation, thus classical critical behavior, is expected to hold as long as $\xi \lesssim \xi_G$, which in terms of ε means

$$|\varepsilon| \lesssim v\xi_0^{-3} v x_0 \beta \qquad \text{(III.60)}$$

where $v = V/N$ is a molecular volume. Such a relation is known as Ginzburg criterion[161] and has been discussed, in the context of chemical instabilities by Nitzan and Ortoleva.[27, 162] It expresses the critical width as a ratio of molecular volume v to the volume of coherence ξ_0^3. As discussed earlier, this is usually very small in chemical systems, and where it is not, because of fast chemical relaxation times, the coupling of reaction-diffusion equations to heat transport can most likely not be neglected.

2. Next we discuss briefly the slowing down as it appears in the correlations when a marginal stability point is approached. For a system like (III.56), (III.16) simply gives

$$\langle \delta X(t)\delta X(0)\rangle = \int d^3\mathbf{r}\int d^3\mathbf{r}'\langle \delta x(\mathbf{r}t)\delta x(\mathbf{r}'0)\rangle = V\sigma(\mathbf{k}=0)e^{-|\lambda_0\varepsilon t|} \qquad \text{(III.61)}$$

The corresponding spectrum is

$$\int_{-\infty}^{\infty} e^{i\omega t}\langle \delta X(t)\delta X(0)\rangle\, dt = V\sigma(\mathbf{k}=0)\,\frac{2|\lambda_0\varepsilon|}{\omega^2+(\lambda_0\varepsilon)^2} \qquad \text{(III.62)}$$

As ε approaches zero this gets sharper, $(\Delta\omega)_{1/2} = |\lambda_0\varepsilon|$, and because we have the relation $\sigma(\mathbf{k}=0) = x_0/\varepsilon$, the structure factor develops a very pronounced peak proportional to ε^{-2} at $\omega = 0$. Regarding the experimental accessibility of these functions, the foregoing considerations suggest doing correlation spectroscopy on the smallest possible systems. Of course the theory predicts that the same relaxation time $|\lambda_0\varepsilon|$ appears in a response function that might be much easier to obtain, but the connection between fluctuations and response behavior, in nonequilibrium systems, is still a matter of debate[124] and should be subject to experimental testing.

3. Cross-correlation functions may display an asymmetry with respect to time reversal. Equation III.53 illustrates some typical features of a two-component system, in a nonequilibrium steady state.[146] The antisymmetric part of $\boldsymbol{\sigma}(\mathbf{k},\omega)$ must be imaginary and odd in ω, which means that it vanishes as $\omega \to 0$. Furthermore, as ω goes to infinity, it goes to zero more rapidly than the normal dissipative terms. Thus it exhibits a peak at some finite frequency and in this way represents an oscillatory tendency. It may be present even if the oscillations are overdamped in the relaxation spectrum. In real time, the existence of this anomalous part of the correlation function means that $\langle \delta x_1(t)\delta x_2(0)\rangle$ is not even in t and not symmetric in 1,2. Near a hard instability, the critical fluctuation enhancement pertains to the anomalous part, whereas the normal dissipative part stays small. Again, all this is probably most easily seen in a response function, but for the reasons mentioned, a direct investigation of the correlation functions is highly desirable.

Acknowledgments

We thank Michael DelleDonne for helpful discussions.

We acknowledge support by the Department of Energy, the National Science Foundation, and the Air Force Office of Scientific Research.

A fellowship from the Deutsche Forschungsgemeinschaft for P.R. is also gratefully acknowledged.

References

1. R. M. Noyes and R. J. Field, *Ann. Rev. Phys. Chem.*, **25**, 95–119 (1974).
2. P. Glansdorff and I. Prigogine, *Thermodynamics of Structure, Stability and Fluctuations*, Wiley-Interscience, New York, 1971.
3. B. Hess and A. Boiteux, *Ann. Rev. Biochem.*, **40**, 237–258 (1971).
4. G. Nicolis and J. Portnow, *Chem. Rev.*, 365–384 (1973).
5. B. Chance, E. K. Pye, A. K. Ghosh, and B. Hess, Eds., *Biological and Biochemical Oscillators*, Academic Press, New York, London, 1973.
6. "Physical chemistry of oscillatory phenomena," *Faraday Symp. Chem. Soc.*, **9** (1974).
7. G. Nicolis and R. Lefever, Eds., *Membranes, Dissipative Structures, and Evolution, Advances in Chemical Physics*, Vol. 29, Wiley-Interscience, New York, 1975.
8. R. Schmitz, in *Chemical Reaction Engineering Reviews 1975, Advances in Chemistry Series*, Vol. 148, American Chemical Society, Washington, D.C., p. 176.
9. H. Haken, *Synergetics—An Introduction*, Springer-Verlag, Berlin, 1977.
10. A. Pacault, P. Hanusse, P. De Kepper, C. Vidal, and J. Boissonade, *Acc. Chem. Res.*, **9**, 438–445 (1976).
11. J. Ross, *Ber. Bunsenges. Physik. Chem.*, **80**, 112–125 (1976).
12. G. Nicolis and I. Prigogine, *Self-Organization in Nonequilibrium Systems*, Wiley-Interscience, New York, 1977.
13. P. Ortoleva, in Ref. 17, pp. 235–286.
14. P. Hanusse, J. Ross, and P. Ortoleva, in Ref. 16, pp. 317–361.
15. "Nonlinear nonequilibrium statistical mechanics," *Suppl. Prog. Theor. Phys.*, **64** (1978).
16. S. Rice, Ed., *Advances in Chemical Physics*, Vol. 38, Wiley-Interscience, New York, 1978.
17. H. Eyring and D. Henderson, Eds., *Theoretical Chemistry*, Vol. 4, *Periodicities in Chemistry and Biology*, Academic Press, New York, 1978.
18. W. Jessen, *Naturwissenschaften*, **65**, 449–455 (1978).
19. P. C. Martin, *J. de Phys.*, **37**:Cl 57–66 (1976).
20. A. M. Zhabotinsky, *Dokl. Akad. Nauk SSSR*, **157**, 392–395 (1964).
21. A. M. Zhabotinsky, in Ref. 5, pp. 89–95.
22. A. T. Winfree, *Sci. Am.*, **230**:6, 82–95 (1974).
23. W. Geiseler and H. H. Föllner, *Biophys. Chem.*, **6**, 107–115 (1977).
24. H. Degn, *Biochim. Biophys. Acta*, **180**, 271–290 (1969).
25. L. F. Olsen and H. Degn, *Nature* (London), **267**, 177–178.
26. A. Nitzan, P. Ortoleva, J. Deutch, and J. Ross, *J. Chem. Phys.*, **61**, 1056–1074 (1974).
27. A. Nitzan, *Phys. Rev. A*, **17**, 1513–1528 (1978).
28. H. Haken, *Z. Phys. B*, **21**, 105–114 (1975).
29. H. E. Stanley, *Introduction to Phase Transitions and Critical Phenomena*, Clarendon Press, Oxford, 1971.
30. S. -k. Ma, *Modern Theory of Critical Phenomena*, Benjamin, Reading, MA, 1976.
31. M. Kubicek and M. Marek, *J. Chem. Phys.*, **67**, 1997–2006 (1977).
32. R. Graham, in T. Riste, Ed., *Fluctuations, Instabilities and Phase Transitions*, Plenum, New York, pp. 215–279, 1975.

33. R. Thom, *Structural Stability and Morphogenesis: An Outline of a General Theory of Models*, Benjamin, Reading, MA, 1975.
34. E. C. Zeeman, *Catastrophe Theory*, Addison-Wesley, Reading, MA, 1977.
35. O. E. Rössler, *Bull. Math. Biol.*, **39**: 275–289 (1977).
36. O. E. Rössler, *Z. Naturforsch.* **31a**, 1168–1172 (1976).
37. T. Y. Li and J. A. Yorke, *Am. Math. Mon.*, **82**, 985–992 (1975).
38. S. Jorna, Ed., *Topics in Nonlinear Dynamics*, American Institute of Physics Conference Proceedings No. 46, AIP, New York, 1978.
39. H. -G. Busse, *Nat. Phys. Sci.*, **233**, 137–138 (1971).
40. E. Körös, M. Orbán, and Zs. Nagy, *Nat. Phys. Sci.*, **242**, 30–31 (1973).
41. I. Lamprecht and B. Schaarschmidt, *Thermochim. Acta*, **22**, 257–266 (1978).
42. A. Einstein, *Sitzungsber. Preuss. Akad. Wiss. Kl. Phys.-Math.*, 380–385 (1920).
43. R. Landauer, *J. Appl. Phys.*, **33**, 2209–2216 (1962).
44. C. L. Creel and J. Ross, *J. Chem. Phys.*, **65**, 3779–3389 (1976).
45. A. Pacault, P. de Kepper, and P. Hanusse, *C.R. Acad. Sci. Ser. B*, **280**, 157–161 (1975).
46. G. Ley and H. Gerrens, *Makromol. Chem.*, **175**, 563–581 (1974).
47. M. H. Bernstein, Dissertation, University of Pennsylvania, Philadelphia, 1976.
48. C. Gabrielli and M. Keddam, *J. Electroanal. Chem. Interfacial Electrochem.*, **45**, 267–277 (1973).
49. P. Ortoleva and J. Ross, *Biophys. Chem.*, **1**, 87–96 (1973).
50. F. Schlögl, *Z. Phys.*, **253**, 147–161 (1972).
51. R. Graham and H. Haken, *Z. Phys.*, **237**, 31–46 (1970).
52. S. Grossmann and P. H. Richter, *Z. Phys.*, **242**, 458–475 (1971).
53. R. J. Field and R. M. Noyes, *Acc. Chem. Res.*, **10**, 214–221 (1977).
54. R. J. Bose, J. Ross, and M. S. Wrighton, *J. Am. Chem. Soc.*, **99**, 6119–6120 (1977).
55. M. Marek and I. Stuchl, *Biophys. Chem.*, **3**, 241–248 (1975).
56. M. Marek and E. Svobodova, *Biophys. Chem.*, **3**, 263–273 (1975).
57. D. Ruelle, *Trans. NY Acad. Sci.*, **35**, 66–71 (1973).
58. L. Onsager, *Phys. Rev.*, **37**, 405–426 (1931).
59. W. Jost, *Z. Naturforschg.*, **2a**, 159–163 (1947).
60. E. Körös, *Termeszettud. Vilaga*, **103**, 109–112 (1972), in Hungarian.
61. A. M. Turing, *Phil. Trans. R. Soc. London Ser. B*, **237**, 37–72 (1952).
62. I. Prigogine and R. Lefever, *J. Chem. Phys.*, **48**, 1695–1700 (1968).
63. J. J. Tyson, *J. Chem. Phys.*, **58**, 3919–3930 (1973).
64. E. Hopf, *Ber. Math. Phys. Kl. Saechs. Akad. Wiss. Leipzig*, **94**, 3–22 (1942).
65. Y. Kuramoto and T. Tsuzuki, *Prog. Theor. Phys.*, **52**, 1399–1401 (1974).
66. H. Mashiyama, A. Ito, and T. Ohta, *Prog. Theor. Phys.*, **54**, 1050–1066 (1975).
67. J. F. G. Auchmuty and G. Nicolis, *Bull. Math. Biol.*, **38**, 325–350 (1976).
68. L. D. Landau and E. M. Lifshitz, *Statistical Physics*, Pergamon Press, Oxford, 1959, Ch. 14.
69. P. H. Richter and S. Grossmann, *Z. Phys.*, **248**, 244–253 (1971).
70. A. J. Lotka, *Proc. Nat. Acad. Sci. (US)*, **6**, 410–415 (1920).
71. V. Volterra, *Leçons sur la Théorie Mathématique de la Lutte pour la Vie*, Gauthier-Villars, Paris, 1936.
72. R. Lefever, G. Nicolis, and I. Prigogine, *J. Chem. Phys.*, **47**, 1045–1047 (1967).
73. H. T. Davis, *Introduction to Nonlinear Differential and Integral Equations*, Dover, New York, 1962.
74. J. S. Frame, *J. Theor. Biol.*, **43**, 73–81 (1974).
75. A. A. Andronov, A. A. Vit, and C. E. Khaikin, *Theory of Oscillators*, Pergamon Press, Oxford, 1966.

76. R. J. Field, E. Körös, and R. M. Noyes, *J. Am. Chem. Soc.*, **94**, 8649–8664 (1972).
77. R. J. Field and R. M. Noyes, *J. Chem. Phys.*, **60**, 1877–1884 (1974).
78. J. J. Tyson, "The Belousov-Zhabotinskii reaction," in *Lecture Notes in Biomathematics*, Vol. 10, Springer, Berlin, 1976.
79. L. F. Olsen and H. Degn, *Biochim. Biophys. Acta*, **523**, 321–334 (1978).
80. A. Goldbeter and R. Lefever, *Biophys. J.*, **12**, 1302–1315 (1972).
81. G. Nicolis, T. Erneux, and M. Herschkowitz-Kaufman, in Ref. 17, pp. 263–315.
82. A. T. Winfree, in Ref. 17, pp. 2–51, and the references cited therein.
83. N. Kopell and L. N. Howard, *Stud. Appl. Math*, **52**, 291–328 (1973).
84. P. Ortoleva and J. Ross, *J. Chem. Phys.*, **60**, 5090–5107 (1974).
85. Y. Kuramoto and T. Yamada, *Prog. Theor. Phys.*, **56**, 679–681 (1976).
86. T. Yamada and Y. Kuramoto, *Prog. Theor. Phys.*, **56**, 681–683 (1976).
87. H. Fujisaka and T. Yamada, *Prog. Theor. Phys.*, **57**, 734–745 (1977).
88. A. Nitzan, P. Ortoleva, and J. Ross, *Faraday Symp. Chem. Soc.*, **9**, 241–253 (1974).
89. E. W. Montroll, in S. A. Rice, K. F. Freed, and J. C. Light, Eds., *Statistical Mechanics*, University of Chicago Press, Chicago, 1972, pp. 69–89.
90. P. Ortoleva and J. Ross, *J. Chem. Phys.*, **63**, 3398–3408 (1975).
91. L. M. Pismen, *J. Chem. Phys.*, **69**, 4149–4158 (1978).
92. P. Ortoleva and J. Ross, *J. Chem. Phys.*, **58**, 5673–5680 (1973).
93. Y. Kuramoto and T. Tsuzuki, *Prog. Theor. Phys.*, **54**, 687–699 (1975).
94. Y. Kuramoto and T. Yamada, *Prog. Theor. Phys.*, **56**, 724–740 (1976).
95. B. A. Huberman, *J. Chem. Phys.*, **65**, 2013–2019 (1976).
96. I. Balslev and H. Degn, *Faraday Symp. Chem. Soc.*, **9**, 233–240 (1974).
97. M. Flicker and J. Ross, *J. Chem. Phys.*, **60**, 3458–3465 (1974).
98. D. Feinn, P. Ortoleva, W. Scalf, S. Schmidt, and M. Wolff, *J. Chem. Phys.*, **69**, 27–39 (1978).
99. R. Lovett, P. Ortoleva, and J. Ross, *J. Chem. Phys.*, **69**, 947–955 (1978).
100. I. M. Lifshitz and V. V. Slyozov, *J. Phys. Chem. Solids*, **19**, 35–50 (1961).
101. B. Hess, A. Boiteux, H. G. Busse, and G. Gerisch, *Adv. Chem. Phys.*, **29**, 137–168 (1975).
102. A. Gierer, and H. Meinhardt, *Kybernetik*, **12**, 30–39 (1972).
103. R. J. Field and R. M. Noyes, *J. Am. Chem. Soc.*, **96**, 2001–2005 (1974).
104. K. Showalter and R. M. Noyes, *J. Am. Chem. Soc.*, **98**, 3730–3731 (1976).
105. R. Gilbert, P. Ortoleva, and J. Ross, *J. Chem. Phys.*, **58**, 3625–3633 (1973).
106. L. D. Landau and E. M. Lifshitz, *Statistical Physics*, Pergamon Press, Oxford, 1959, Ch. 12.
107. R. Landauer, *Phys. Today*, **31**: 11, 23–30 (1978).
108. D. Forster, *Hydrodynamic Fluctuations, Broken Symmetry and Correlation Functions*, Benjamin, Reading, MA, 1975.
109. W. Pauli, in P. Debye, Ed., *Probleme der Modernen Physik*, Leipzig, Hirzel, 1928, pp. 30–45.
110. D. A. McQuarrie, *J. Appl. Prob.*, **4**, 413–478 (1967).
111. H. Haken, *Rev. Mod. Phys.*, **47**, 67–121 (1975).
112. C. W. Gardiner and S. Chaturvedi, *J. Stat. Phys.*, **17**, 429–468 (1977).
113. R. Görtz, *J. Phys., A, Math. Gen.*, **9**, 1089–1092 (1976).
114. J. Schnakenberg, *Rev. Mod. Phys.*, **48**, 571–585 (1976).
115. N. G. van Kampen, *Can. J. Phys.*, **39**, 551–567 (1961).
116. N. G. van Kampen, *Adv. Chem. Phys.*, **34**, 245–309 (1976).
117. R. Kubo, K. Matsuo, and K. Kitahara, *J. Stat. Phys.*, **9**, 51–96 (1973).
118. M. Suzuki, *Prog. Theor. Phys.*, **55**, 383–399 (1976).

119. C. Blomberg, *J. Stat. Phys.*, to be published.
120. C. W. Gardiner, *J. Stat. Phys.*, **15**, 451–454 (1976).
121. S. Grossmann, *J. Chem. Phys.*, **65**, 2007–2012 (1976).
122. W. Horsthemke and R. Lefever, *Phys. Lett. A*, **64**, 19–21 (1977).
123. M. Lax, *Rev. Mod. Phys.*, **32**, 25–64 (1960).
124. F. Jähnig and P. H. Richter, *J. Chem. Phys.*, **64**, 4645–4656 (1976).
125. J. Keizer, *J. Chem. Phys.*, **63**, 5037–5043 (1975).
126. L. D. Landau and E. M. Lifshitz, *Fluid Mechanics*, Pergamon Press, Oxford, 1959, Ch. 17.
127. R. Graham, *Phys. Rev., A*, **10**, 1762–1784 (1974).
128. S. Grossmann and R. Schranner, *Z. Phys. B*, **30**, 325–337 (1978).
129. K. J. McNeil and D. F. Walls, *J. Stat. Phys.*, **10**, 439–448 (1974).
130. H. K. Janssen, *Z. Phys.*, **270**, 67–73 (1974).
131. F. Jähnig and P. H. Richter, *Ber. Bunsenges. Phys. Chem.*, **80**, 1132–1136 (1976).
132. G. Nicolis and J. W. Turner, *Physica, A*, **89**, 326–338 (1977).
133. G. Nicolis and R. Lefever, *Phys. Lett., A*, **62**, 469–471 (1977).
134. G. H. Weiss, *Adv. Chem. Phys.*, **13**, 1–18 (1967).
135. I. Procaccia and J. Ross, *J. Chem. Phys.*, **67**, 5565–5571 (1977).
136. I. Oppenheim, K. E. Shuler, and G. H. Weiss, *Physica, A*, **88**, 191–214 (1977).
137. J. S. Turner, *Adv. Chem. Phys.*, **29**, 63–83 (1975).
138. B. N. Belintsev, M. A. Livshits, and M. V. Volkenstein, *Z. Phys., B*, **30**, 211–218 (1978).
139. C. Y. Mou, *J. Chem. Phys.*, **68**, 1385–1390 (1978).
140. K. Binder, C. Billolet, and P. Mirold, *Z. Phys., B*, **30**, 183–195 (1978).
141. G. Nicolis and M. Malek-Mansour, in Ref. 15, pp. 249–268.
142. C. W. Gardiner, K. J. McNeil, D. F. Walls, and I. S. Matheson, *J. Stat. Phys.*, **14**, 307–331 (1976).
143. K. Tomita, T. Ohta, and H. Tomita, *Prog. Theor. Phys.*, **52**, 1744–1765 (1974).
144. Y. Kuramoto and T. Tsuzuki, *Prog. Theor. Phys.*, **54**, 60–71 (1975).
145. S. Chaturvedi, C. W. Gardiner, I. S. Matheson, and D. F. Walls, *J. Stat. Phys.*, **17**, 469–489 (1977).
146. P. H. Richter and F. Jähnig, to be published.
147. D. T. Gillespie, *J. Phys. Chem.*, **81**, 2340–2361 (1977).
148. R. Schranner, S. Grossmann, and P. H. Richter, *Z. Phys. B.*, to be published.
149. K. Tomita and H. Tomita, *Prog. Theor. Phys.*, **51**, 1731–1749 (1974).
150. G. Nicolis, *J. Stat. Phys.*, **6**, 195–222 (1972).
151. J. Keizer, *J. Stat. Phys.*, **15**, 477–483 (1976).
152. B. J. Berne and R. Pecora, *Dynamic Light Scattering with Applications to Chemistry, Biology, and Physics*, Wiley, New York, 1976.
153. D. Magde, E. Elson, and W. W. Webb, *Phys. Rev. Lett.*, **29**, 705–708 (1972).
154. D. Magde, E. Elson, and W. W. Webb, *Biopolymers*, **13**, 29–61 (1974).
155. G. Feher and M. Weissman, *Proc. Nat. Acad. Sci. (US)*, **70**, 870–875 (1973).
156. H. P. Zingsheim and E. Neher, *Biophys. Chem.*, **2**, 197–207 (1974).
157. M. Eigen and L. DeMaeyer, *Theoretical Basis of Relaxation Spectrometry*, in A. Weissberger and G. Hammes, Eds., *Techniques of Chemistry*, Vol. 6/2, Wiley, New York, 1973, pp. 63–146.
158. C. Allain, H. Z. Cummins, and P. Lallemand, to be published.
159. R. A. Ferrell, in *Contemporary Physics*, Vol. 1, pp. 129–156, Trieste Symposium, 1968, International Atomic Energy Agency IAEA, Vienna, 1969.
160. S. Grossmann and P. H. Richter, *Phys. Lett., A*, **33**, 39–40 (1970).
161. V. L. Ginzburg, *Sov. Phys.—Solid State*, **2**, 1824–1834 (1961).
162. A. Nitzan and P. Ortoleva, *Phys. Rev. A*, to be published.

AUTHOR INDEX

Numbers in parentheses are reference numbers and show that an author's work is referred to although his name is not mentioned in the text. Numbers in *italics* indicate pages on which the full references appear.

SUBJECT INDEX